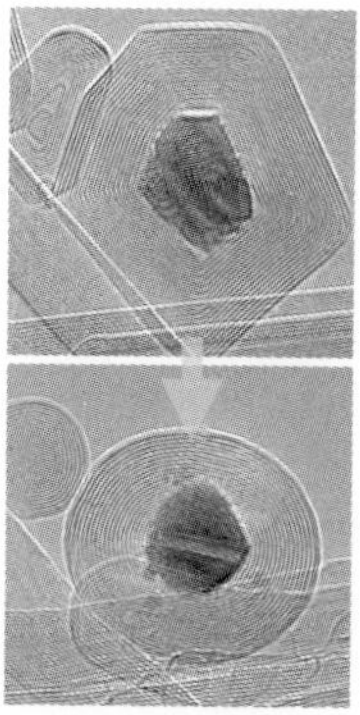

In-situ Electron Microscopy at High Resolution

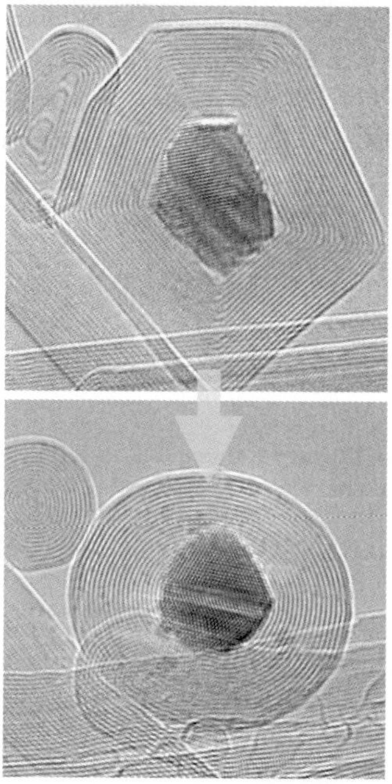

In-situ Electron Microscopy at High Resolution

Editor

Florian Banhart
Université de Strasbourg, France

World Scientific

NEW JERSEY • LONDON • SINGAPORE • BEIJING • SHANGHAI • HONG KONG • TAIPEI • CHENNAI

Published by

World Scientific Publishing Co. Pte. Ltd.
5 Toh Tuck Link, Singapore 596224
USA office: 27 Warren Street, Suite 401-402, Hackensack, NJ 07601
UK office: 57 Shelton Street, Covent Garden, London WC2H 9HE

British Library Cataloguing-in-Publication Data
A catalogue record for this book is available from the British Library.

IN-SITU **ELECTRON MICROSCOPY AT HIGH RESOLUTION**

ISBN-13 978-981-279-733-9
ISBN-10 981-279-733-5

Editor: Tjan Kwang Wei

Typeset by Stallion Press
Email: enquiries@stallionpress.com

Printed by FuIsland Offset Printing (S) Pte Ltd, Singapore

CONTENTS

CHAPTER 1

INTRODUCTION TO IN-SITU ELECTRON MICROSCOPY

Florian Banhart

Institut de Physique et Chimie des Matériaux, Université de Strasbourg
23 rue du Loess, 67034 Strasbourg, France
banhart@ipcms.u-strasbg.fr

This chapter gives an introduction to *in-situ* electron microscopy. The historical background, the achievements, and modern techniques of *in-situ* electron microscopy are briefly reviewed, and the limitations of the technique as well as the prospects for future developments are discussed.

1. Definition and History of *In-Situ* Electron Microscopy

The interest in structures with sizes on the micron, nanometer, and eventually atom cluster or molecule scale has resulted in the development of sophisticated tools of microscopy during the past decades. Today, materials or biological science cannot be imagined without the characterization techniques of modern transmission electron microscopy.[1,2] Transmission electron microscopes permit a view into the interior of small objects and are complementary to scanning tip microscopies that provide images from specimen surfaces. Due to continuous efforts in electron optics, the lateral resolution of electron microscopes has now dropped below 0.1 nanometers, and this scale is already smaller than the distance between atoms in densely packed crystals or in molecules. Nowadays, images with the resolution of crystal lattice spacings are recorded as a matter of routine and even individual atoms have been observed, though in special systems only. With the ability of forming strongly focused electron probes, an electron microscope is also able to provide the platform for analytical techniques with high lateral resolution such as energy-dispersive X-ray or electron energy loss spectroscopy.

By looking through textbooks of electron microscopy, it appears that microscopy is just able to provide information from the space of the objects. However, time is another parameter and, indeed, it has been demonstrated since decades that experiments can be conducted in real time inside the specimen chambers of electron microscopes. This field of experimentation became known as *in-situ* electron microscopy. The monitoring of the image in real time allows us to study structural changes in the specimens and, thus, to use the electron microscope as a 'nanolaboratory' for carrying out experiments on a small spatial scale.

In-situ electron microscopy dates back to the 1960s when serious problems in materials science, for example, the fatigue of metals for applications in aviation, had to be solved. The need to design spacious experimental setups in the specimen chamber of the electron microscope resulted in the development of high-voltage instruments operating at or above 1 MV and with large gaps between the objective pole pieces. The lateral resolution of these machines was hardly below 1 nm but it was possible to introduce specimen stages with dimensions of several centimetres. Electron-transparent metal sheets have been strained and monitored at the same time so that the movement of defects, e.g. dislocations, was accessible to direct observation.[3] Dedicated stages were designed that allowed to heat specimens up to high temperatures in the microscope. In such a way, phase transformations were observable, though not on a very small scale. Imaging was in most cases carried out in diffraction contrast (bright or dark field imaging of specimens under Bragg conditions) whereas electron diffraction gave information about the crystallography during transformations of the material.

Some *in-situ* experiments were based on the accidental observation of dynamic processes in the specimen during normal inspection in standard stages. An important field has been electron irradiation of specimens which is generally unavoidable during electron microscopy. The energetic electron beams in high-voltage electron microscopes were used to generate and study radiation damage and to simulate the behavior of materials for applications in nuclear reactors or in space.

2. Modern *In-Situ* Electron Microscopy

Numerous advancements have been achieved in electron microscopy in the past decades. Not only has the spatial resolution been improved, image recording techniques have been revolutionized by the application of CCD cameras, dedicated specimen stages have been made with micromechanical tools, and analytical spectroscopies with almost atomic resolution have been integrated into standard electron microscopes. All these improvements had their impact in the development of modern *in-situ* electron microscopy. Systems with dimensions on the nanometer scale are in the focus of interest at this time and particularly well suited to experiments in the electron microscope. These experiments have also led to the discovery of new phenomena on the scale of nanoparticles or atom clusters.

Many *in-situ* experiments in the last years have been carried out with a spatial resolution of better than 0.3 nanometers. On this scale, the atom columns in well aligned crystals become visible. In some studies, the monitoring of even single heavy atoms within light materials has been achieved, though with considerable image noise in the recordings.[4–6] But it still remains a fascinating goal of *in-situ* studies to 'see the atoms moving'. An advantage of transmission electron microscopy is the parallel recording of the whole image. Scanning tip microscopies, on the other hand, need a certain time to scan the image and can 'see the atoms', but only one at a time so that dynamic processes are difficult to monitor. Dynamic processes where many atoms in a crystal lattice are involved and lattice planes change their position are ideally suited for high-resolution *in-situ* electron microscopy as will be demonstrated in the following chapters of this book.

However, some difficulties remained a challenge to the experimentalist. Lattice resolution of a crystal is only achieved when a low-indexed zone axis of a crystal is precisely aligned parallel to the electron beam. This condition is often difficult to fulfill in nanosystems that are subjected to mechanical, thermal, or electrical influence during the experiments. Another difficulty is the signal-noise ratio which is often quite high in high-resolution images that have to be taken with short exposure times. The image formation in high-resolution electron microscopy is based on phase contrast which has to be converted to amplitude contrast by the optical system to make the object details visible. The contrast in high-resolution images is generally much lower than in conventional mass-, thickness- or diffraction contrast images. Time-resolved *in-situ* electron microscopy does not allow long exposure times of single image frames, therefore the problem of image noise cannot be overcome.

The output of *in-situ* electron microscopy is generally a series of images that is either taken manually frame by frame or recorded as a real time video. Image intensifiers with attached TV cameras have been used since decades although they provide noisy images when used at high magnification of the microscope. Nowadays, CCD cameras with high sensitivity and low noise are replacing photographic films or TV cameras. Multi-scan cameras enable us to record single frames or video sequences in the electron microscope, just like in digital consumer cameras or camcorders. With the steadily increasing computing power, online image or video processing is now possible. Offline processing allows the selection of suitable frames from the videos and to extract the whole information from the recordings.

In-situ experimentation needs specially designed specimen stages that fit into the objective lens and contain the whole setup around the specimen. Due to advancements in miniaturization techniques, specimen stages for many experiments can now be made small enough to fit even into the narrow gap of objective pole pieces for high resolution microscopy. Nowadays, high-voltage microscopes with large specimen chambers are only needed for very special setups or for irradiation experiments. Specimen stages for several applications are now available commercially, for example for heating, cooling, electrical probing, straining, or indentation of the specimen. Even scanning tunneling or atomic force microscopy tips have been integrated into these specimen stages so that mechanical manipulations of the specimen can be carried out by piezo drivers with highest precision and the simultaneous imaging of the secimen by TEM and AFM resp. STM became feasible.[7,8] These stages allow *in-situ* experiments in amost every standard electron microscope. On the other hand, the columns of some microscopes have been modified for special experiments, for example, microscopy in a gas atmosphere,[9,10] crystal growth in ultra-high vacuum,[11] ion irradiation of the specimen,[12] or the application of pulsed laser beams for nanosecond microscopy.[13] Complicated setups have been attached to the columns of the microscopes that sometimes needed more space in the laboratory than the microscope itself.

The demands on the stability of the setup are particularly high in *in-situ* electron microscopy at high resolution because the specimen is not allowed to drift or vibrate by more than the desired image resolution. Moving parts of the specimen stages may cause mechanical vibrations, and local magnetic or electric fields may deteriorate the image formation in the objective lens. Another difficulty is the thermal expansion of the specimen stages which is unavoidable in heating experiments. Specially designed electronic image drift compensation systems have helped to overcome some of these problems. Furthermore, specimen preparation techniques had to be developed and adapted to the specific requirements of *in-situ* experimentation. Standard preparation techniques that are applied as a matter of routine for inspection in the electron microscope are aften not suitable for *in-situ* experiments.

3. The Techniques of *In-Situ* Electron Microscopy

A great variety of *in-situ* experiments has been carried out in the past decades. Many special setups have been designed and built in the electron microscopy labs. Only the most common types of experiments, or those where specimen stages or special setups are available commercially, will be summarized in the following. Figure 1 shows the principle of *in-situ* experiments in a schematic drawing.

The response of materials to mechanical stress was one of the first applications of *in-situ* electron microscopy.[3] The straining of specimens in specially designed stages has been carried out at ambient or high specimen temperature.[14] The nucleation, glide, or pileup of dislocations or the operation of dislocation sources has been made visible in impressive video sequences. In more recent experiments, nanoindentation of materials by tiny diamond tips have been applied to study deformation mechanisms on a small scale.[15] The integration of an AFM into the TEM specimen holder[7] allows one to measure small forces resulting from the elastic response of the specimen.

The variation of the specimen temperature by resistive heating of the holder has also been carried out since a long time. Modern heating stages allow imaging with lattice resolution at specimen temperatures up to more than 1000°C. Phase transformations such as solid-solid or solid-liquid

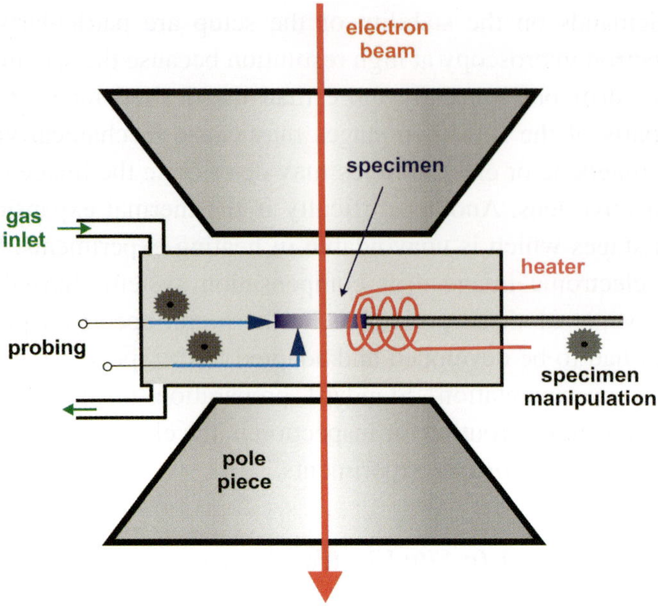

Fig. 1. Principle of *in-situ* experimentation in the transmission electron microscope. The specimen can be manipulated by heating, straining, electrical probing, or reaction in a gas atmosphere during inspection at high magnification. (In a real experiment, not all manipulations as shown here will be carried out in a single setup.) The space between the pole pieces of the objective lens limits the dimensions of the "lab inside the microscope".

transitions or chemical reactions can be studied *in-situ* by varying the specimen temperature. *In-situ* microscopy of thermal effects is particularly interesting in nanomaterials because macroscopic characterization or analytical tools do not provide the information needed for understanding transformations of clusters or particles that, themselves, have dimensions which are only accessible by techniques of high-resolution microscopy. Other applications of high-temperature microscopy are, for example, crystal growth or epitaxy in specially designed ultra-high vacuum electron miroscopes[11] or irradiation studies of materials.

The observation of chemical reactions of solids is a particular challenge of *in-situ* electron microscopy. Moving reaction fronts or the transformation of nanoparticles can be observed with lattice resolution.[16] There are several examples of solid state reactions that were initiated by heating the specimens and observed in real time. Not much detailed information exists about the course of reactions in nanoparticles, therefore *in-situ* observation of reactions promises very valuable information, for example in the technically most important field of nanoparticles in catalysis.

The study of reactions between solids and gases by applying environmental cells in the electron microscopes has been developed in the past years and resulted in a new field of electron microscopy[9,10] (see Chapters 2 and 3). Specially designed gas flow and differential pumping systems have been attached to high-resolution electron microscopes and allow the observation of chemical reactions in a gas atmosphere at low pressure.

It has already been realized in the early days of electron microscopy that the energetic electron beam may alter the structure of the specimen. This has been applied in numerous studies of radiation damage of materials,[17] for example, for applications in nuclear technology. More recently, interesting transformations of materials, in particular of nanoparticles, have been observed to occur under electron irradiation.[18] The experiments are often straightforward because the same beam is used for both imaging and modifying the specimen.

A related topic is ion irradiation of specimens which has been realized by attaching the beam tube of an ion accelerator to the specimen chamber of an electron microscope.[12,18] In such a setup, the ion beam is directed onto the specimen so that observation at high resolution is possible during ion irradiation. However, large setups outside the microscope are needed, and these experiments have only been carried out in a few specially designed microscopes. The modification of specimens by electron or ion beams has also been used for pre-defined structuring of the specimens. By using focused beams, different techniques of lithography have been developed and later applied in systems outside the microscope on a larger scale.

When an electrical current passes through a conducting specimen, the current-voltage characteristics can be measured. This has been carried out by electrically probing microscopic structures within specimens in special stages so that the relationship between structure and electrical properties could be investigated.[19,20] Furthermore, the structure and properties of the specimen material may be modified by applying an electrical current. A specimen stage has recently been made available where a STM is integrated into the TEM stage so that the advantages of both techniques can be combined. Electrical currents in the specimen can be measured with high precision through the STM tip.

The application of a magnetic field to ferromagnetic specimens causes a certain magnetisation and has been studied on a small scale by Lorentz electron microscopy or by phase contrast imaging with a biprism in the electron beam. The micromagnetic behavior of small magnetic structures, for example, hysteresis loops, magnetoresistive signals, or thermal effects have been observed while changing the local magnetic field.[21]

In-situ electron microscopy is normally limited to time scales above the minimum exposure time of an image. Experiments on much smaller time scales have been carried out by applying pulsed electron beams from photocathodes that are illuminated by pulsed laser beams. Triggering the image recording with the same pulses permits the study of dynamic processes on a time scale down to 10^{-13} seconds.[13,22,23] However, these experiments need a specially designed electron microscope and an extensive external setup.

The principles of image formation that are applied for *in-situ* experimentation are basically the same as for usual static characterization of specimens. Most chapters of this book were written with focus on lattice resolution electron microscopy in the imaging mode because this technique is rather new and has shown many spectacular phenomena on the nanoscale. However, imaging in diffraction contrast remains important for many problems of *in-situ* electron microscopy as shown in Chapter 4.

4. Limitations of *in-situ* Electron Microscopy and Future Demands

In-situ electron microscopy observes dynamic processes on a small spatial scale. Of course, the time scale of the processes may span over many orders of magnitude. Thermally activated processes depend exponentially on the temperature; for example, the diffusivity of atoms may vary by six orders of magnitude when the temperature is varied by only 300 K. Hence, it is obvious that just a narrow time window is accessible to dynamic observation. The lower limit is given by the exposure time of one video frame which is approximately 0.05 seconds. An upper time limit is normally set by the regular start-up and shut-down procedures of the microscopes which is in most labs done every day. Of course, this time could be expanded in some exceptional experiments, but a typical upper time limit of 50,000 seconds appears realistic. This gives us an accessible time scale over six orders of magnitude. However, as mentioned above, it has already been demonstrated that in a few special systems much shorter time scales can be explored by pulsed electron beams.

Another factor which limits the applicability of *in-situ* electron microscopy towards long observation times is radiation damage of the specimens. If irradiation is not the purpose of the study, it is important to minimize the electron dose on the specimen during an experiment and to work at low acceleration voltages to avoid ballistic atom displacements. Contamination of the specimen with organic molecules is another limiting factor when working close to room temperature.

The lateral resolution of *in-situ* electron microscopy depends on the specimen stage but is nowadays close to the specified resolution limit of the microscope when modern miniaturized *in-situ* stages are used. However, the space that is needed for the experiment inside the objective lens is crucial. With increasing gap width in the pole piece, the resolution of the lens decreases due to increasing spherical aberration. It is to be expected that this problem can be partly solved when aberration correctors are applied.[24]

The applicability of electron microscopy is often limited due to the concern whether the results on thin specimens are representative for bulk materials. If the behavior of macroscopic bulk materials is of interest, special care has to be taken that artefacts due to thinning or small-particle effects are avoided. However, nanoparticles which are in the focus of current interest are small systems and do not have to be thinned for electron microscopy experiments and observation.

Of course, every experimental setup has its own limits but there are some common problems that always appear. The mechanical and thermal stability of the setup has to be optimized so as to minimize vibrations or drift during observation. Improvements due to the development of new *in-situ* stages can be expected in the near future. The enormous efforts and achievements in the mechanics of tip microscopies (STM, AFM) can also be applied in TEM stages to move the specimen with almost atomic precision by piezo drivers. Everything is facilitated when such stages are available commercially because micromechanical engineering cannot be done in most laboratories of electron microscopy. Some developments are already on the market and have been used in novel experimentation techniques (see Chapter 6). The affordability of such stages is, as usual, another important point. Improvements of other *in-situ* setups are to be expected, for example in heating stages with less thermal drift or applicability at higher temperature[25] or in gas reaction cells with higher flexibility for studying different types of chemical reactions.[10]

The spatial resolution of modern TEMs has already been extended below the 1 Å level by applying aberration correctors.[24] Another advantage of aberration correction is the easier interpretation of phase contrast images when the coefficient of spherical aberration is close to zero. As an example, delocalization artefacts in the images can be eliminated. As stated above, *in-situ* electron microscopy profits from high resolution imaging at lower voltage of the electron microscope and a larger gap in the objective pole piece. Both can be achieved with correction of the spherical aberration of the objective lens. The application of aberration-corrected condensor systems will enable us to focus electron beams onto spots in the 0.1 nm range and so manipulate specimens on the atomic scale. Nevertheless, high-voltage electron microscopy will continue to have its justification (Chapters 7 and 8). Many specimen materials or special setups for *in-situ* experimentation do not allow the preparation of specimens with thickness in the 10 nm range. Due to the lower inelastic scattering of energetic electrons, high-voltage microscopy remains the only useful technique for obtaining images from thicker specimens.

For many reasons, considerable further improvements in spatial resolution of electron microscopes is not expected in the near future, and the usefulness of resolutions towards the 10 pm range in materials science is not undisputed. But even if there is not too much 'room at the bottom', there is plenty of time at the bottom, and *in-situ* electron microscopy should be considerably extended towards time scales below 0.1 seconds. Many processes on the atomic scale happen within femtoseconds, so there remain 14 orders of magnitude almost unexplored. Modern electron detectors are quite sensitive, but exposure times below 0.01 seconds appear unrealistic with the present beam current densities. Due to radiation damage of the specimens, brighter continuous beams are not desirable, but pulsed electron beams promise to open new windows. A few very exciting advances by the application of pulsed electron beams have already been reported,[13,22,23] but an *in-situ* technique that is able to cover the whole remaining time scale is far from feasible today.

5. Concept of this Book

One of the motivations for this book was the fact that *in-situ* experimentation is hardly treated in textbooks about electron microscopy. In this book recent technical developments and advancements of *in-situ* electron microscopy are presented in a collection of articles. The main focus is on transmission electron microscopy with high resolution. Although the book was edited with the intention to give a concise overview, not all aspects of *in-situ* transmission electron microscopy were treated, for example, *in-situ* microscopy of magnetic materials[21] or the application of pulsed electron beams.[13]

It was not the purpose of this introductory *Chapter 1* to provide an overview of the extensive literature about *in-situ* electron microscopy. Only reference to a few review or milestone papers is given here. Collections of papers about recent work in *in-situ* electron microscopy can be found in some special issues of journals or conference proceedings[26–28] and in the following chapters of this book.

Chapter 2 gives an overview of environmental electron microscopy as carried out in specially designed microscopes where the specimen is in a gas atmosphere under the electron beam. Chemical reactions are studied *in-situ* with lattice resolution and gas-solid or liquid-solid interactions become visible. Processes of highest technical importance, e.g. the CVD technique, can now be studied in a reaction cell inside an electron microscope. Environmental electron microscopy is meanwhile indispensable in the chemistry of nanomaterials.

In *Chapter 3* a technical alternative to the extensive setup of dedicated environmental electron microscopes is shown. Specially designed heating stages and gas nozzles for the exposure of specimens to gases allow the study of reactions in a standard electron microscope. Chemical reactions and transformations between different phases of nanomaterials, e.g. solid-solid, solid-liquid, or solid-gas reactions are investigated. Technically important reactions such as the solid-state formation of SiC from Si and C are imaged at high temperature in real time and at lattice resolution. It is also shown how heating stages can be applied to study the melting behaviour of embedded nanoparticles and nucleation phenomena at solid-liquid interfaces.

In *Chapter 4* it is shown how mechanical deformation of specimens can be carried out *in-situ* during inspection in the electron microscope. Nanoindentation of the specimen under the beam allows the study of the dynamics of dislocations and grain boundaries or superplasticity in crystalline materials. Straining of specimens at high temperature gives further insight into the plastic behaviour of crystal grains and the evolution of substructures.

Chapter 5 shows the application of *in-situ* hot stage electron microscopy to the study of interphase boundaries. The collective motion of atoms at interfaces is made visible with lattice resolution. These observations are indispensable in the understanding of phase transformations at the atomic scale. Examples for order-disorder and precipitation phenomena in metallic alloys are given.

The electrical and mechanical manipulation of specimens is presented in *Chapter 6*. By using a dedicated *in-situ* specimen holder, electrical probing experiments, e.g. of carbon nanotubes filled with ferromagnets, are carried out. Current-voltage characteristics of nanoparticles are measured with such a device. Piezo drivers in the holder also allow the mechanical deformation which is shown here in the example of bending of nanoparticles. This chapter also shows the application of cooling and heating holders for the construction of nanodevices, e.g. a thermometer on the basis of a filled nanotube.

Chapter 7 is devoted on the one hand to ion irradiation of specimens which has been realized by connecting an ion beam line to the specimen chamber of a high-voltage electron microscope. With such a setup, ion implantation processes can be studied *in-situ* with high spatial resolution. This is shown here in the example of the implantation of Xe atoms into a metal matrix which enables us to monitor the growth and behaviour of Xe crystals with lattice resolution. As a second subject, this chapter treats the fabrication of nanostructured materials by electron beam-induced deposition of metals. This is realized by the decomposition of metal-organic gases on a substrate under the electron beam in the microscope. *In-situ* observation allows the monitoring of the growth of pre-defined nanostructures with high resolution.

Chapter 8 treats the effects of electron irradiation on the specimens in the electron microscope. The displacement of atoms by knocks from the energetic electrons in the beam leads to a restructuring of the materials and, in certain systems, to new morphologies and phases of nanoparticles. The combination of heating and electron irradiation allows us to control the defect dynamics in the systems under the beam and to generate new structures by self-organization processes. Examples are shown for structures based on graphite such as carbon nanotubes.

Chapter 9 shows the detection limits of modern *in-situ* high-resolution electron microscopy. Individual point defects such as vacancies or interstitial atoms as self-interstitials or foreign atoms are observed in real time. Electron irradiation is also used here to create point defects in graphitic nanostructures.

References

1. L. Reimer, *Transmission Electron Microscopy* (Springer, Berlin, 1989).
2. D. B. Williams and C. B. Carter, *Transmission Electron Microscopy* (Plenum Press, New York, 1996).
3. E. P. Butler and K. F. Hale, *Dynamic Experiments in the Electron Microscope, in Practical Methods in Electron Microscopy*, Vol. 9, (ed.) A. M. Glauert (Elsevier, Amsterdam, 1981).
4. S. Iijima, *Optik* **48**, 193 (1977).
5. N. Tanaka, H. Kimata, and T. Kizuka, *J. Electron Microsc.* **45**, 113 (1996).
6. K. Suenaga, R. Taniguchi, T. Shimada, T. Okazaki, H. Shinohara, and S. Iijima, *Nano Lett.* **3**, 1395 (2003).
7. D. Erts, A. Lohmus, R. Lohmus, H. Olin, A. V. Prokopivny, L. Ryen, and K. Svensson, *Appl. Surf. Sci.* **188**, 460 (2002).
8. T. Kizuka, *Phys. Rev. Lett.* **81**, 4448 (1998).
9. E. D. Boyes and P. L. Gai, *Ultramicroscopy* **67**, 219 (1997).
10. T. W. Hansen, J. B. Wagner, P. L. Hansen, S. Dahl, H. Topsøe, and C. J. H. Jacobsen, *Science* **294**, 1508 (2001).
11. K. Takayanagi, K. Yagi, K. Kobayashi and G. Honjo, *J. Phys.* **E 11**, 441 (1978).
12. C. W. Allen, *Ultramicroscopy* **56**, 200 (1994).
13. O. Bostanjoglo, R. Elschner, Z. Mao, T. Zink, and M. Weingärtner, *Ultramicroscopy* **81**, 141 (2000).
14. U. Messerschmidt and M. Bartsch, *Ultramicroscopy* **56**, 163 (1994).
15. A. M. Minor, J. W. Morris Jr., and E. A. Stach, *Appl. Phys. Lett.* **79**, 1625 (2001).
16. R. Sinclair, *Mater. Res. Soc. Bull.* **19/6**, 26 (1994).

17. R. C. Birtcher, M. A. Kirk, K. Furuya, G. R. Lumpkin, and M. O. Ruault, *J. Mat. Res.* **20**, 1654 (2005).
18. F. Banhart, *Rep. Progr. Phys.* **62**, 1181 (1999).
19. Z. L. Wang, P. Poncharal, and W. A. de Heer, *J. Phys. Chem. Sol.* **61**, 1025 (2000).
20. C. M. Grimaud and O. R. Lourie, *Microsc. Microanal.* **10**, 1112 (2004).
21. J. N. Chapman and M. R. Scheinfein, *J.* Magn. Magn. *Mater.* **200**, 729 (1999).
22. W. E. King, G. H. Campbell, A. Frank, B. Reed, J. F. Schmerge, B. J. Siwick, B. C. Stuart, and P. M. Weber, *J. Appl. Phys.* **97**, 111101 (2005).
23. M. S. Grinolds, V. A. Lobastov, J. Weissenrieder, and A. H. Zewail, *Proc. Nat. Acad. Sci.* **103**, 18427 (2006).
24. M. Haider, H. Rose, S. Uhlemann, B. Kabius, and K. Urban, *J. Electron Microsc.* **47**, 395 (1998).
25. T. Kamino, T. Yaguchi, T. Sato, and T. Hashimoto, *J. Electron Microsc.* **54**, 505 (2005).
26. H. Saka (ed.), *Proc. of the Int. Symp. on In-Situ Electron Microscopy*, Nagoya, 2003, *Phil. Mag.* **84,** 25/26 (2004).
27. I. M. Robertson, M. Kirk, U. Messerschmidt, J. Yang, and R. Hull (eds.), *J. Mater. Res.* **20**, 7 (2005).
28. P. J. Ferreira, I. M. Robertson, G. Dehm, and H. Saka (eds.), *Mater. Res. Soc. Symp. Proc.* **907E** (2006).

CHAPTER 2

OBSERVATION OF DYNAMIC PROCESSES USING ENVIRONMENTAL TRANSMISSION OR SCANNING TRANSMISSION ELECTRON MICROSCOPY

Renu Sharma

LeRoy Eyring Center for Solid Sate Science, School of Materials
Arizona State University, Tempe, AZ 85287-1704, USA
renu.sharma@asu.edu

Transmission electron microscopy (TEM) is one of the preferred techniques to obtain nano scale information of morphology, structure and chemistry of nanomaterials. Such in depth characterization of reactants and products is generally enough to deduce possible reaction mechanism. However, direct observation of the dynamic processes is needed to understand the fundamental properties of the reaction. Recent developments in the instrument and holder designs have made it possible to obtain atomic level structural and spatial resolution during gas-solid, liquid-solid and liquid-liquid interactions with a time resolution approaching 1/60th of a second using environmental transmission or scanning transmission electron microscopes (ETEM and ESTEM). Such combination of time, temperature, pressure, structural and chemical resolution has made ESTEM a versatile instrument that may be considered as a nano-scale laboratory for *in situ* synthesis and characterization. This chapter starts with a brief description of various approaches for making *in situ* TEM observations in gaseous and/or liquid environments including the design of ETEM or ESTEM. Special considerations for designing ETEM or ESTEM experiments and data collection are also provided. Its applications to obtain nano-scale information of the morphological, structural and chemical changes occurring during synthesis and/or functioning of nanomaterials are described. Such information is crucial in assisting us to understand and improve the synthesis and functioning of nanoscale processes.

1. Introduction

Environmental transmission or scanning transmission electron microscope (ETEM or ESTEM) is a modified instrument that permits us to make *in situ* observations of dynamic processes under reaction conditions (at temperature under gaseous or liquid environment). Transmission electron microscopy (TEM) related techniques are usually preferred for nanoscale characterization. The combination of diffraction, atomic level imaging and elemental analysis is a powerful approach to obtain relationship and/or local variations between the morphology, structure and chemistry of nanomaterials. Moreover, modern TEM, especially aberration corrected TEM can be used to obtain atomic level structural and chemical information from individual nano-sized particles before and after synthesis. However, such pre- and post-synthesis characterization fails to provide us the information about the synthesis routes and intermediate reaction steps. This drawback has been rectified by developing techniques to make *in situ* observations of the dynamic reaction processes.

During early days, electron beam radiation was often used to initiate the structural and morphological changes. For example, coalescence of gold particles and reorganization of atomic layers within small gold particles (dancing atom) were initiated by electron beam radiation.[1] Similarly the electron beam was used to decompose praseodymium and neodymium carbonate hydrate and hydroxyl carbonates.[2,3] A combination of electron diffraction and high resolution electron microscopy images (HREM) were used to obtain an atomic level structural transformation of the decomposition process to form praseodymium and neodymium oxides. However, such TEM observations failed to provide thermodynamic data which was obtained from *in situ* x-ray diffraction and thermogravimetric analysis. It is obvious that the knowledge of reaction conditions in the TEM column is required to understand the chemistry, thermodynamics and kinetics of a reaction. A modified specimen holder and/or the TEM column enable us to observe the effects of stress, strain, electrical bias, temperature, and gaseous environment on solids during synthesis and/or during operation in a controlled manner.[4]

Currently *in situ* TEM observations are applied to understand the synthesis as well as functioning of nanomaterials and nanodevices. Selective synthesis of nanomaterials is often required to effectively integrate various nano-sized components into a functional device for nanotechnology applications. Components of such systems also require high structural and spatial resolution for characterization due to their small size. Therefore, a nano-level understanding of the relationship between the synthesis, structure and properties has become vital to improve the quality and function of nanoscale devices. We can achieve this by direct (*in situ*) observations of the dynamic processes occurring during synthesis and functioning of the individual components of a nano system. *In situ* TEM observations can also be used to obtain atomic level understanding of the effects of environment on synthesis and properties of nano materials by using a modified TEM. Although such modifications for *in situ* observations have a long history of development,[4] the advent of nanotechnology has made *in situ* TEM techniques indispensable during last decade due to following reasons.

(1) Same area or same nanoparticle is observed before, after and during the reaction therefore all of the steps, including intermediate steps (if any), are identified.
(2) A careful design of the experiments leads to observation and understanding of structural, morphological and chemical changes simultaneously.
(3) Both the thermodynamic and the kinetic data of the reaction process from individual nanometer sized particles are obtained.
(4) *In situ* observations result in considerable time savings as the synthesis, property measurement and characterization can be performed simultaneously.

Modified holders and microscopes are now commercially available which reduces time and effort involved in developing a holder design for a particular application. Developments in the design and performance of the microscopes have made it possible to obtain a host of information simultaneously. For example, high resolution TEM and STEM images along with X-ray dispersive and electron energy-loss spectra can be collected from the same area of the sample or same nanoparticle using the same instrument.[5]

The possibility to make real time dynamic observations of the changes in the structure and/or chemistry of nano materials has revitalized the interest of researchers and manufactures in the area of both the application as well as the instrumentation. Most of the modifications have been made to sample holders with specific applications in mind. Following is list of some of the holders currently available and their applications:

- **Heating holders:** Phase transformation, coalescence, sintering process etc.
- **Cooling holders:** Low temperature phase transformation, observation of beam sensitive materials etc.
- **Piezo Drives:** bonding at the nanometer level, manipulation of nano structures such as nanotubes.
- **Straining holders:** To measure the effect of stress and follow the structural changes and structural failure due to strain.
- **Nano-indentation holder:** Follow the effect of indentation at nanometer level.
- **Biasing holders:** Effect of electric field on properties, measuring properties such as emission and conductivity.
- **Liquid holders:** Observation of biological materials, electro-chemistry, solid-liquid interactions, solution chemistry etc.
- **Gas reaction holders:** Gas-solid interactions to understand various chemical interactions, synthesis of nano-materials.

These holders are often used in conjunction with the ESTEM to monitor the effect of ambient on various processes and some of them are covered in other chapters of this book. In this chapter we will discuss the design, functioning and the applications of ETEM or ESTEM for *in situ* observations of gas-solid interactions.

2. Environmental Scanning/Transmission Electron Microscope

In a TEM high energy electrons (100–1500 KeV) are used to obtain images and elemental analysis from thin (>30 nm) sections of a sample. In order to avoid scattering from the gas molecules and increase the life of the electron source, the column vacuum is kept in 10^{-6}–10^{-10} Torr

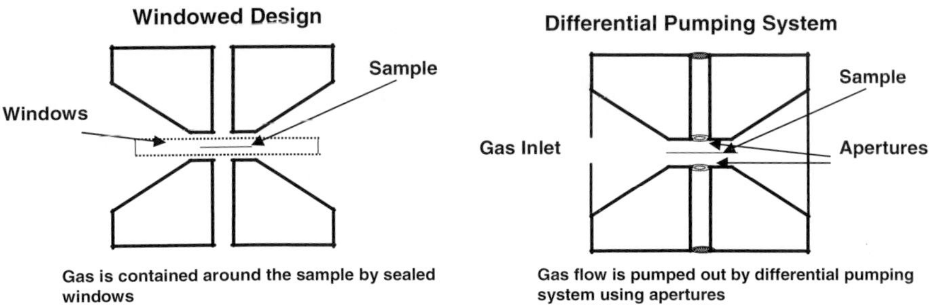

Fig. 1. Schematic showing the sample region within the objective pole piece for windowed (left) and differential pumping (right) systems. See text for details.

range. Moreover, the gun chamber must be kept at 10^{-10} Torr level when a field emission gun is used as an electron source. Therefore, in order to follow the gas-solid or gas-liquid interactions, we need to confine the gas or the liquid to the sample area and the TEM/STEM instrument with this capability is known as an environmental transmission or scanning/transmission electron microscope (ETEM or ESTEM). The modified part, used to confine gas or liquid around sample, is also known as environmental cell or E-cell. Such environment control around the specimen area in a TEM column can be achieved by (a) using vacuum sealed electron transparent windows above and below the specimen (Fig. 1(a)) and/or (b) using a combination of small apertures (Fig. 1(b)) and extra pumps (not shown in Fig. 1) in the TEM column which is also known as differential pumping system.[4–7] The first can be achieved by modifying specimen holder while the latter requires modifications to the TEM column as described below.

2.1. *Windowed cell*

In a windowed cell the sample is sandwiched between two electron transparent thin films, such as amorphous carbon, using specially modified environmental cell (EC) holders and was first employed by Marton in 1935.[6] The gas or liquid is introduced through fine tubes running inside the specimen holder rod and their volume around the sample is controlled by the height of the spacers used between the

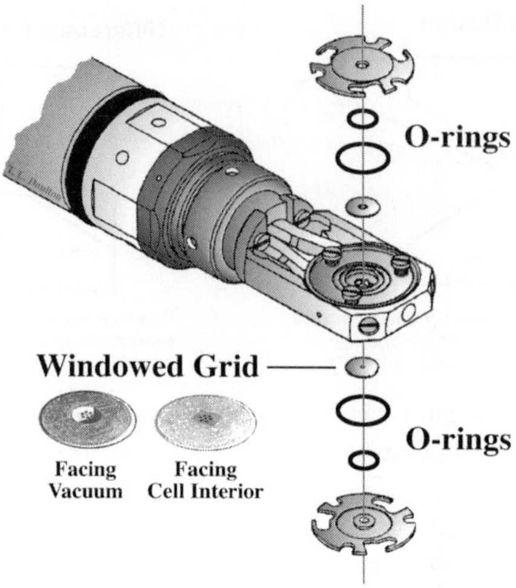

Fig. 2. Schematic illustration of the sample holder tip showing various parts of the windowed E-cell with high magnification pictures of the top and bottom grids with window grids.[7]

sample and the windows. Figure 2 shows a schematic of such a holder commercially available for JEOL microscopes.[7] Windowed grids are sealed to the cap plates on top and bottom using o-ring seals. A flow system for the gas can be maintained using a two-line EC holder while a four-line EC holder is used for flowing gas and liquid over the sample (Fig. 3). The flow rates are controlled using a computer system and a schematic diagram of the external plumbing and its relationship to the TEM column is shown in Fig. 3. Such a system is particularly useful to observe samples in a water saturated environment with variable humidity.

Windowed EC holders have also been employed for *in situ* observations of liquid-solid interactions and to observe the samples under wet or highly humid conditions at moderate imaging resolutions.[8] Image resolution of these systems is generally lower than the TEM specifications in which they are used, due to the scattering contributions from the window material as well as from the gas/liquid to the image formation. Parkinson has reportedly resolved graphite fringes (0.335 nm) by reducing the gas path length to 30 μm in an EC holder.[9] Sample heating is also a problem

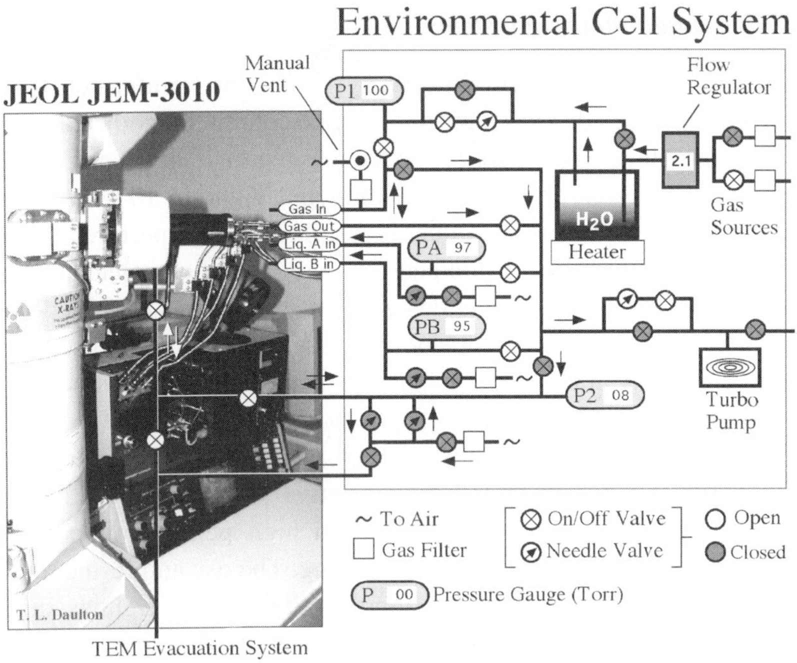

Fig. 3. A schematic of gas/liquid handling system and its relationship to the JEOL TEM column. Flow rates and pressure inside the cell is controlled by the pumping speeds, diameter of the tubes and thickness of the window. Daulton *et al.*[7] have reported that the EC system designed by JEOL is capable of achieving 0–15 L min^{-1} flow rates and water saturated gas pressure up to 200 Torr in the specimen area.[7]

due to constraints imposed by the difference in thermal expansion of windows, sample, gas and other parts of the holder. Recently, Giorgio *et al.* have employed an EC holder, with slightly different design than described above, to heat powder catalyst samples up to 350°C in H_2 and O_2 at a maximum pressure of 7.6 Torr.[10]

The advantage of the windowed design is that a holder can be built for and used in any microscope and a dedicated instrument need not be purchased with this specific application. However, EC holders have limited tilt capability due to increased thickness of the holder tip. Moreover, perfect seal of the windows is difficult to achieve as they are prone to vacuum failure, and a limited range of temperature is achievable.

2.2. *Differential pumping systems*

A differentially pumped environmental system is often preferred over the windowed design. In this case, the microscope column is modified to constrain the gaseous environment to the sample region. The basic principle is based on the design proposed by Swann and Tighe in 1972 where the gas leak rate from specimen chamber is restricted using apertures, and the area beyond the apertures is pumped out using a turbo molecular pump (TMP).[11] The advantage of this design is that regular TEM holders or modified holders such as heating, cooling, straining etc. can be used. High angle tilt capability allows us to orient crystal particles along zone axis or obtain 3D information using tomography holders.

Early modifications were incorporated in high voltage microscopes (1000–1500 KeV) due to their large objective pole-piece gap that provided space to incorporate the E-cell, and high penetration power to reduce the loss of intensity from gas scattering. The use and further development of ETEM diminished considerably during the eighties due to problems associated with high voltage microscopes. Moreover, vibrations from regular TMP affected the resolution limit making it difficult to use the modified microscope for general TEM applications.[12–15]

In the 1990s, improvements in objective pole-piece design for medium-voltage (200–500 kV) microscopes with large enough pole-piece gaps (7–9 mm), and 0.2 to 0.35 nm point-to-point resolution renewed the interest in E-cell technology.[13,16–18] Incorporating multiple pumping levels and vibration free Mag-Lev TMP made it possible to use a TEM with a field-emission gun as an ESTEM.[15] A schematic of a three level differential pumping system, used in a Tecnai F-20 ETEM, is shown in Fig. 4. It consist of two sets of apertures, marked a:a′ and b:b′ (Fig. 4), fitted in the bores of the upper and the lower objective-lens pole-pieces. The region between the two sets of apertures is pumped out by a Mag-Lev turbomolecular pump so that there is negligible gas leakage past the second set of apertures (marked b and b′ in Fig. 4). The vacuum in the electron gun area can be further improved by adding another pump after the first level of pumping. Multiple levels of pumping combined with multiple sets of apertures have been successfully used to obtain up to 50 Torr of gas pressures in the sample region while maintaining high-resolution capability.[5,12]

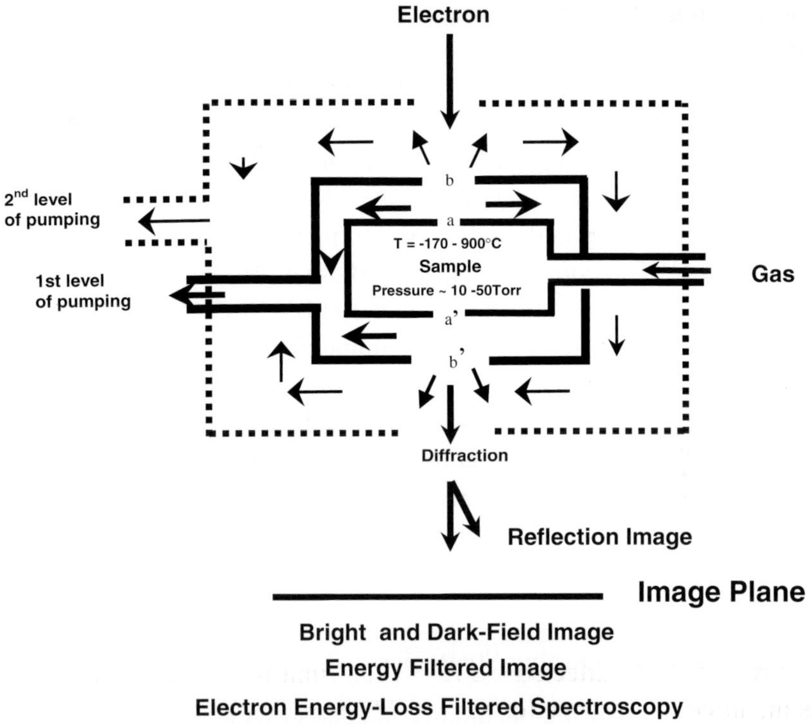

Fig. 4. Schematic of a three-stage differential pumping system that can be used to convert a TEM to an ETEM. Gas is introduced in the sample area and the leak rate into the microscope column is restricted by a set of small apertures (e.g. 100 μm), placed above and below the sample. Gas leaked through these apertures is pumped out using a Mag-Lev turbo-molecular pump (1st level of pumping). The space between the condenser aperture and viewing chamber is pumped using a molecular drag pump (MDP) (2nd level of pumping). The region between condenser aperture and gun chamber is pumped by an ion pump (3rd level of pumping).

The analytical capability in this configuration is limited to EELS, since the EDS detector is not compatible due the possible contamination of its window by the gaseous environment. Moreover, as the lower differential aperture blocks high-angle diffracted beams, HAADF images can not be obtained using the current ETEM. The nanoprobe size of a FEG microscope makes it possible to obtain high resolution STEM images and analyze nanoscale regions. In the future, a redesign of the differential pumping system can allow us to use large differential pumping aperture to obtain HAADF images.

The gas pressure inside the cell is limited by the choice of aperture size, pumping speed, and distance between apertures. These parameters have been discussed in detail by several researchers who have been involved in designing and modifying the microscope in the seventies and the nineties.[4,13] The combined effect of these efforts has made it possible to obtain 0.14 nm lattice resolution in a 200 KV FEG (Tecnai F20), better than 1 eV energy resolution and it is commercially available.[15] Hitachi has recently combined differential pumping with a modified heating holder design equipped with a needle dozer to introduce gas in the sample region to image gas-solid interaction at high temperature.[19] In the future, aberration corrected ESTEM will be readily available and will further improve structural and spatial resolution as well as energy resolution for spectroscopy applications.

3. Experimental Planning Strategies

In situ TEM is a versatile technique and can provide us answers to questions that could not be answered otherwise. But in order to fully exploit the capabilities, a carefully thought out plan of action is crucial. First and foremost, it must be emphasized that ESTEM is not just a characterization tool but an experimental set-up, where we perform the experiments in a TEM column, and instead of making *ex situ* characterization of reactants and products after the fact, each experimental step is characterized *in situ*. Therefore choice of experimental conditions, specimen holder, grid material, and microscope environment must be considered before performing an experiment in the ESTEM. **A wrong choice can not only destroy that particular experiment but even the microscope.**

Following are some of the important parameter to take in to account:

- **Sample preparation:** Sample must be prepared according to the geometry of the specimen holder to be used.
- **Choice of grid material:** If the sample is to be loaded on a TEM grid, the grid materials should not (a) react with your sample or specimen

holder material under the conditions to be employed for *in situ* observations, for example the Tamman temperature (half the melting temperature in K) of the grid material should be higher than the reaction temperature to be used.

- **Choice of heating holder:** The heating holder not only controls the temperature limit that can be achieved; its stability directly affects our ability to obtain atomic level structural or spectroscopic information. Also, the heating holders should not react with the gas to be used for reaction, for example Ta or W holders will be corroded in oxygen rich ambient (used to study oxidation reactions). Moreover, the composition of the area directly in contact with sample (furnace body) should be compatible with the grid material and there should be no reaction between the sample, grid and the furnace body at the conditions to be used for observations.
- **Choice of gas:** For a windowed cell, gas temperature, pressure and gasses must comply with the requirements specified by the manufacturer. For example, the gas should not react with the window, holder, and gas inlet and outlet tubes. In case of differential pumping apertures, the gas (to be used) should not react with the materials used inside the microscope column or leave harmful residue in the gas delivery lines or inside the microscope column.
- **Choice of data collection system:** The choice of data collection system, e.g. CCD, film, or video (tape or digital) depends on the reaction rates. Currently the fastest rate for data collection is using digital video (1/30 s). Some CCD cameras can record data at the rate of 15 frames/s for 256×256 images. On the other hand, chemical information and STEM images can only be acquired at a much lower time resolution. This is an area where major improvement is highly desirable.

Gas reaction systems of ETEM or ESTEM have an added requirement of keeping the external and internal plumbing clean after each use to avoid cross contamination of the gasses. For example, presence of water in the system will increase the oxygen partial pressure in the column and thereby affect the reduction experiments. Similarly, carbonaceous vapor may leave residual hydrocarbons that could be a source of contamination for the next user.

5. Data Collection

For most of the *in situ* observations data is collected using regular imaging and spectroscopy detectors i.e. film, CCD or TV rate camera for imaging, CCD for EELS and EFTEM and annular bright and dark-field detectors for STEM imaging. Choice of recording media is governed by the temporal resolution required to follow various reaction steps. For example STEM imaging is not suitable to follow a process with reaction rate of less than one second. On the other hand, slow processes, such oxidation and reduction reaction can be followed using analytical methods such electron energy-loss spectroscopy.

Some of the reaction processes require fast detectors in order to reduce the effect of long and multiple exposures to the same area in order to avoid radiation effects to the sample. The time resolution for TV rate (30 fps) video cameras, currently being used to record bright field images in TEM mode, may not be enough to record fast reaction process such as nucleation events or spinoidal decomposition. On the other hand a fast detection system will require high beam intensity. For example, the intensity required to record images is directly proportional to the speed. Commonly employed recording media in NTSC format with 640×480 image size and 8 bit depth requires a beam intensity of 1.4 nA for 30 fps recording speed. If the speed is increased to 100 fps, we will need to increase the intensity to 5 nA in order to maintain the image quality. Moreover, the amount of data recorded per second will increase from 6 MB s^{-1} to 20.1 MB s^{-1}. Therefore data storage and data handling becomes a tedious process and we need to plan properly for it. Fortunately, the computer disc space and speed has become quite cheap and with careful planning the data handling and data reduction problem can be solved.

6. Applications

ETEM and lately ESTEM has been successfully employed to understand a number of reaction process in the area of synthesis and functioning of catalysts as well as for nanomaterials such as oxidation,[12,20] reduction,[21–23] nitridration,[24,25] polymerization,[26] chemical vapor deposition,[27] electron beam induced deposition,[28,29] hydroxylation,[30] dehydroxylation,[31] sintering[32,33] etc. Most of these achievements have also been summarized in various review articles, books and book chapters.[4,5,12,34] Some selective examples are given here.

6.1. *Nanoscale characterization during synthesis*

In principle it is not practical to use ESTEM for synthesis. But the capability to make *in situ* observations of dynamic process is immensely valuable to understand the synthesis process and effect of various synthesis parameters on structure and morphology of the products. Such information can be applied for selective synthesis of nanostructured materials and thereby improve their quality. Moreover, *in situ* observations can also reveal intermediate phases that can later be synthesized using *ex situ* techniques in large quantities. For example, Sayaguès *et al.* observed formation of a new type of structure as defects during oxidation of niobium tungsten oxide $(NbW)_{12}O_{32}$).[20] Careful structural analyses revealed the chemical composition of the defect structure to be $Nb_7W_{10}O_{47}$. The chemical composition and reaction conditions obtained from *in situ* observations were successfully used to synthesize a single phase stable compound with the same structure.[35]

Another example is the decomposition process of vanadyl hydrogen phosphate hydrate (VHPO) to prepare vanadyl pyrophosphate (VPO), an important commercial catalyst. In this case *in situ* observations of the morphological and structural transformation associated with the decomposition of VHPO have provided the insight necessary for choosing appropriate temperature regions for selective catalysis.[17] Recently, the morphological changes observed, *in situ*, during the polymerization reaction of propylene over Ziegler-Natta catalyst have been used to obtain the growth rates and the growth mechanism of polypropylene.[26,36]

ESTEM can also be employed to follow the nucleation and growth mechanisms of nanoparticles as well as the effect of environment and support on their morphology. Such information can be related to the properties and thereby designing experimental conditions for selective synthesis of nanomaterials with desired properties. Some recent examples are described in the following sections.

6.1.1. *Effect of the environment on nanoparticle morphology*

Nanostructured materials such as catalyst nanoparticles are often synthesized by precipitating their salts, decomposing them to metal oxide, followed by reduction to metals. The decomposition process and reducing environment can affect the size and morphology of the resulting nanoparticles.

For example, model catalysts were prepared by impregnating Cu-acetate on ZnO or silica followed by reduction of CuO to Cu. Reduction was performed at 220°C under three different reducing environments using a CM300 ETEM.[37] *In situ* observations of the process revealed that the shape of the particles depended on the reducing environment. For example, the particles were observed to be faceted with their (111) planes in contact with the support in H_2 atmosphere (Figs. 5(a) and (b)). Addition of water to H_2 (mild reducing environment) resulted in more round shaped particles terminated by (110) and (100) planes but the contact area with

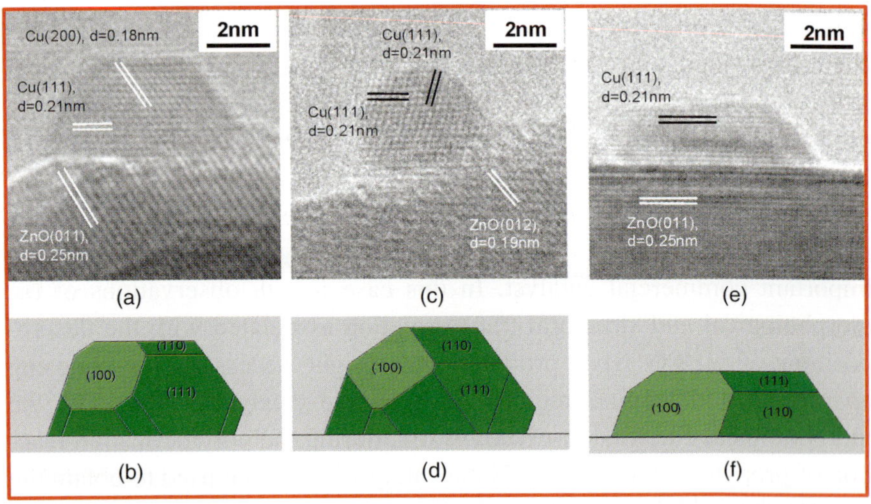

Fig. 5. *In situ* TEM images (a, c and e) of a Cu/ZnO catalyst in various gas environments together with the corresponding Wuff construction of the Cu nanocrystals (b, d, and f). (a) The image was recorded at a pressue of 1.5 mbar of H_2 at 220°C. The electron beam was parallel to [011] zone axis of Cu. (c) obtained in a gas mixture of H_2 and H_2O, H_2:H_20 = 3;1 at a total pressure of 1.5 mbar at 220°C. (e) Obtained in a gas mixture of H_2 (95%) and CO (5%) at a total pressure of 5 mbar at 220°C. (Hansen *et al.* Science, 2002, reproduced with permission.)[22]

the support remained unchanged (Figs. 5(c) and (d)). On the other hand a more reducing environment obtained by replacing H_2 by CO, had a more pronounced effect on the morphology and increased the wetting area to the support (Figs. 5(e) and (f)). A detailed analysis of the structure and morphology was used to obtain surface energies and work of adhesion. These observations were directly related to the activity of catalysts prepared under various reducing environments.

Such studies provide direct information of the micro-kinetics and activity of various surfaces for selective catalytic processes.

6.1.2. *Effect of support on nanoparticle morphology*

As mentioned in the previous section, reduction is often the final step for the synthesis of nano structured metal particles. *In situ* observations of the process can help us understand the nucleation and growth mechanisms, the growth morphology, the role of support, the choice of precursor etc. Dynamic observation of the nucleation and growth of Ni nanoparticles were made by heating $Ni(NO_3)_2 \cdot 6H_2O$ in 0.2 Torr of CO at 350°C on two types of TiO_2 supports (rutile and anatase).[38–40]

The HREM image of the sample, before heating, shows the spacing of 0.35 nm of precursor structure on the TiO_2 support (Fig. 6(a)). Most of Ni particles were observed to nucleate from the precursor particles present on the surface of the titania support. Ni metal particles had non-wetting morphology on the anatase TiO_2 (Fig. 6(b)) but nucleated with wetting morphology on rutile TiO_2 (Fig. 7). Moreover, rutile TiO_2 was observed to migrate and grow by layer-by-layer mechanism on the surface of Ni metal particles.

Thus *in situ* observations were not only able to reveal the nucleation behavior of nanoparticles but also show that the wetting and non wetting morphology may not always depend upon reducing environment as observed for Cu/ZnO but the structure of the support (anatase versus rutile) can also affect the wetting behavior.

6.1.3. *Nanoparticle synthesis by de-hydroxylation*

Metal oxides, such as copper, iron and magnesium oxides, with high surface area can be synthesized by decomposing their hydroxides.

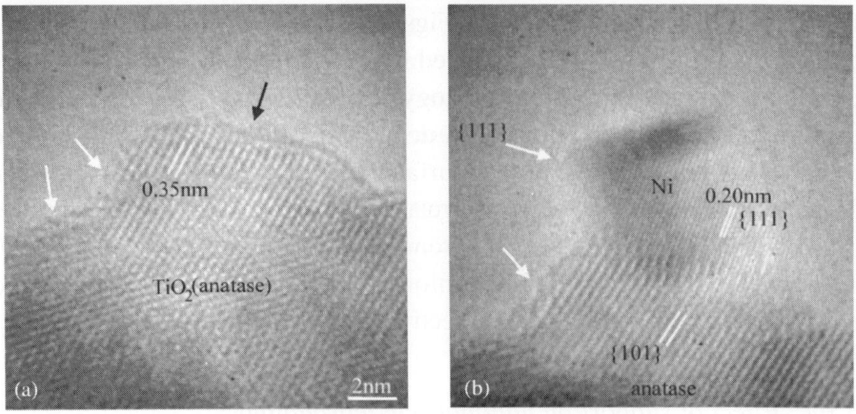

Fig. 6. Nucleation of Ni particles during reduction of precursor. (a) Before reduction, precursor particles are spread on the TiO_2 (anatase) support that (b) converts in to Ni particle after 20 minutes of heating in H_2 at 350°C. Note that the particle changes from wetting to non-wetting morphology upon reduction.

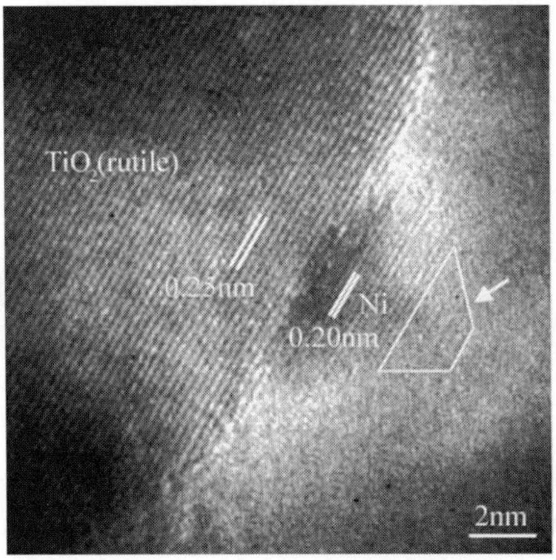

Fig. 7. Ni particles formed on TiO_2 (rutile) support under similar reduction conditions as shown in Fig. 6. Note the wetting morphology and rutile (101) surface.

Magnesium oxide is specifically of interest as it is used as catalyst support as well as is the key component for mineral sequestration of CO_2 to reduce the green house problem. Brucite is a naturally occurring mineral form of magnesium hydroxide ($Mg(OH)_2$ and decomposes easily in the vacuum of a regular TEM. Therefore HREM images of pristine mineral could not be obtained using a TEM.

The overpressure of water vapor around the sample in an ETEM (Phillips EM430) not only made it possible to image the layered structure of hydroxide crystal fragments (Fig. 8(a)) but the de-hydroxylation mechanism could also be followed by reducing the water vapor pressure in a controlled manner (Fig. 8(b)–(d)).[30,31] The structure of $Mg(OH)_2$ can be described as MgO layers intercalated by water molecules. Therefore it was hypothesized that removal of water will collapse the layers to form the oxide structure. But de-hydroxylation was observed to proceed by nucleation and growth of nanoscale oxide particles instead of layer by layer compression

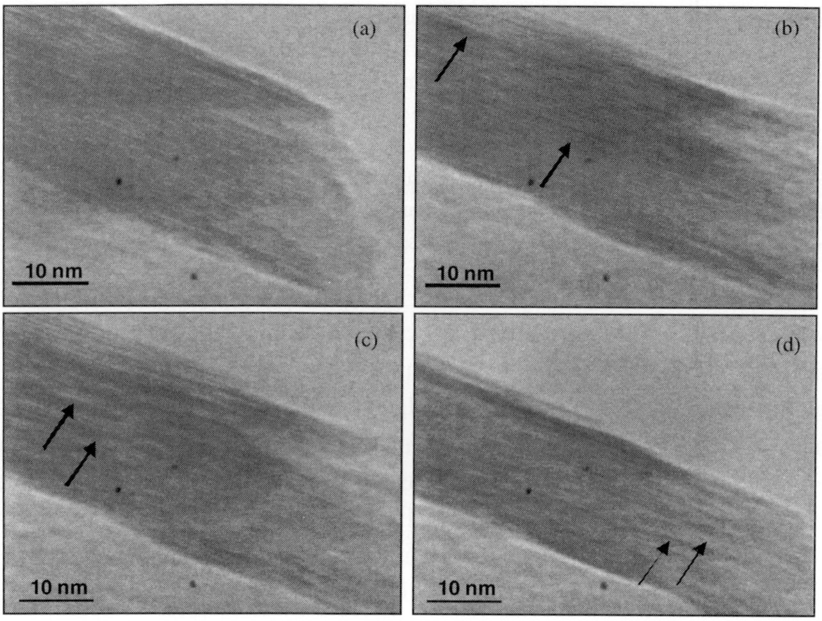

Fig. 8. Time resolved sequence of digitized images showing the nucleation and growth of lamellar oxide/oxy-hydroxide regions as the water vapor pressure in the E-TEM was reduced from 1 to 0 Torr; (a) a brucite crystal fragment prior to dehydroxylation via electron beam heating and (b–d) dehydroxylation via the formation of oxide/oxyhydroxide regions (e.g. select regions marked by arrows). Note: the dramatic one-dimensional shrinking of the fragment in the direction perpendicular to the hydroxide layers. (Sharma *et al.*)[30]

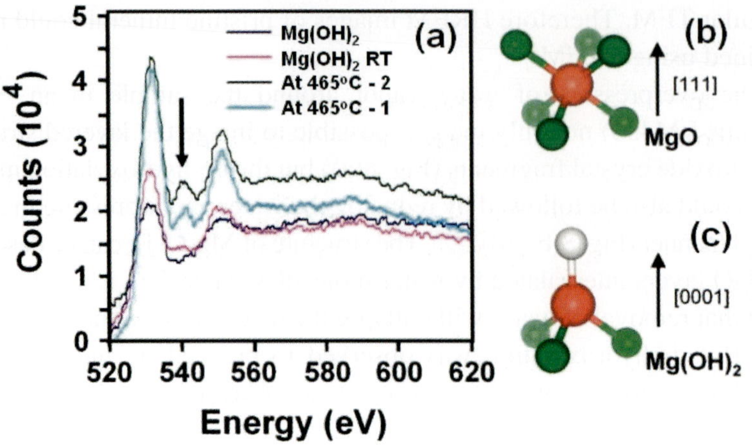

Fig. 9. (a) Oxygen *K*-edge EELS spectra for $Mg(OH)_2$ standard brucite at room temperature (RT; dark blue curve), after 20 minutes of observation (pink curve) and from two different areas after *in situ* annealing in vacuum at 465°C. Note the appearance of a central peak in the ambient temperature spectra after 20 minutes (marked by arrow at 542 eV) that became more prominent at 465°C. The vertical scale (intensity) is arbitrarily shifted for the sake of clarity; (b) schematic illustrating octahedral coordination in MgO; the arrow illustrates the ⟨111⟩ direction relative to this bonding geometry; (c) schematic showing the four-coordinated oxygen site in brucite and the associated ⟨0001⟩ direction. (Bearat *et al.*)[42]

as expected from general structural considerations. The resolution of this microscope was not enough to obtain HREM images of MgO (0.24 lattice resolution), but change in the near-edge structure of O-K edge and appearance of a 3rd peak due to change in the nearest neighbor configuration (Fig. 9)[41] during de-hydroxylation was used to identify MgO nanoparticles.

Such nucleation and growth resulted in fracturing large single crystals of brucite into nano-sized MgO particles with high surface area and high reactivity. Therefore the de-hydroxylated material was found to be more reactive than MgO powder and reacted with CO_2 at room temperature to form the mineral carbonate.[42]

6.1.3. *Chemical vapor deposition (CVD)*

An ETEM or ESTEM can also be used as a cold wall CVD reactor and permits us not only to observe the nucleation and growth of nanoparticles but also to obtain thermodynamic and kinetic data.[27] The CVD process involves the deposition of solid species on a substrate by the dissociation of the vapors of organic or organometallic precursors. The energy needed to decompose or dissociate the precursor molecules can be provided either by the electron beam or by thermal heating of the substrate. High-temperature CVD may or may not require an active catalyst as substrate. Deposited materials nucleate and grow to form nano-particles, thin films, nanowires or nanotubes.

Electron beam assisted CVD

In principle this technique is similar to using UV or laser beam for patterning surfaces. When the electron beam is used for dissociation of precursor molecules, it is also called electron beam induced deposition (EBID) or electron beam lithography.

The effect of the electron beam on the nucleation and growth of gold particles on Si (100) surfaces has been evaluated.[43] Total coverage of the surface by gold particles was measured during deposition as a function of time at different temperatures. Deposition was also performed without electron beam irradiation and time-resolved images were recorded after evacuating the precursor from the microscope column. The rate of gold deposition obtained with electron beam irradiation was higher than the rate without the electron beam but the difference reduced as a function of temperature. This result provided an important conclusion that electron-beam-induced decomposition of the precursor is more prominent at room temperature than at high temperature. Therefore, electron-beam-induced decomposition (EBID) can be used for electron lithography at room temperature while pure thermal decomposition of the precursors can be obtained at high temperatures.

There has been great interest in using EBID techniques to fabricate periodic arrays of nanostructures for various applications. Until recently scanning electron microscope (SEM) and focused ion beam (FIB) have been employed to fabricate patterns for semiconductor applications. But the nanoprobe formation capability of a FEG-ESTEM has provided an opportunity to use nanolithography to form and immediately characterize nanostructures. In principle, the beam (electron probe) size should dictate the diameter of the deposited structure, but in practice it is larger than the electron probe size due to secondary ionization effects.

Nanosized structures can be formed by introducing the precursor vapors into the microscope column with the electron gun-valve closed, and then high-energy electrons are used to locally decompose and deposit the material. This technique has been used to deposit an array of GaN dots on SiO_x using highly reactive hydride, D_2GaN_3 that dissociated exothermally under electron irradiation forming stoichiometric GaN.[28] These arrays are small enough to manifest true quantum effects and are likely to possess unique electronic and optical properties. The smallest size reported is 1 nm for arrays deposited using tungsten carbonyl as precursor.[29]

Similar nanometer-sized arrays, lines or boxes can be generated for other materials using appropriate precursors.[44]

Thermally assisted CVD

As mentioned in the previous section, CVD can also be performed by decomposing precursor molecules adsorbed on a heated substrate. The process consists of four steps; (1) the precursor molecules adsorb on the surface, (2) they decompose to form a solid and gaseous species, (3) the solid phase nucleates and (4) grows to form nanostructures with various morphologies. *In situ* observations provide us the knowledge of the intermediate steps of the process. For example, specially modified UHV microscopes, with low pressure gas introduction capability, have been used to understand the nucleation and growth process of Ge islands on clean surfaces.[45] The observed Ostwald ripening was found to be dependent on the size and the shape of the islands. These observations have provided a fundamental understanding of the role of surfaces in the growth process.

Another approach to CVD is to deposit material on active sites such as catalyst surfaces via thermal decomposition. This procedure is routinely used to synthesize nanowires and nanotubes for nanotechnology applications. For example, nanowires are natural candidates for a number of applications such as optoelectronics and sensors. A modified UHV TEM has been used to understand the effect of environment on the nucleation and growth mechanism of Si nanowires. Gold (Au) nanoparticles are generally used as catalyst and silane (SiH_4) or disilane (Si_2H_6) is used as the Si source.[46,47] It has been proposed that Si is dissolved in Au at high temperatures (above 600°C) and is released after nanoprticles are saturated, nucleates and grows at the tail end of the catalyst particle via vapor-liquid-solid (VLS) growth mechanism. *In situ* observations reveal that after a period of growth Au particles start to shrink and disappear, and the growth of the nanowire stops at this point. But introduction of low amounts of oxygen nucleates the Au particles and the growth of nanowires resumes.[47]

Figure 10 shows time resolved images extracted from a video sequence of the growth of Si nanowires. The nanowire under observation grew steadily in a mixture of disilane and oxygen up to 1.5 μm in length and 40 nm in diameter with a gold droplet at the top. The droplet volume of Au particle and hence wire diameter decreased very slowly over several hours of growth, perhaps because of very slow Au diffusion, but at

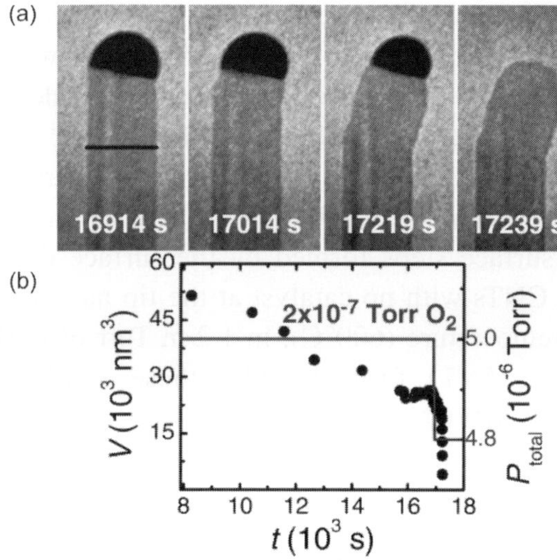

Fig. 10. (a) Series of images extracted from a video sequence showing the effect of reducing the oxygen pressure. Wire growth was carried out for 282 min at 600°C using 1 √←10^{-6} Torr disilane and 2 √←10^{-7} Torr O_2, after which the oxygen was switched off but the disilane pressure remained steady. The time since growth began is shown in each image. The scale bar is 40 nm. (b) Volume *V* of the droplet as a function of time *t*. (Kodambaka *et al.*)[47]

$t = 16914$ s, when the oxygen pressure was decreased, the Au particle volume started to shrink rapidly and the Au droplet disappeared within 325 s. It appears that the gold diffusion away from the growing tip is responsible for stopping the growth of nanowires and oxygen can be used to reduce the rate of gold diffusion and increase the growth time and thereby length of the nanowire.[47]

Similarly the nucleation and growth of carbon nanotubes (CNT) has been investigated using modified UHV TEM or ESTEM.[48–50] CNTs, especially single-walled carbon nanotubes (SWCNTs) have interesting electronic and mechanical properties making them one of the most sought after nanomaterials. But full potential of their applications has not been realized as a number of issues, concerning synthesis of CNTs with desired structure and morphology, and the role of the catalyst in their synthesis are still unresolved. Baker and co-workers were the first to report dynamic observations of the growth of graphitic fibers at elevated temperature during the decomposition of acetylene on Ni catalyst.[51,52] Growth of carbon filaments on various transition-metal catalysts remained a central theme of their research for several years.[53] This group also worked on understanding the influence of gaseous environment on a wide range of transition metal catalysts (Cu, Fe, Co, Ni, Cr, V, Mo etc.).[51,54]

Recently, Helveg *et al.* have shown that the Ni particles are mobile during the carbon nanofibre growth and graphene sheets nucleate and grow from the surface steps formed by the surface diffusion of Ni.[48] Almost straight CNTs with no catalyst at the tip have been reported to form at higher temperature (600°C), in 1.2 m Torr of C_2H_2.[56] Although mostly SWCNTs (66%) formed under these synthesis conditions, growth of straight MWCNT was also observed (Fig. 11). As an example, a time resolved digital image sequence extracted from a video recorded during the growth of a multiwall CNT is shown in Fig. 11. Such sequences have been used to obtain linear growth rates and were often observed to be discontinuous. This tube grew with an average linear growth rate of 4 nm/s. HREM images confirmed that this tube had a large diameter (12.2 nm) and 9 walls. Therefore, the slow growth rate can be explained, in part, in terms of the rate of arrival of carbon atoms compared with the amount of carbon required to construct a unit length of the nanotube.[56]

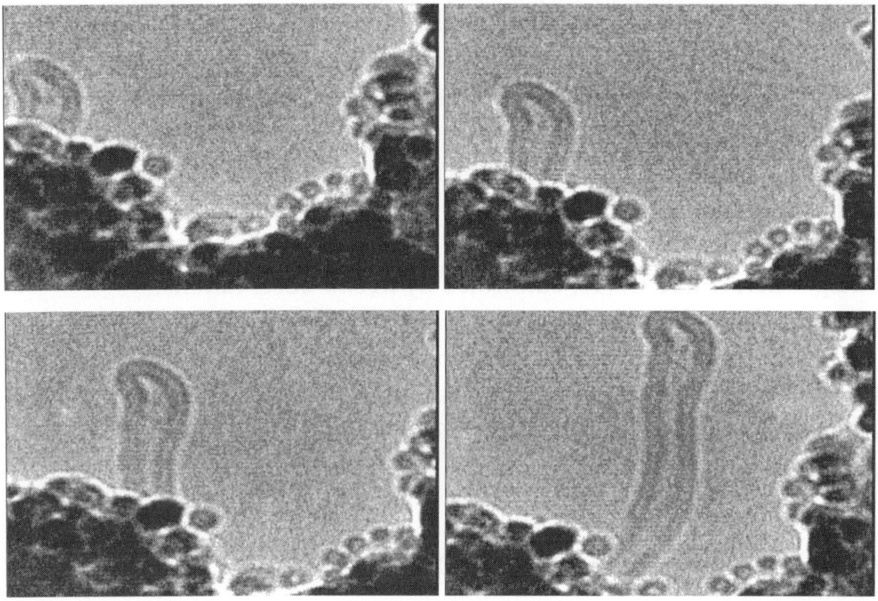

Fig. 11. Individual frames digitized from a video sequence recorded at 600°C and 1.2 m Torr acetylene pressure showing the growth of multi-walled CNT. Individual frames shown here were extracted after (a) 0.03, (b) 1.67, (c) 3.33 and (d) 14.8 seconds from the start of the video recording. (Sharma *et al.*)[56]

Effect of synthesis conditions, such as temperature and pressure, on growth rates, structure and morphology of CNTs formed, has also been investigated using ESTEM.[55,56] *In situ* observations of the CNT growth reveal that both the number of walls and the diameter of the tubes decreased with increasing temperature and decreasing precursor pressure (carbon flux). Although SWCNTs were observed to form at temperatures as low as 480°C, their yield increased to 90% at 650°C at low precursor pressure. The kinetic measurements show that a very small fraction of C atoms available are used to form CNTs indicating that high precursor pressure is not required for SWCNT synthesis.[56]

Most interestingly the catalyst particles are observed to change shape[48] and such movements may be responsible for providing new steps for nucleation of new graphene layers.[57] Preliminary *in situ* observations also show that catalyst particles often changed their form and size to fit inside the diameter of the tube.[58] The general morphology of the CNTs was also observed to depend upon the synthesis conditions. Bent and zigzag multi-walled nanotubes were observed to form at low temperature and/or high pressure, most probably due to formation of 5- and 7-ring defects. It is easy to perceive that high nucleation and growth rates will reduce the time and/or energy required for annealing such defects. On the other hand high temperature and low C flux (pressure) will encourage the formation of straight defect free CNTs, confirmed by *in situ* observations.[55]

Time, temperature and pressure resolved observations of the CNT growth process have revealed that the morphology and the diameter can be controlled by synthesis conditions for a given catalyst/support and precursor. Such information can be used for selective synthesis of CNTs for device fabrication.

6.2. *Effect of environment on catalytic activity*

Catalysts by nature are nano structured materials and function at high temperature under gas or liquid environments. Although the exact role of a catalyst is not completely understood, they are responsible for lowering the activation energy of a chemical reaction. Catalysts generally do not get consumed during the reaction process but their activity often decreases with time. *In situ* observations of the catalyst under reaction conditions can aid us to comprehend their functioning and deactivation process. ETEM has played a crucial role in helping us understand a number of catalytic processes over the years. Therefore, ETEM/ESTEM has been used to characterize the nanoscale behavior of various catalytic processes. Baker *et al.*[51–54] and Gai *et al.*[12,17,34] have pioneered both the application and instrumentation aspect of catalysis and ETEM. Some of the recent examples from other research groups are described below:

Regeneration of hydrogenation catalyst

Pd/Al_2O_3 is industrial catalyst, widely used for hydrogenation of unsaturated hydrocarbons. But its activity deteriorates after extended periods of use due to carbon build up from the dissociation of hydrocarbons (a process that under selective conditions is used to form CNTs). The carbon build up can be oxidized by heating the used catalyst in steam and in principle the catalyst should regain its original activity. Unfortunately, the activity of the regenerated catalyst is always lower than fresh catalyst. Liu *et al.* have compared the sintering behavior of fresh catalyst with used catalyst in oxidizing environment (air and steam) from *in situ* obseravtions.[32,33]

Negligible sintering was observed for fresh catalyst particles up to 700°C while used catalyst particles started to sinter at 350°C. Low temperature sintering was found to be due to high surface mobility of catalyst particles during regeneration and the particle coalesce as they come in contact with each other to form larger particles (Fig. 12). The difference in the mobility of the catalyst particles in these two samples can be explained on the basis of metal support interactions. Fresh catalyst particles have strong metal support interactions and therefore are not mobile. But as they are pushed out of the support by the coke deposits on the surface during the hydrogenation process, the metal support interactions are weakened after removal of the coke layer. As a result the particles become mobile and the physical contact with other particles results in coalescence, reducing the surface area and the activity of the catalyst.

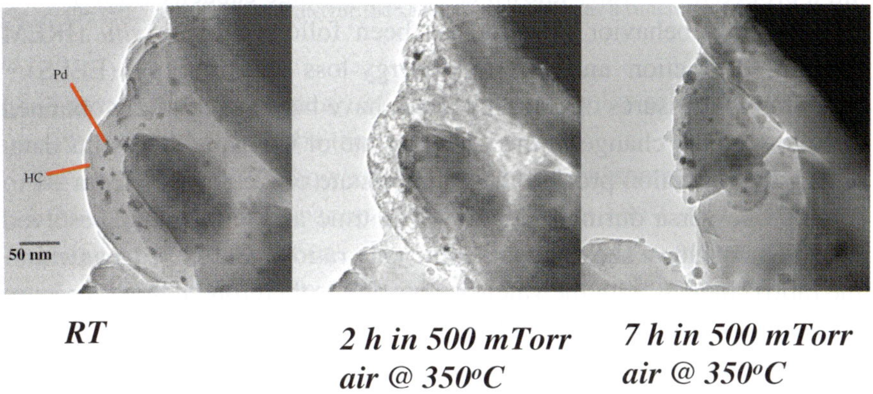

Fig. 12. TEM images showing the regeneration process of Pd/Al2O3 catalyst. Used catalyst as received at RT (left) shows Pd particles on hydrocarbon (HC) instead od Al2O3. After heating in 500 mTorr of air at 350°C for 2 hours HC region becomes patchy as Pd particles become mobile (center) and after 7 hours the HC is burnt off but Pd particles have increased in size due to sintering (right). (Liu *et al.*)[33]

concluded that the nanoparticles which have not oxidized back to +4 valence state after redox cycle become inactive. This is most probably the reason for a drop in catalytic activity after each redox cycle.[63]

The white-line ratio for most of the transition metal atoms is a function of their valance state. Therefore similar experiments can be used to understand the redox behavior for transition metal oxides.

6.2. *Effect of humidity on aerosol particles*

Aerosol particles in the atmosphere are responsible for radiation shielding effects by absorbing or scattering sunlight and also act as the nuclei for cloud formation. These particles are salts with various composition (e.g. NaCl, NaBr, $(NH_4)_2SO_4$, KBr etc.) and can vary in size and shape. They can reversibly absorb and desorb moisture from the air with changing relative humidity. Recently, the effect of relative humidity (RH) on various salts has been investigated using ESTEM.[64] Figure 14 shows the change in the shape and size of the particle as relative humidity in TEM column was increased (deliquescence) and then decreased (effervescence). It is interesting to note that the particle resumed its original shape and size as the relative humidity dropped to about 13%.

Similar studies were also performed on the aerosol particles collected from the young smoke of flaming and smoldering fires during SAFARI-2000, a comprehensive air quality campaign in southern Africa.[65] The aerosol particles collected could be divided in six representative carbonaceous particle categories such as soot, tar balls, and heterogeneously internally mixed particles containing C with S-, K-, Mg- or Na rich inorganic phases. It was found that the soot and tar balls did not take up water, whereas the mixed organic–inorganic particles took up water between 55 and 100% RH. The inorganic phase appeared to determine the hygroscopic properties of all mixed organic–inorganic particles with exact value of RH depending on the composition of their water-soluble phases. Thus, incorporation of inorganic plant material or reactions with inorganic atmospheric components can dramatically alter the hygroscopic properties of carbonaceous particles in smoke plumes. The fraction of these mixed organic–inorganic particles plausibly increases with time, which will modulate the effects of smoke on radiative budgets.

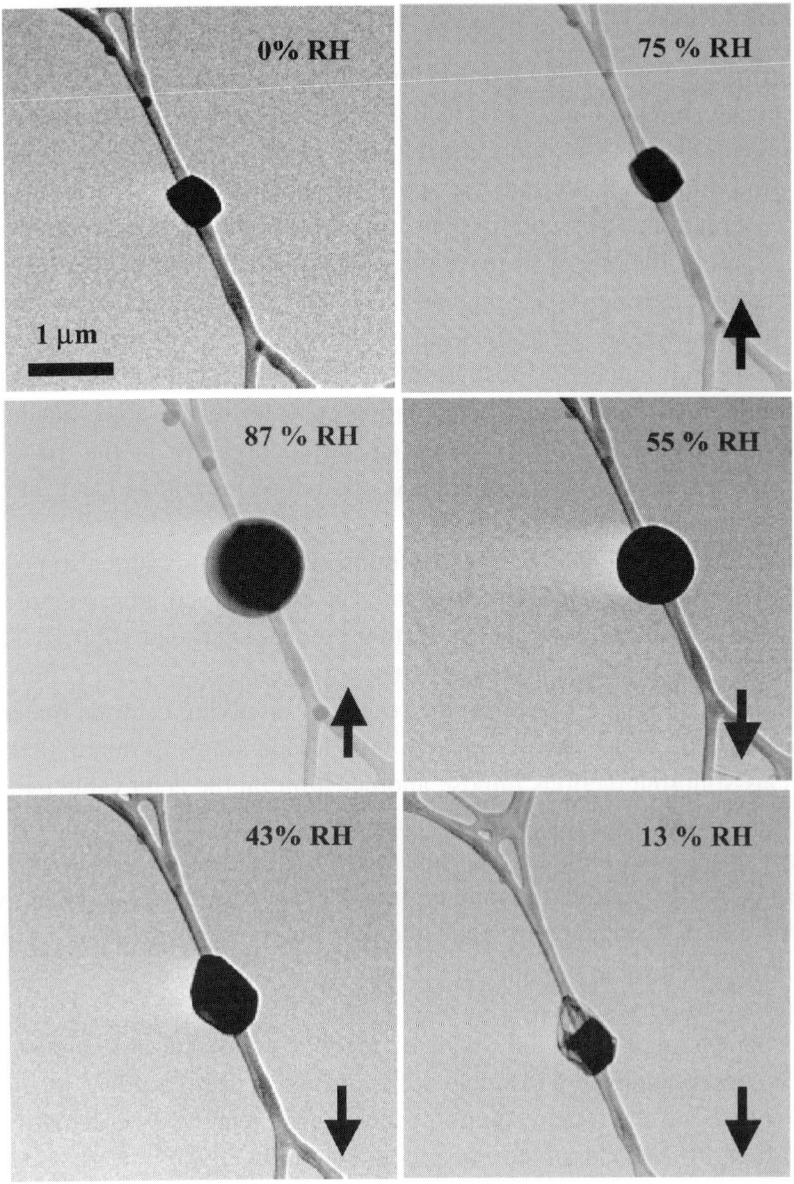

Fig. 14. Images of a NaCl particle as the RH is increased from 0 to 87% and then decreased from 87 to 13% at ~279 K. The up arrows indicate increasing water vapor pressure, and the down arrows indicate decreasing water vapor pressure in the environmental cell of the ETEM. (Wise *et al.*)[64]

7. Limitations

Unfortunately, as for any other technique *in situ* ESTEM has limitations and disadvantages. One of the main disadvantages is the effect of the electron beam that is also a concern for most of other TEM techniques. It is very important to make sure that the electron beam is not interfering with our observation, especially for collecting kinetic and thermodynamic data. For example, the rate of hydroxylation of MgO changes by electron beam irradiation.[66] Similarly, decomposition of many metal organic compounds can be achieved by electron beam radiation, a property that is used for electron beam induced deposition of metals. On the other hand CeO_2 can be reduced by strong electron radiation at room temperature but not at temperatures above 500°C.[49] Oxygen partial pressure in the 10^{-6} Torr vacuum of the microscope column is enough to re-oxidize CeO_2 at high temperatures.

The limited range of achievable temperature and pressure also poses a limitation on the type of processes that can be observed. For example the upper temperature limit of most heating holders is around 1000°C. There are some holders (Saka's design)[67] that can be used up to 1500°C but only for powder samples. Therefore the reactions involving ceramic materials can not be studied as they happen at high temperatures (around 2000°C).

Gas-solid interactions impose another limit on the achievable temperature as high temperatures can damage the internal parts, such as o-rings, of the differential pumping aperture assembly. Moreover, increased electrical power is required to attain high temperature as thermal conductivity of the gasses is a source of a considerable amount of heat loss. Similarly currently achievable pressure in the microscope for dynamic observations is limited to 50 Torr.

There is also a limited range of reaction rates that are suitable for *in situ* observations. For example if the reaction happens too fast, it can not be recorded as the time resolution is only 1/30s. On the other hand if it is too slow, it is not possible to keep the conditions in the microscope constant for long periods of time. Therefore, changes happening within a few seconds to a few hours are most suitable for *in situ* TEM observations.

For most of the heating holders, the thermocouple is attached to the body of the furnace that is in direct contact with the grid. Therefore the temperature measured by the thermocouple may not be the temperature of the nanoparticle under observation. The temperature variation depends upon the thermal conductivity of the grid material as well as that of the sample and/or sample support. For gas-solid interactions, it also depends upon the thermal conductivity of the gas being used. It is therefore necessary to calibrate the thermocouple using melting points of metals and alloys if knowledge of exact reaction temperature is required for thermodynamic calculations.

Another challenge is data reduction and data processing. We obtain an enormous amount of data from each set of *in situ* observations and handling of this data could be Herculean task. All of the data must be analyzed manually and is a tedious and lengthy process. Computer programs for automated measurements could solve some of the problem.

Conclusions

ESTEM can be a very powerful technique to obtain atomic-level information of the gas-solid or liquid-solid interactions at elevated temperatures. A combination of HREM images, electron diffraction and chemical analysis can be used to establish relationships between synthesis, morphology, structure and chemistry of nanomaterials. There is also a time advantage as synthesis and characterization can be performed simultaneously. Moreover, intermediate steps or metastable phases formed during the reaction can be easily identified.

ESTEM has been successfully used to understand oxidation, reduction, nitridation, de-hydroxylation, hydroxylation, polymerization, and sintering processes. Moreover, *in situ* observations of the CVD process have been used to understand and optimize synthesis mechanisms for nanosized structures such as quantum dots, nanowires or nanotubes.

In situ observations using ESTEM are made from full set of experiments performed within the TEM column and should be planned very carefully. The possible reaction of the TEM grid material with gas and sample, or sample with grid material at the temperature or gas with part of E-cell or TEM column should be considered. As always, the effect of the electron beam should be evaluated by making observations in the areas not previously irradiated.

Acknowledgments

I am thankful to Dr. Peter Crozier, Prof. Jeff Drucker, Prof. Ray Carpenter, Prof. David Smith, Dr. Michael McKelvy, Prof Marija Gajdardziska, and Mr. Karl Weiss for their support and advice. The ESTEM is a part of John Cowley Center for High Resolution Electron Microscopy within LeRoy Eyring Center for Solid State Science. The support from Department of Energy (DE-FG0395TE00068 and DE-AC36-99GO10331) and National Science Foundation (DMR-0210023, CBET-0625340, CBET 0612940 and CBET 0306688) is gratefully acknowledged.

References

1. D. J. Smith, A. K. Petford-Long, L. R. Wallenberg and J.-O, and Bovin, *Science* **233**, 872 (1986).
2. H. Hinode, R. Sharma, and L. Eyring, *Journal of Solid State Chemistry* **84**, 102 (1990).
3. R. Sharma, H. Hinode, and L. Eyring, *Journal of Solid State Chemistry* **92**, 401 (1991).
4. P. Butler and K. Hale, *In Situ Gas-Solid Reactions, Practical Methods in Electron Microscopy*, Experimental Microscopy (North Holland Co., 1981), pp. 239 and 309.
5. R. Sharma and P. A. Crozier, *Transmission Electron Microscopy for nanotechnology* N. Y. Z. L. Wang (ed.) (Springer-Verlag and Tsinghua University Press, 2005), pp. 531–565.
6. L. Marton, *Nature* **133**, 911 (1935).
7. T. L. L. Daulton, B. J. Lowe, and J. Jones-Meehan, *Microscopy and Microanalysis* **7**, 470 (2001).
8. J. W. Kim, Y. Furukawa, T. L. Daulton, D. Lavoie, and S. W. Newell, *Clays and Clay Minerals* **51**, 382 (2003).
9. G. M. Parkinson, *Institute of Physics Conference Series* **119**, 151 (1991).
10. S. Giorgio, S. S. Jao, S. Nitsche, D. Chaudanson, G. Sitja, and C. R. Henry, *Ultramicroscopy* **106**, 503 (2006).
11. P. R. Swann and N. J. Tighe, paper presented at the *Proc. 5th Eur. Reg. Cong. Electron Microscopy* (1972).
12. P. L. Gai and E. D. Boyes, *In Situ Microscopy in Materials Research* P. L. Gai (ed.) (Kluwer Academic Publishers, 1997) pp. 123–146.
13. I. M. Robertson and D. Teter, *Microscopy Research & Technique* **42**, 260 (1998).
14. R. Sharma, *Microscopy and Microanalysis* **7**, 494 (2001).
15. R. Sharma, *Journal of Materials Research* **20**, 1695 (2005).
16. R. C. Doole, G. M. Parkinson, and J. M. Stead, *Institute of Physics Conference Series* **119**, 157 (1991).
17. P. L. Gai and K. Kourtakis, *Science* **267**, 661 (1995).
18. R. Sharma and K. Weiss, *Microscopy Research & Technique* **42**, 270 (1998).
19. T. Kamino, T. Yaguchi, M. Konno, A. Watabe, T. Marukawa, T. Mima, K. Kuroda, H. Saka, S. Arai, H. Makino, Y. Suzuki, and K. Kishita, *Journal of Electron Microscopy* **54**, 497 (2005).

20. M. J. Sayagués and J. L. Hutchison, *Journal of Solid State Chemistry* **146**, 202 (1999).
21. P. A. Crozier, V. P. Oleshko, A. D. Weswood, and R. D. Cantrell, *Institute of Physics Conference Series* **168**, 393 (2001).
22. T. W. Hansen, J. B. Wagner, P. L. Hansen, S. Dahl, H. Topsoe, and J. H. Jacobsen, *Science* **294**, 1508 (2001).
23. R. Sharma, P. A. Crozier, Z. C. Kang, and L. Eyring, *Philosophical Magzine* **84**, 2731 (2004).
24. Z. Atzmon, R. Sharma, S. W. Russell, and J. W. Mayer, *Proceedings of Materials Research Society Symposium* **337**, 619 (1994).
25. R. Sharma, E. Schweda, and D. Naedele, *Chemistry of Materials* **13**, 4014 (2001).
26. V. P. Oleshko, P. A. Crozier, R. D. Cantrell, and A. D. Westwood, *Journal of Electron Microscopy* **51**, S27 (2002).
27. J. Drucker, R. Sharma, J. Kouvetakis, and K. Weiss, *Journal of Appied Physics* **77**, 2846 (1995).
28. P. A. Crozier, J. Tolle, J. Kouvetakis, and C. Ritter, *Applied Physics Letters* **84**, 3441 (2004).
29. W. F. van Dorp, B. van Someren, W. Cornelis, P. Kruit, and P. A. Crozier, *Nano Letters* **5**, 1303 (2005).
30. R. Sharma, M. J. McKelvy, H. Béarat, A. V. G. Chizmeshya, and R. W. Carpenter, *Philosophical Magzine* **84**, 2711 (2004).
31. M. J. McKelvy, R. Sharma, A. V. G. Chizmeshya, R. W. Carpenter, and K. Streib, *Chemistry of Materials* **13**, 921 (2001).
32. R.-J. Liu, P. A. Crozier, C. M. Smith, D. A. Hucul, J. Blackson, and G. Salaita, *Microscopy and Microanalanalysis* **10**, 77 (2004).
33. R.-J. Liu, P. A. Crozier, C. M. Smith, D. A. Hucul, J. Blackson, and G. Salaita, *Applied Catalysis* **A282**, 111 (2005).
34. P. L. Gai and E. D. Boyes, *Electron Microscopy of Heterogeneous Catalysis. Series in Microscopy and Materials Science* (Institute of Physics Publishing, Bristol, Philadelphia, 2003).
35. M. J. Sayagués and J. L. Hutchison, *Journal of Solid State Chemistry* **143**, 33 (1999).
36. V. P. Oleshko, P. A. Crozier, R. D. Cantrell, and A. D. Westwood, *Studies in Surface Science and Catalysis* **130**, 935 (2000).
37. P. L. Hansen, J. B. Wagner, S. Helveg, B. S. Calusen, and H. Topsoe, *Science* **295**, 2053 (2002).
38. P. Li, J. Y. Liu, N. Nag, and P. A. Crozier, *Journal of Physical Chemistry B* **109**, 13883 (2005).
39. P. Li, J. Y. Liu, N. Nag, and P. A. Crozier, *Surface Science* **600**, 693 (2006).
40. P. Li, J. Liu, N. Nag, and P. A. Crozier, *Applied Catalysis A* **307**, 212 (2006).
41. P. Rez, J. Bruley, P. Brohan, M. Payne, and L. A. J. Garvie, *Ultramicroscopy* **59**, 159 (1995).
42. H. Bearat, M. J. McKelvy, A. V. G. Chizmeshya, R. Sharma, and R. W. Carpenter, *Journal of the American Ceramic Society* **85**, 742 (2002).
43. J. Drucker, R. Sharma, J. Kouvetaki, and K. Weiss, *Proceedings of Materials Research Society Symposium* **404**, 75 (1996).
44. M. Shimojo, M. Takeguchi, and K. Furuya, *Nanothecnology* **17**, 3637 (2006).

45. F. M. Ross, M. Kammler, M. C. Reuter, and R. Hull, *Philosophical Magzine* **84**, 2687 (2004).
46. M. J. Williamson, R. M. Tromp, P. M. Vereecken, R. Hull, and F. M. Ross, *Nature Materials* **2**, 532 (2003).
47. S. Kodambaka, J. B. Hannon, R. M. Tromp, and F. R. Ross, *Nano Letters* **6**, 1292 (2006).
48. S. Helveg, C. Lopez-Cartes, J. Sehested, P. L. Hansen, B. S. Clausen, J. R. Rostrup-Nielsen, F. Abild-Pedersen, and J. Norskov, *Nature* **427**, 426 (2004).
49. R. Sharma and I. Zafar, *Applied Physics Letters* **84**, 990 (2004).
50. M. Lin *et al.*, *Nano Letters* **6**, 449 (2006).
51. R. T. K. Baker, M. A. Barber, P. S. Harris, F. S. Feates, and R. J. Waite, *Journal of Catalysis* **26**, 51 (1972).
52. R. T. K. Baker, P. S. Harris, R. B. Thomas, and R. J. Waite, *Journal of Catalysis* **30**, 86 (1973).
53. R. T. K. Baker, *Carbon* **34**, 715 (1986).
54. R. T. K. Baker, J. J. Chludzinski, Jr., N. S. Dudash, and A. J. Simoens, *Carbon* **21**, 463 (1983).
55. R. Sharma, P. Rez, M. M. J. Treacy, and S. J. Stuart, *Journal of Electron Microscopy* **54**, 231 (2005).
56. R. Sharma, P. Rez, M. Brown, G. Du, and M. M. J. Treacy, *Nanothecnology* **18**, 125602 (2007).
57. F. Abild-Pedersen, J. K. NørskovJens, R. Rostrup-Nielsen, J. Sehested, and S. Helveg, *Physical Review B* **73**, 115419 (2006).
58. S. Hofmann, R. Sharma, C. Ducati, G. Du, C. Mattevi, C. Cepek, M. Cantoro, S. Pisana, A. Parvez, F. Cervantes-Sodi, A. C. Ferrari, R. Dunin-Borkowski, S. Lizzit, L. Petaccia, A. Goldoni, and J. Robertson, *Nano Letters* **7**, 602 (2007).
59. P. L. Gai and E. D. Boyes, *Catalysis Review. Science and Engineering* **34**, 1 (1992).
60. P. Fornasiero, G. Balducci, R. Di Monte, J. Kaspar, V. Sergo, G. Gubitosa, A. Ferrero, and M. Graziani, *Journal of Catalysis* **164**, 173 (1996).
61. P. Fornasiero, G. Balducci, J. Kaspar, S. Meriani, R. Di Monte, and M. Graziani, *Catalysis Today* **29**, 47 (1996).
62. R.J. Gorte, *AICHE Journal* **51**, 2377 (2005).
63. P.A.C. Ruigang, R. Wang, and J.B.A. Sharma, *Journal of Physical Chemistry B* **110**, 18278 (2006).
64. M.E. Wise, G. Biskos, S.T. Martin, L.M. Russell, and P.R. Buseck, *Aerosol Scienec and Technology* **39**, 849 (2005).
65. T.A. Semeniuk, M.E. Wise, S.T. Martin, L.M. Russell, and P.R. Buseck, *Journal of Atmospheric Chemistry* **56**, 259 (2007).
66. M. Gajdardziska-Josifovska and R. Sharma, *Microscopy and Microanalysis* **11**, 524 (2005).
67. T. Kamino and H. Saka, *Microscopy, Microanalanalysis and Miccrostructure* **4**, 127 (1993).

CHAPTER 3

IN-SITU HIGH-RESOLUTION OBSERVATION OF SOLID-SOLID, SOLID-LIQUID AND SOLID-GAS REACTIONS

Hiroyasu Saka

Department of Quantum Engineering, Nagoya University
Nagoya 464-8603, Japan
saka@nagoya-u.jp

The achievements of *in-situ* HREM observation carried out over the past decade by the author's group are reviewed. The subjects include solid-solid reaction (formation of SiC via solid-state reaction between Si and C, formation of void in SiC during sintering, vibration of a grain boundary and an interface), melting of metals with small dimensions, solid-liquid interfaces, wetting of non-metallic substrates with liquid metals and solid-gas reaction (oxidation of Si and catalyst). Analytical methods and holographic methods were also applied.

1. Introduction

In-situ observation in a transmission electron microscope (TEM) has rendered a powerful tool to characterize materials and material processing. Indeed, the first *in-situ* experiment can be traced back to 1956, when Hirsch, Horne and Whelan[1] first succeeded in observing dislocations motion. The achievements obtained in 1970s have been reviewed by Butler,[2] Imura and Saka.[3] In the late 1970s, the weak-beam technique[4] was applied to the *in-situ* experiments.[5–8] Furthermore, over the past decade or two, high resolution techniques that allow observations of lattice images have been applied to *in-situ* experiments and information obtained by the *in-situ* experiments has increased drastically. This is particularly true for the *in-situ* heating experiment. This chapter deals with *in-situ* heating experiments, carried out by the author and his colleagues, using high-resolution technique, including near-atomic resolutions, as well as analytical techniques such as EDX and EELS. The content of this chapter is essentially based on the reviews[9–13] by the author and his group.

2. Specimen-Heating Holders

Needless to say, in order to carry out an *in-situ* heating experiment specimen-heating holders are indispensable. The requirements for the heating holders can be summarized as follows:

(1) The maximum achievable temperature;
(2) Thermal stability;
(3) Temperature measurement;
(4) Easy operation.

The heating holders developed hitherto can be classified into two categories. One has an indirect heater and the other a direct heater. In the former, a miniature furnace is installed into the specimen holder and the specimen heated indirectly. The advantage of this type is that the temperature of the furnace (and hence the specimen) can be measured precisely with a thermocouple. Also, the conventional 3mmφ specimen can be used. On the other hand, the disadvantage of this type is that the maximum temperature is limited to a rather low temperature range. To operate at a high temperature range it is often necessary to cool the holder with circulating water, which certainly makes the operation much more difficult.

In the direct type a fine wire or a mesh is directly heated by direct electric current. The specimens are mounted directly onto the heater. Thus, the geometry of the specimens is limited to either powder or flake; the conventional 3mmφ specimen cannot be used. Furthermore, it is usually impossible to measure the temperature of the heater (and hence specimens) directly. Temperature is to be estimated from a calibration curve of temperature versus current, prepared beforehand. However, the largest advantages of this type are the high maximum temperature and the high thermal stability.

One example of such a direct-type heating holder is Kamino holder which was developed by Kamino and Saka.[14] Figure 1(a) shows a schematic diagram of the Kamino holder. A fine filament made of tungsten (W) of 25 μm in diameter, which is bridged across two electrodes, is heated by direct electric current from a battery. It is of vital necessity to use a dry battery as an electric source to obtain thermal stability.

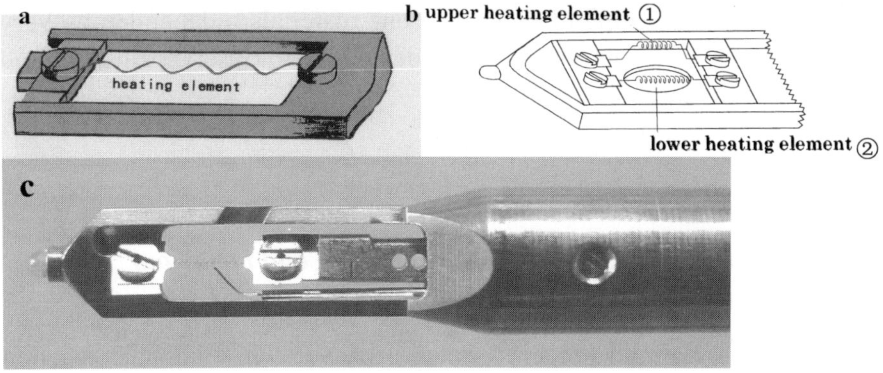

Fig. 1. Kamino holders (a) One-wire type, (b) Two-wire type, (c) Gas-injection type.

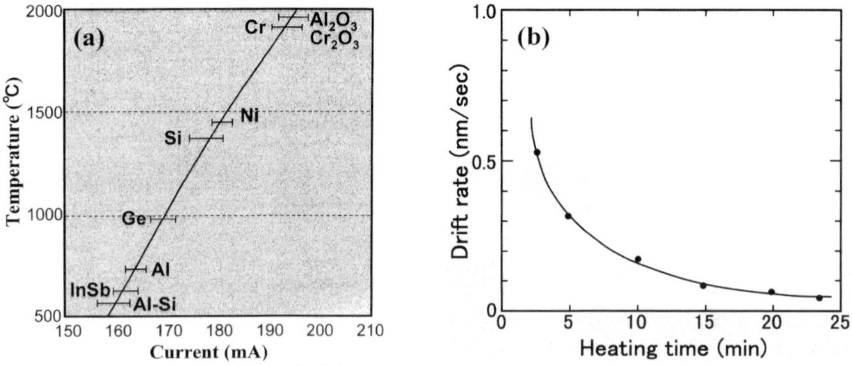

Fig. 2. (a) Temperature versus electric current, (b) drift rate as a function of time.

Temperature is estimated in two manners, that is, either by using an optical pyrometer outside a TEM or observing *in situ* in a TEM melting of known materials as a function of the electric current. An example is shown in Fig. 2(a). Temperature as high as 2000°C can be achieved with an electric current as low as 195 mA. This small thermal mass ensures the thermal stability of the heater, leading to a very small drift rate as shown in Fig. 2(b). After 15 min, the drift rate becomes as small as 0.1 nm/s.

Following this prototype, a variety of versions of the Kamino holder have been developed. Two-wire or three wire type has been developed,[15] which facilitates heating of more than one materials independently. Very recently a gas-injection type has been developed to observe solid-gas and liquid-gas reactions.[16]

3. Solid-Solid Reactions

3.1. *Formation of SiC via solid-state reaction and behaviour of grain boundary in SiC*

By using the Kamino holder, formation of SiC *via* solid-state reaction between Si and graphite was successfully observed.[10,17] Mixtures of particles of Si and graphite were mounted on the heating wire of Kamino holder, and then heated at 1400°C. Initially Si particle was single crystalline (Fig. 3(a)): Graphite was poly-crystalline, as can be seen in Fig. 3(e).

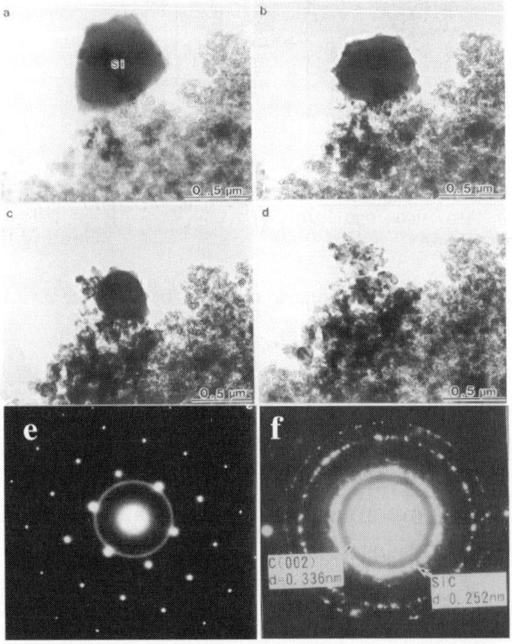

Fig. 3. A sequence of formation of SiC via reaction between Si and graphite (a–d). (e) Diffraction pattern from graphite and Si before reaction. (f) Diffraction pattern of SiC.

On heating, Si shrank while keeping the crystalline form with Si atoms penetrating into graphite (Fig. 3(b),(c)) and eventually disappeared. At the same time, the contrast of that region of graphite which had lain just underneath the Si particle darkened (Fig. 3(d)). The diffraction pattern taken from this darkened region showed definitely that this region became now SiC (Fig. 3(f)).

Figure 4 reproduces a sequence of HREM micrographs showing the process of formation of SiC. In Fig. 4(a) the Si particle lay at the left bottom corner of the micrograph, and partially had reacted with graphite to

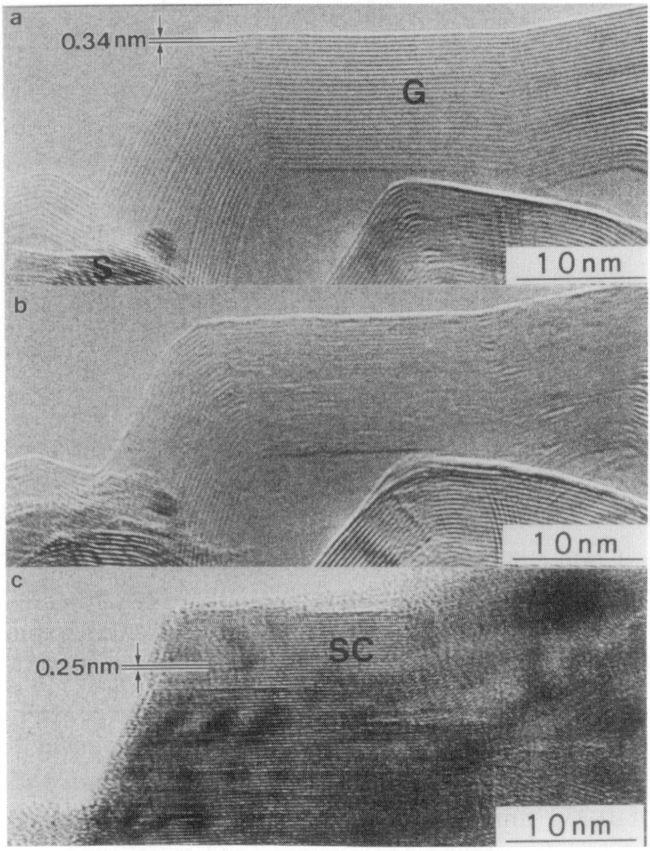

Fig. 4. HREM showing the process by which SiC is formed via solid-state reaction between Si and graphite at 1400°C. These micrographs were taken on photographic films.

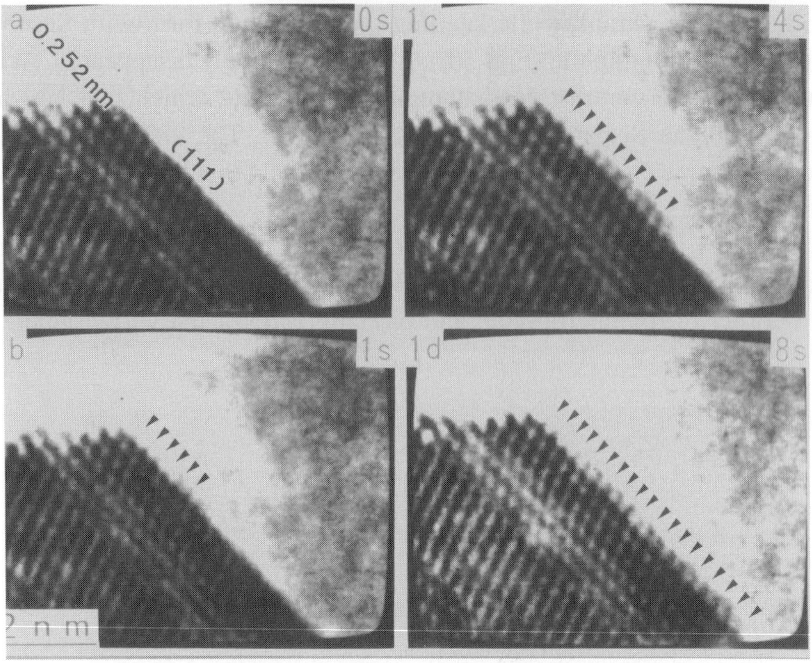

Fig. 5. A sequence of growth of a SiC crystal at 1500°C.

form SiC. However, most of the graphite remained unchanged. In Fig. 4(c), which was taken 2 minutes after Fig. 4(a), almost the whole region of the graphite reacted with Si, and SiC crystals were formed. Lattice fringes with a spacing of 0.252 nm of cubic β-SiC are evident. Figure 4(b), which was taken 1 minute after Fig. 4(a), is one example of the stage between these two extremes. The lattice fringes of the graphite became very faint, suggesting penetration of Si into the graphite lattice.

On further heating at 1500°C, SiC continued to grow.[16] Figure 5 shows an example of dynamic observation of a sequence of crystal growth at 1500°C. The crystal is viewed along the <110> direction and the flat surface corresponds to the (111) plane of cubic-SiC. The lattice fringes with spacing of 0.252 nm parallel to the surface are evident. Black dots appeared at the edge of the surface (arrows in Fig. 5(b)). The number and intensity of the contrast of the dots increased with time (Fig. 5(c–d)).

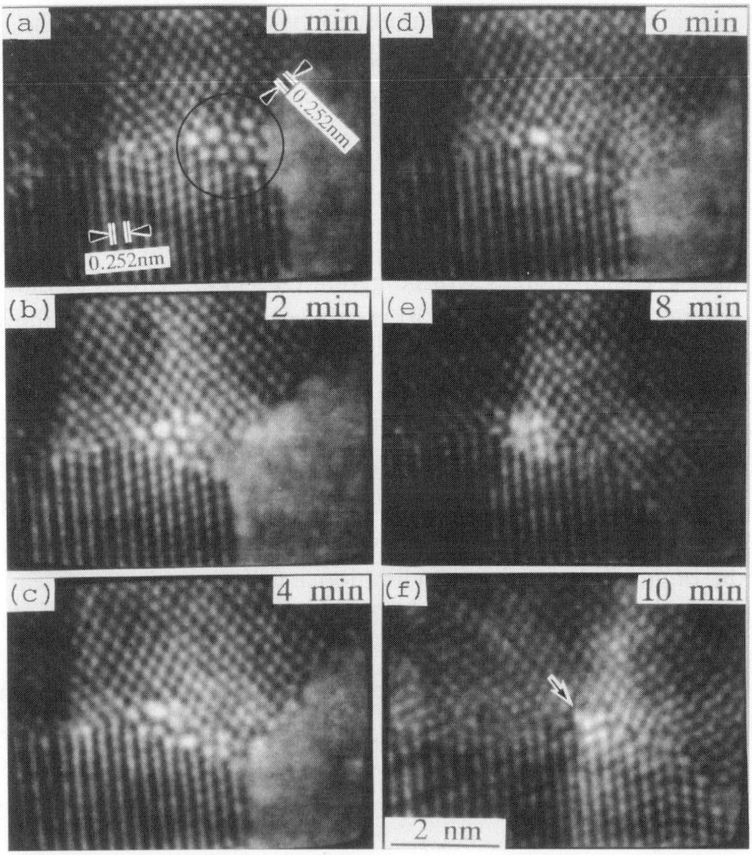

Fig. 6. Formation of voids in a grain boundary in SiC during sintering.

From the size, contrast and behavior during dynamic observation, the dot is considered to be a column of a pair of Si and C atoms, that is, a SiC molecule. The growth of the single monolayer was completed in 8 seconds in this case (Fig. 5 (d)).

Figure 6 shows an example of dynamic observation of the formation of a grain boundary during sintering. Two grains were growing together from left to right. At the boundary, the (111) planes of the lower grain faced to the (111) planes of the upper grain at an angle of ~140°. Near the triple point between the grain boundary and the vacuum, an electron-transparent network appeared (Fig. 6(a)). While the two grains continued to grow, this network was surmounted by the matrices of the grains and deformed, and eventually, a few voids were left behind in the grain boundary (indicated by arrow in Fig. 6(a)).

3.2. *Vibration of a grain boundary and an interface*

A grain boundary and an interface can be very unstable under some circumstances. Two examples will be given. One is the vibratory motion of a grain boundary under electron irradiation,[19,20] and the other is the vibratory motion of an interface between Si and SiO_2 during reduction of SiO_2 to Si.[21]

When an intermetallic compound $CuGa_2$ (tetragonal) was observed in TEM at an accelerating voltage ranging from 200 to 1000 kV, a grain boundary (GB) vibrates around its equilibrium position. Figure 7 shows an example.[22] The GB repeated to-and-fro motion around its equilibrium position. Figure 8 shows the GB vibration at a high resolution. Clear (001) lattice fringes are observed in both of the upper and lower grains. The GB is clean and shows no evidence of a grain boundary phase.

Effects of the accelerating voltage on the GB vibration were examined. The frequency and the amplitude of the vibration depended on the accelerating voltage and the electron flux. For instance, at a flux of electron beam of 2×10^{23} e/m^{2}s, the frequency decreased with decreasing the

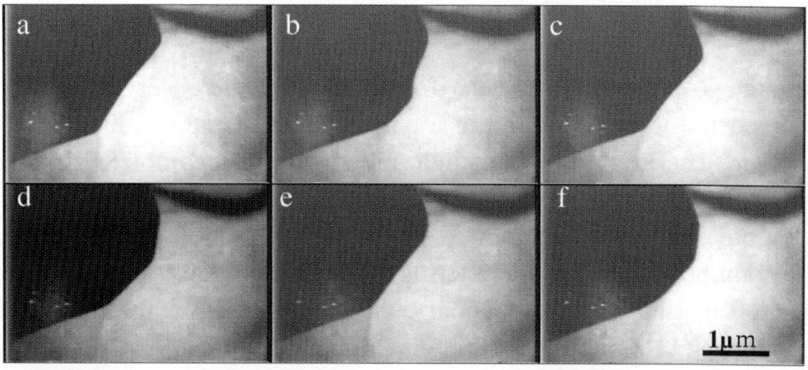

Fig. 7. Vibration of a grain boundary in $CuGa_2$ observed at room temperature at an accelerating voltage of 400 kV.

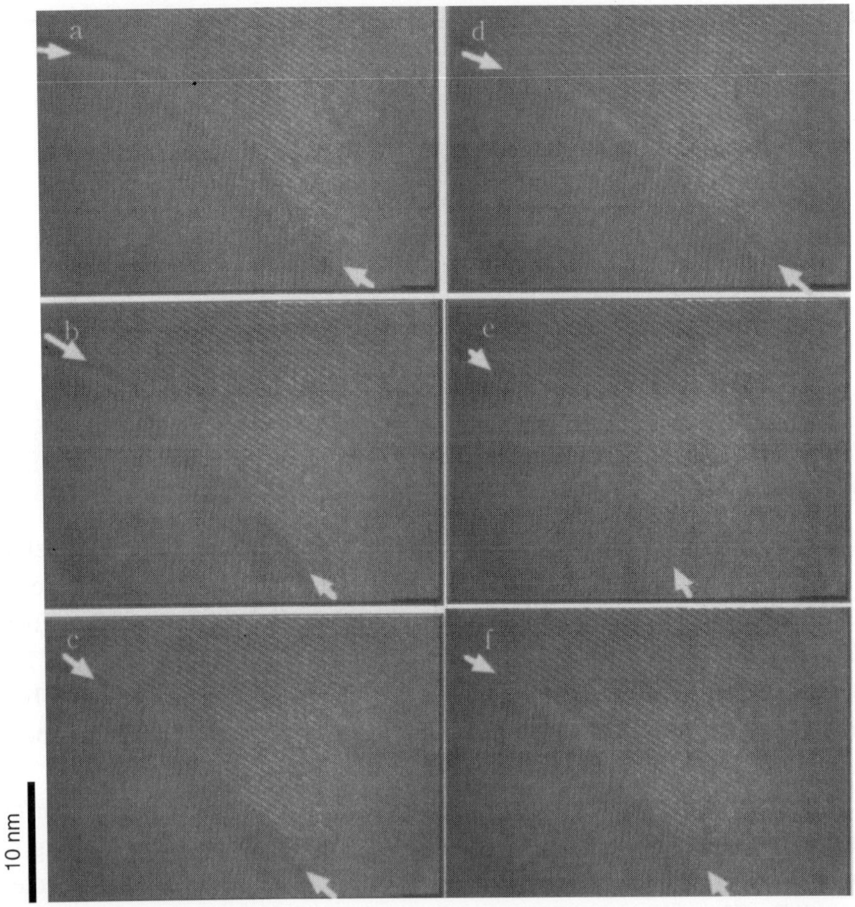

Fig. 8. HREM micrographs of GB vibration observed at room temperature at 400 kV.

accelerating voltage, as shown in Fig. 9(a). However, the vibration was still observed at 200 kV when the flux was increased to 8×10^{23}e/m^{2}s. Thus, it appeared that a critical electron flux exists above which the GB vibration took place. Effects of temperature on GB vibration are more surprising. As shown in Fig. 9(b), the GB vibration diminished with increasing temperature: The GB vibration was observed between –70 and 55°C. Above 55°C neither vibration nor motion of the GB was observed. Not all the GBs examined showed vibration. About 20 GBs were observed to

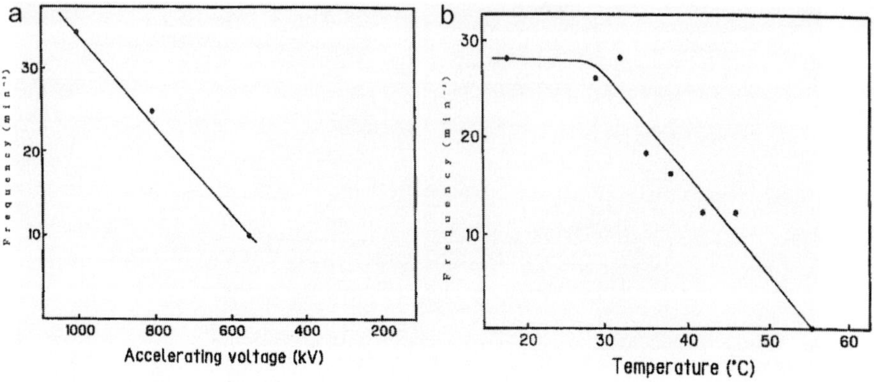

Fig. 9. Effects of accelerating voltage (a) and temperature (b) on GB vibration.

show GB vibration. The orientation relationships of those GB's were classified into three types: Type 1; $(10\bar{1})$//(101), <010>//<010>, Type 2; $(\bar{1}01)$//(110), <101>//<001>, Type 3; (110)//(101), <001>//<010>. From these results, it is suggested that point defects which are introduced by fast electrons during observation are most likely responsible for the GB vibration. A model for the GB vibration due to excess point defects has been proposed.[20]

Next example is vibration of an interface between Si and SiO_2 during the reduction of SiO_2.[21] Our original idea was to observe the melting of pure Si in a TEM, by encapsulating Si with SiO_2 (Fig. 10(a)). We expected the SiO_2 film suppressed evaporation of Si. However, what happened actually was that, when a Si particle coated with a SiO_2 film (formed by thermal oxidation) was heated between 1373 K and 1473 K, SiO_2 was reduced to Si, as can be seen from Fig. 10(b). The inner core of Si increased its volume at the expense of the surface layer of SiO_2. Surprisingly again, during the reduction the interface between Si and SiO_2 vibrated violently (Fig. 11). In Fig. 11, on the right-hand side of the Si particle the SiO_2 did not exist: it was reduced completely and/or evaporated. On the left-hand side the SiO_2 layer still persisted and was being reduced. Figure 11(k) shows the superimposition of traces of the interfaces shown in Fig. 11(a–j). It is evident that the interface between the SiO_2 layer and the core Si vibrated violently. The amplitude was as large as several tens

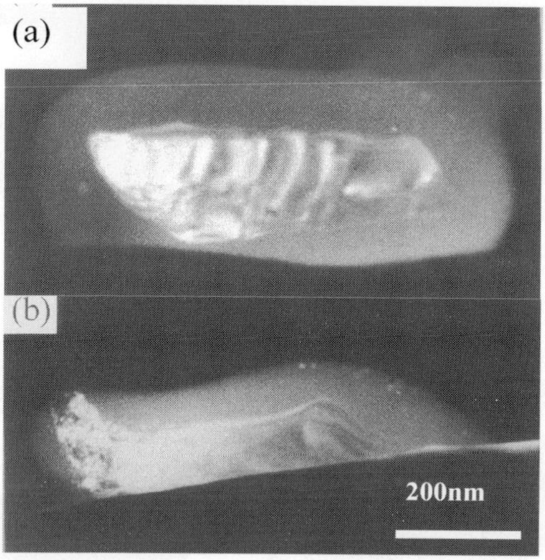

Fig. 10. Si particle covered with a rather thick layer of SiO_2 before (a) and (b) after heating.

nanometers and the frequency was an order of a few Hz. The Si particle behaves as if it were a liquid droplet, but it remains crystalline during the motion. The free surface on the right-hand side did not vibrate very much. This suggests that the existence of the SiO_2 layer is essential for the vibration to take place. There are many possible explanations for the observed vibration of the Si-SiO_2 interface.

(1) Temperature rise due to irradiation by electrons;
(2) Radiation damage due to irradiation by electrons;
(3) Electric charging;
(4) Stress due to volume change during reduction of SiO_2;
(5) Effects caused by point defects produced by reduction.

Among them, (1) and (2) were ruled out because similar vibration was observed even under a much lower flux of electron beam. With regards to (3), if electric charging exists, it should be visualized by electron holography.[23] Figures 12(a) and (b) show the electron holographs taken in a Hitachi HF-2000 microscope equipped with a cold FEG at room temperature

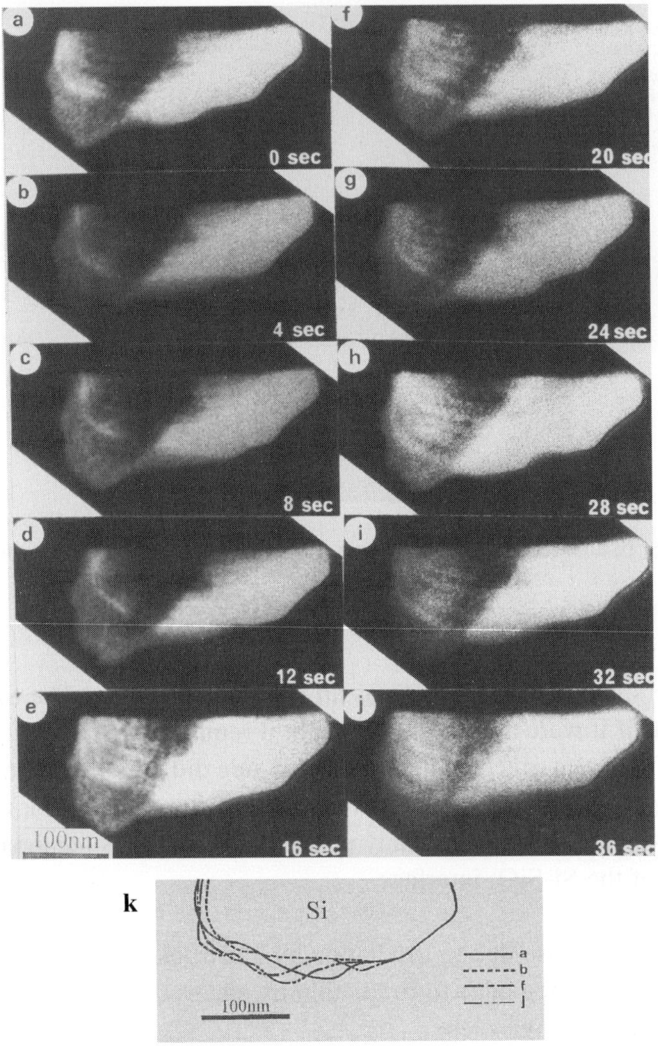

Fig. 11. (a–j) Behaviour of a Si particle covered with SiO_2 layer at 1473 K. (k) Superimposition of the traces of the interface.

and 1273 K, respectively. In Fig. 12(a) the interference fringes avoid the specimen and are curved around it; this indicates that electric charge indeed takes place at the specimen. On the other hand, at 1273 K the interference fringes penetrate the specimen and are not disturbed by the specimen. This

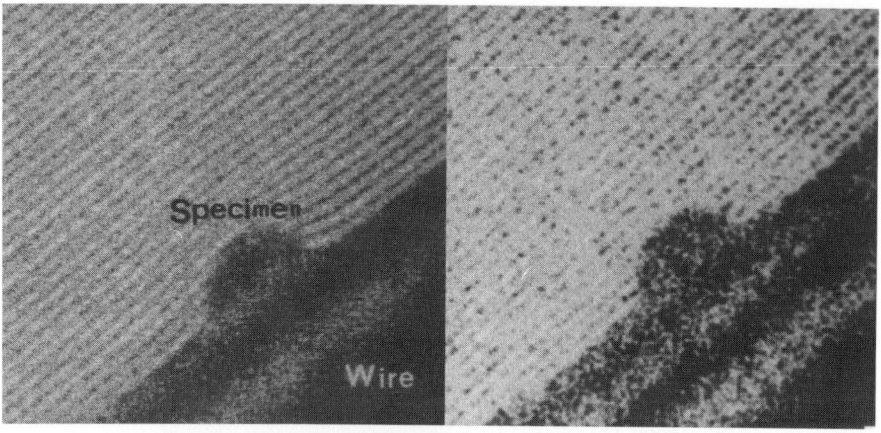

Fig. 12. Interference micrographs of a Si particle covered with SiO_2 layer obtained (a) at room temperature and (b) at 1273 K.

indicates that the electric charging does not take place at high temperatures. This may be rationalized by the fact that the resistivity of SiO_2 decreases with increasing temperature. Also, this result demonstrates effectiveness of an *in-situ* holography experiment in studying resistivity change of an insulator at high temperatures.

During oxidation and reduction a large volume change occurs and this inevitably develops stress at the Si-SiO_2 interface. According to EerNisse,[24] however, this intrinsic interfacial stress is very small above 1123 K where viscous flow of the oxide relieves the stress. Thus, (4) can be ruled out. During reduction a chemical reaction such as $SiO_2 \rightarrow Si + 2O$ should take place. O eventually will escape from the specimen in the form of molecules. However, immediately after O is produced by the aforementioned reduction, it most probably exists as a point defect near the interface, and this may cause the vibration of the interface.

Contribution of O to the vibration was confirmed by carrying out, by EELS, *in-situ* chemical mapping of a vibrating interface[25]. Figure 13 shows plasmon loss spectra of a Si particle covered with SiO_2. The solid and the broken lines were obtained from Si and SiO_2, respectively; Si and SiO_2 have the peaks at 16.7 and 22.4 eV, respectively. Figures 14(a) and (b) show typical images obtained by placing the slit at 16.7 ± 1.5 eV (Si) and 22.4 ± 1.5 eV

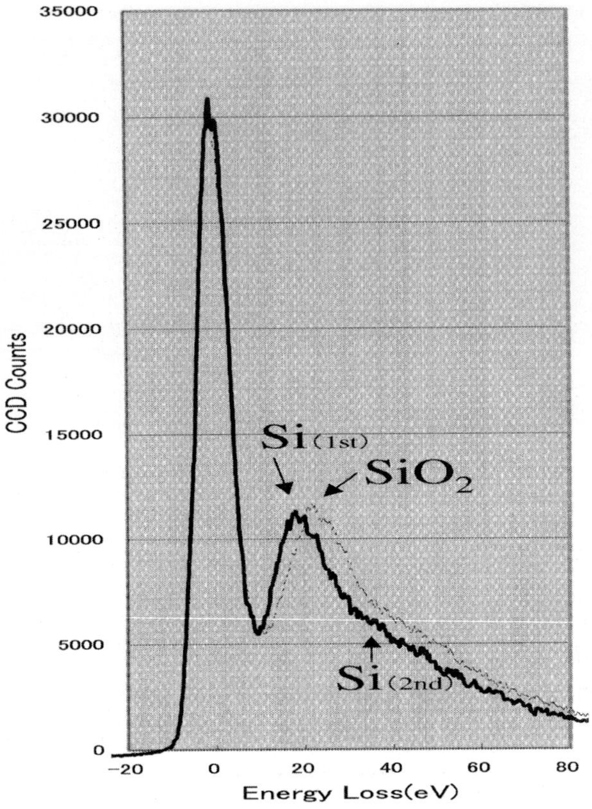

Fig. 13. EELS taken from Si (solid line) and SiO_2 (broken line).

(SiO_2), respectively. Figure 14(c) shows the intensity profiles of the plasmon loss images of Si along A-B and of SiO_2 along A'-B'. Figure 15 shows an example of the vibration of a Si-SiO_2 interface observed at 1473 K, imaged by the first plasmon loss of SiO_2. It is evident that the interface is vibrating in terms of not only the position but also composition. In other words, the vibration is accompanied with to-and-fro motion of oxygen. In the case of GB vibration in $CuGa_2$ already mentioned, the unit cell of $CuGa_2$ changes from one equilibrium position in one grain to the equilibrium position in the other grain without changing its composition, that is, the motion is short-range and is not accompanied with a long range diffusion of the species involved. By contrast, in the case of vibration of the Si-SiO_2

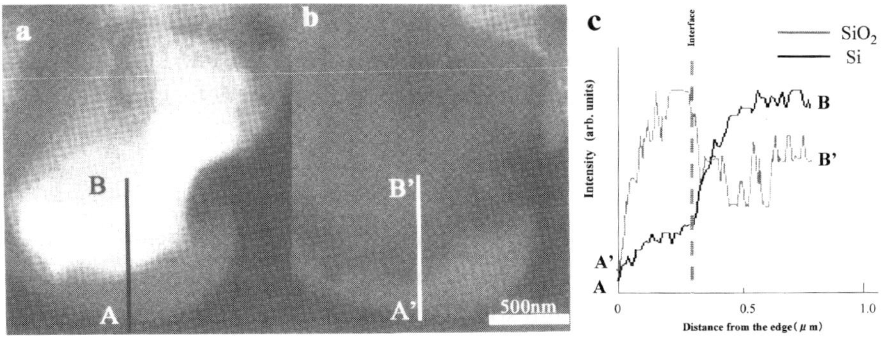

Fig. 14. Chemical mapping of (a) Si and (b) and SiO_2. (c) Intensity profiles along A-B and A'-B'.

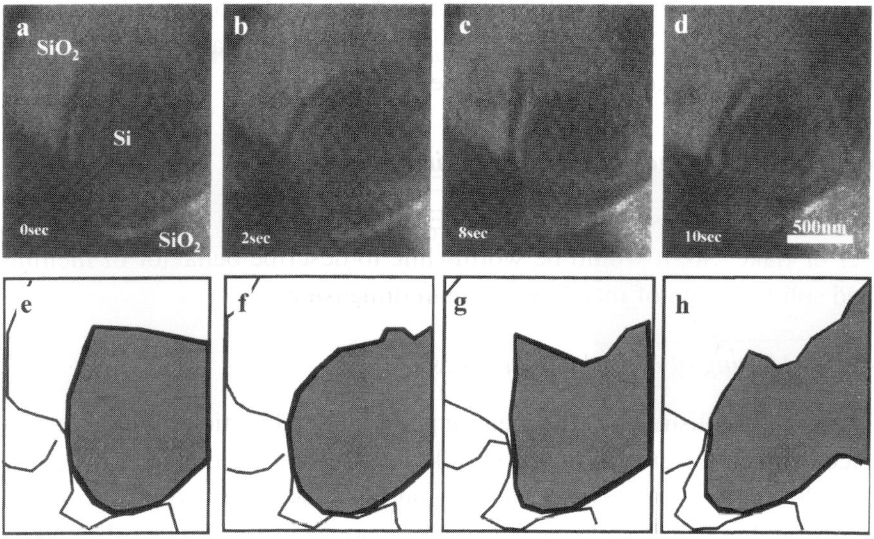

Fig. 15. The plasmon loss images (a–d) of SiO_2 (22.4eV) and the schematic drawings (e–h) showing the vibration of Si-SiO_2 interface.[25]

interface, long range diffusion of oxygen is involved. Another point to be noted is that, during the vibration of the interface, the Si crystal is deforming *plastically*. However, no evidence was obtained for the motion of dislocations during the *plastic deformation* of Si. One possibility is that the

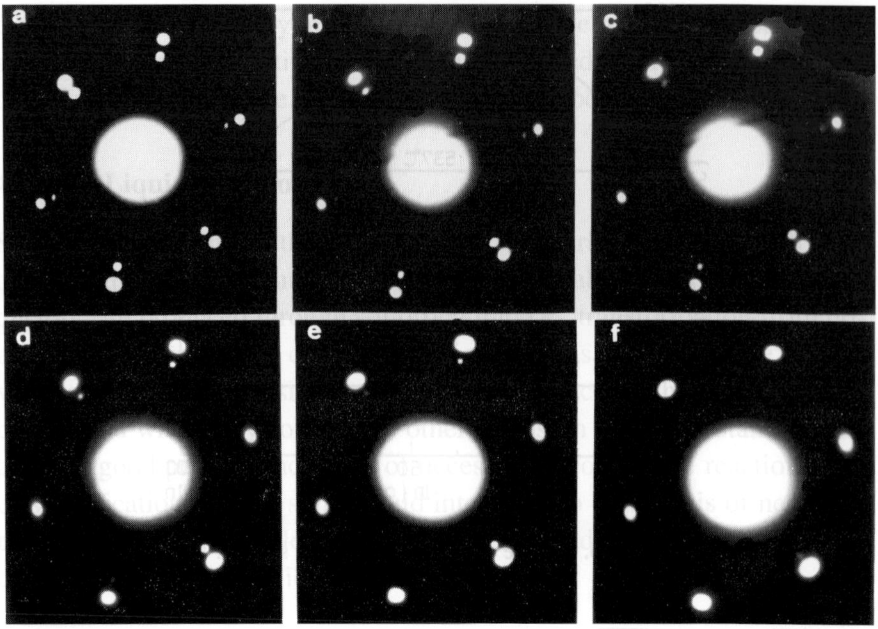

Fig. 18. Diffraction patterns taken at various temperatures during heating. (a) 25°C, (b) 162°C, (c) 175°C, (d) 184°C, (e) 189°C, and (f) 193°C.

not until the temperature reached 193°C that the spots from In crystals disappeared completely. Since the bulk melting point of In is 155°C, this indicates that In particles were superheated by as large as ~38°C.

On cooling, spots from crystalline In appeared only after the specimen was cooled down to 129°C, that is, supercooling as large as ~26°C took place. Examination of the Moire fringes allows determination on individual particles whether they are in solid or liquid states. The melting and freezing temperatures thus obtained are plotted as functions of particle radius (r) in Fig. 20(a). Upper and lower hatched bands indicate the melting point and freezing point, respectively; the region bounded by two broken lines shows melting point of In particles with free surface. The freezing point of In particles embedded in the matrix of Al showed a minimum at around $r = 13$ nm. Furthermore, the melting point of In particles embedded in the Al matrix increases remarkably with decreasing r. This is in sharp contrast to the free particles of In.

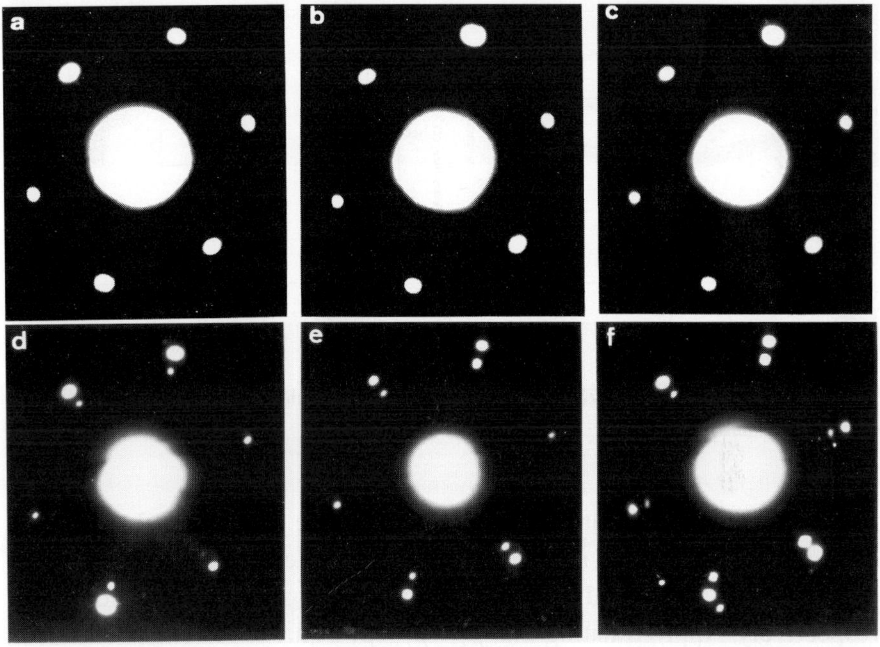

Fig. 19. Diffraction patterns taken at various temperatures during cooling from the liquid state. (a) 156°C, (b) 146°C, (c) 136°C, (d) 129°C, (e) 124°C and (f) 119°C.

One possible explanation would be that the elevated melting temperature is to be attributed to the pressure from the Al matrix. In order to see if this is the case, similar experiment was carried out on In particles but embedded in Fe matrix.[29] The result is shown in Fig. 20(b), where closed circles show melting temperature and open circles show freezing temperature, respectively. Both the freezing and the melting temperatures decrease with decreasing the particle radius (r). This is essentially similar to the behaviour of free particles.

The variation in melting points in the embedded In particles can be explained as follows:

Thermodynamic models suggest that melting point of fine particles is inversely the particle radius r and proportional to the difference in surface energies between the solid and liquid state, i.e.,

$$T_0 - \frac{T}{T_0} = k\,\frac{(\gamma_{sv}/\rho_s - \gamma_{lv}/\rho_l)}{rL}, \tag{1}$$

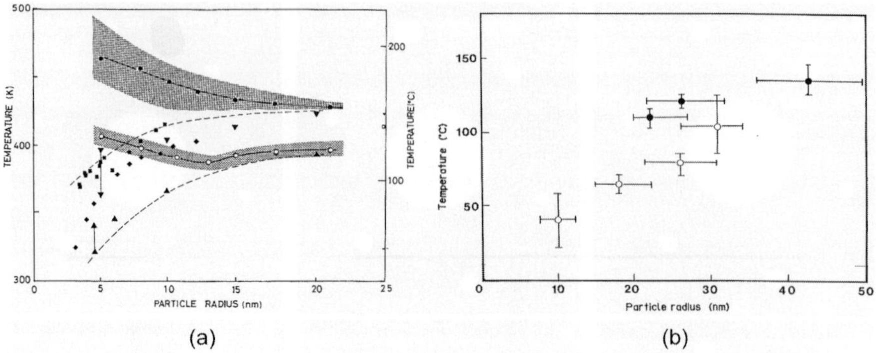

Fig. 20. Melting and freezing temperature of In particles in (a) In-Al system and (b) In-Fe system.

where T_o is the bulk melting point, k is a positive constant, L is the latent heat of melting per unit mass, ρ_s and ρ_l are the densities of solid and liquid, respectively. γ_{sv} and γ_{lv} are the surface energies of solid and liquid, respectively. Approximating that $\rho_s = \rho_l = \rho$, Eqn.1 is rewritten

$$T_0 - \frac{T}{T_0} = k\,\frac{(\gamma_{sv} - \gamma_{vl})}{rL\rho}. \tag{2}$$

For particles with free surface, $\gamma_{sv} - \gamma_{lv} > 0$, $T_0 > T$; depression of melting point is observed.

For embedded particles, γ_{sv} and γ_{lv} should be replaced by γ_{sm} and γ_{lm}, respectively, where γ_{sm} and γ_{lm} are interfacial energy between solid and matrix and interfacial energy between liquid and matrix, respectively.

γ_{sm} and γ_{lm} are related with γ_{sl} and the contact angle θ through the Young's equation:

$$\gamma_{sm} - \gamma_{lm} = \gamma_{sl}\cos\theta, \tag{3}$$

where γ_{sl} is the interfacial energyof the S-L interface.

Substituting Eq. (3) into Eq. (1) we obtain

$$T_0 - \frac{T}{T_0} = k\,\frac{(\gamma_{sl}\cos\theta)}{rL\rho}. \tag{4}$$

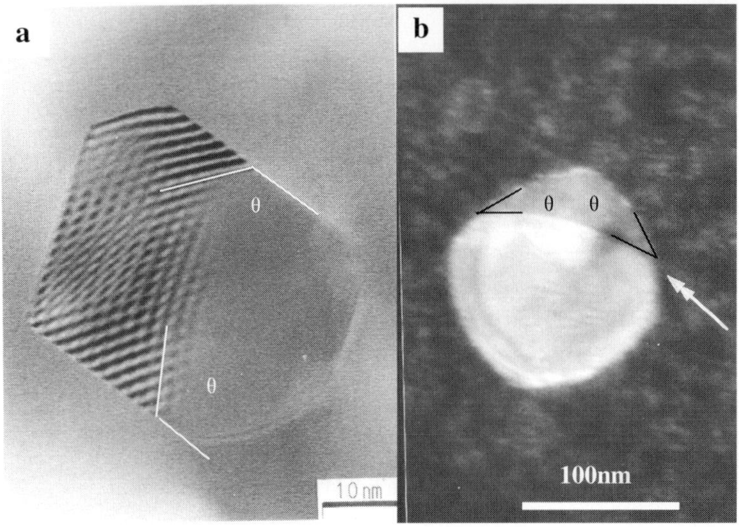

Fig. 21. Partially molten In particles in (a) the In-Al system and (b) the In-Fe system.

In the In-Al system there is a cube-on-cube orientation relationship, while in the In-Fe system no such a simple or rational relationship exists between In and Fe. In other words, in the In-Al system the interfacial energy between the solid In and the solid Al matrix is likely quite low, while this is not the case for In-Fe. This can be visualized by observing the contact angles in the In-Al and In-Fe systems. Figures 21(a) and (b) show TEM micrographs of partially molten In particles embedded in Al and Fe, respectively. For In-Al $\theta > 90°$, while for In-Fe $\theta < 90°$, indicating that elevation and depression of melting temperature take place for In-Al and In-Fe cases, respectively.

It was also possible to study the melting processes of an individual In particle in detail. Figure 22 shows the early stage of the nucleation of a liquid phase in an In particle embedded in the Al matrix.[30] The liquid droplet was nucleated at one of the {100} facets (at the bottom of the particle in this case). The liquid was lenticular in shape at the very beginning (Fig. 22(a)) and then it grew to a spherical shape with much larger volume (Fig. 22(b)). The liquid droplet assumed these two configurations alternately and the time spent in this stage was much longer than that spent in the rest of the melting process.

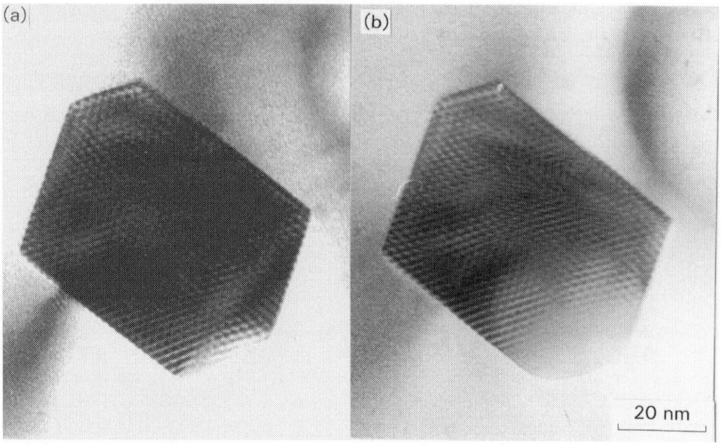

Fig. 22. Nucleation of a liquid droplet at {100} facet.

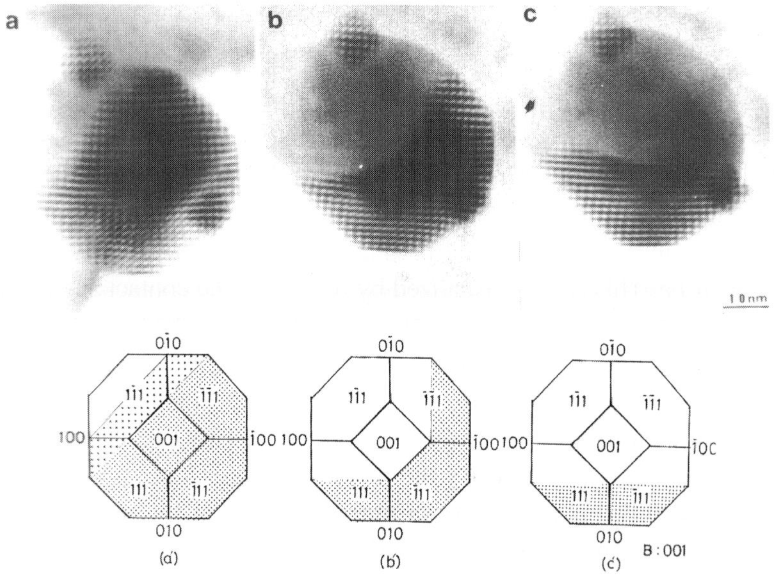

Fig. 23. Propagation of the liquid phase into the interior of In particle.

Figure 23 shows a typical sequence showing the melting process, viewed along [001] direction. In (a), (100) and (0$\bar{1}$0) facets are covered by the liquid phase. In (b) (001) and (00$\bar{1}$) facets are covered by the liquid phase additionally. In (c) the ($\bar{1}$00) facet is also covered by the liquid phase. Figure 24

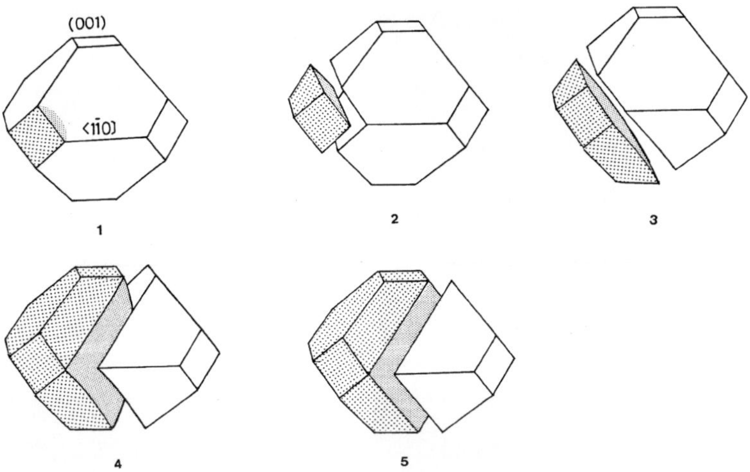

Fig. 24. Schematic illustration of melting processes of an In particle embedded in an Al matrix.

shows schematically the melting processes of an individual In particle embedded in an Al matrix. Melting started at one of the {100} facets and proceeded into the interior of the In particles in 6 separate stages in such a way that, at each of the stages, one of the {100} facets became covered with the liquid phase. Detailed calculations on the thermodynamics confirmed that this is indeed the path of minimization of the interfacial energy.[31,32]

4.1.2. *Melting of a wedge-shaped crystal*

Since Takagi[26] reported depressed melting temperature for fine metal particles with free surface, it has been well established that metallic particles have size-dependent melting temperatures.[27] A fine particle is reduced three-dimensionally. On the other hand, a thin film is reduced one-dimensionally and a needle two-dimensionally. Here, melting behaviour of a thin film, with the shape of a wedge, was studied.[33]

Discs 3 mm in diameter were cut from 7 μm plates of Sn. The central part was thinned by ion milling until a small hole was formed. Thus, the cross section of the specimen had a wedge shape. Whole surfaces of the specimens were coated with a hydrocarbon-polymerized amorphous film 80 μm thick.[34] This hydrocarbon film was strong enough to keep the shape of the initial thin specimen after the crystal was molten. Indeed, this technique was successfully applied to the *in-situ* observation of solder reaction

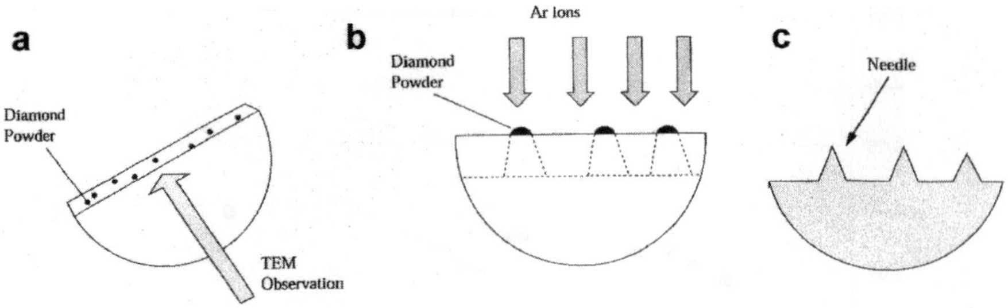

Fig. 27. Preparation method of a conical needle.

40 kV along the direction normal to the section of the half-cut for ~1 hour resulted in the formation of small needles (Fig. 27). The radius of curvature of the tip ranged from 15 to 40 nm and the needle angle α ranged from 20 to 55 degrees. The specimen was coated again with a 80 μm thick hydrocarbon-polymerized amorphous film.[34]

Figure 28 shows a typical Sn needle at six different temperatures. As the temperature increased, the S-L interface moved towards the thicker part. On cooling the S-L interface moved towards the thinner part. The motion of the S-L interface was quite reversible. Thus, it is evident that the melting point depends on the local radius of the needle (Fig. 29). Furthermore, the S-L interface is convex toward the liquid phase. Thermodynamic calculations lead to the following equation:

$$\frac{T_0 - T}{T_0} = \frac{3\tan(\alpha/2)}{L\rho R_2}\left\{\frac{\gamma_{sc}^{Sn-C} - \gamma_{lc}^{Sn-C}}{\sin\alpha} - (1+4\beta^2)\gamma_{sl}^{Sn}\right\}, \quad (6)$$

where β (= 0.32–0.37) is a geometrical factor which depends on α, and R_2 is the local radius of the needle. This equation is indicated by a straight line in Fig. 29.

4.2. *Solid-liquid interfaces*

Solid-liquid (S-L) interfaces, which are part of everyday life, are difficult to study. Most of the experiments rely on indirect measurement of the surface properties.[38] It is only very recently that S-L interfaces can be successfully observed at near atomic level by TEM.[11]

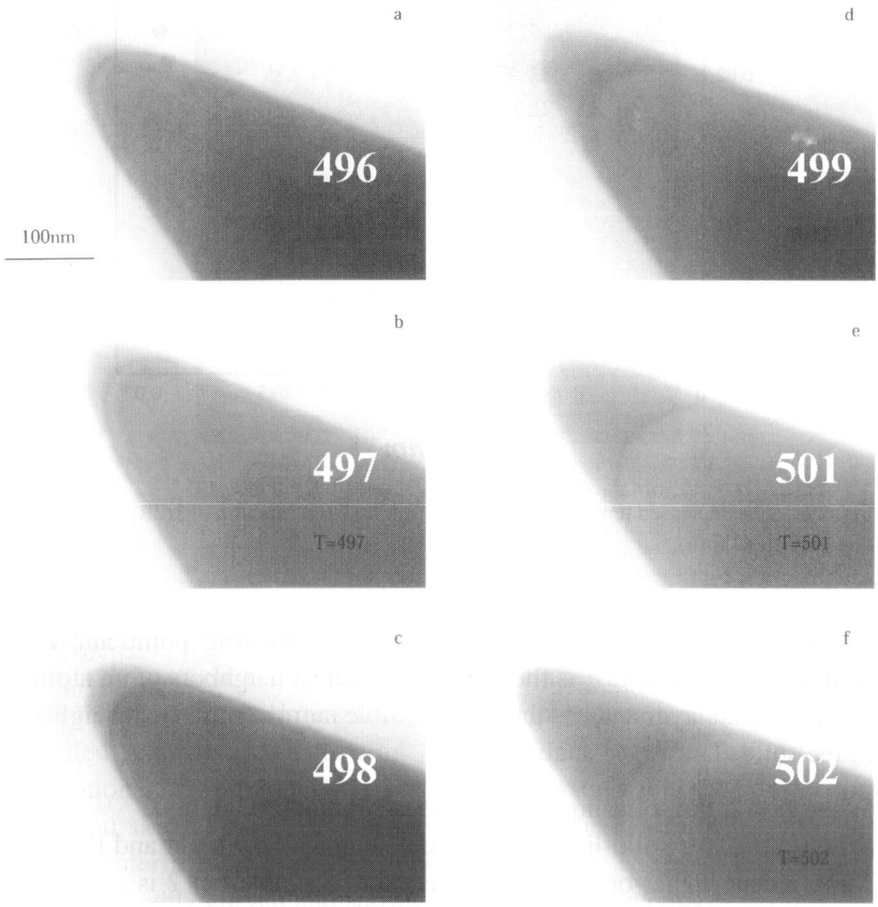

Fig. 28. Sn needle with a S-L interface. Numbers refer to temperatures in K.

The atomic structure and dynamics of a S-L interface are believed to play an important role in crystal growth.[38] Jackson,[39] based on the Ising lattice model, pointed out that a solid-liquid interface is either atomically rough or smooth, depending on a parameter defined by

$$\alpha = \left(\frac{\Delta H_m}{k_B T_m} \right) \left(\frac{z_l}{z} \right), \tag{7}$$

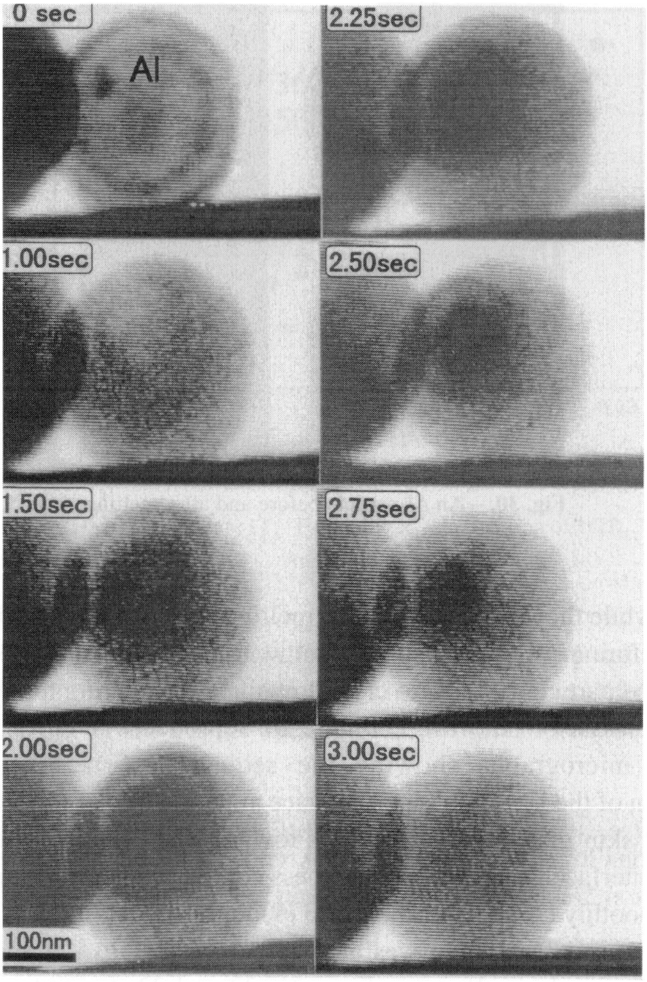

Fig. 31. Sequence of propagation of liquid phase from the surface into the interior of an Al particle.

holder.[42] Powders of alumina were heated up to 2000 K. When the temperature approached 2000 K, a large number of whiskers, with horsetail shape, were formed as shown in Fig. 32. On top of the whiskers sit hemspherical droplets. In the stem of the whisker the $(1\bar{1}02)$ and $(10\bar{1}2)$ lattice fringes were clearly observed as shown in HREM images in Fig. 33.

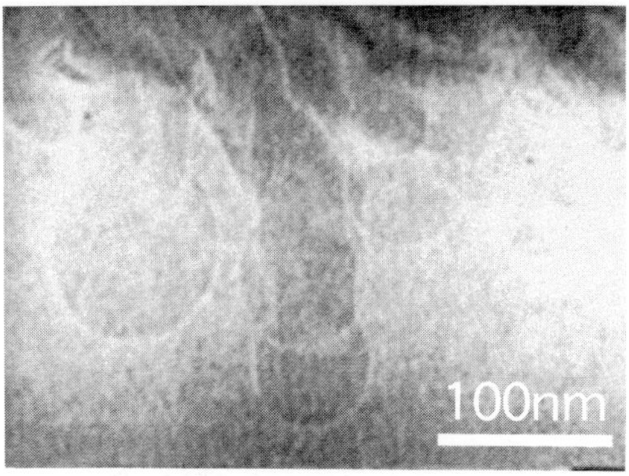

Fig. 32. Formation of horsetail-like whiskers on the surface of Al_2O_3 at 2000 K.

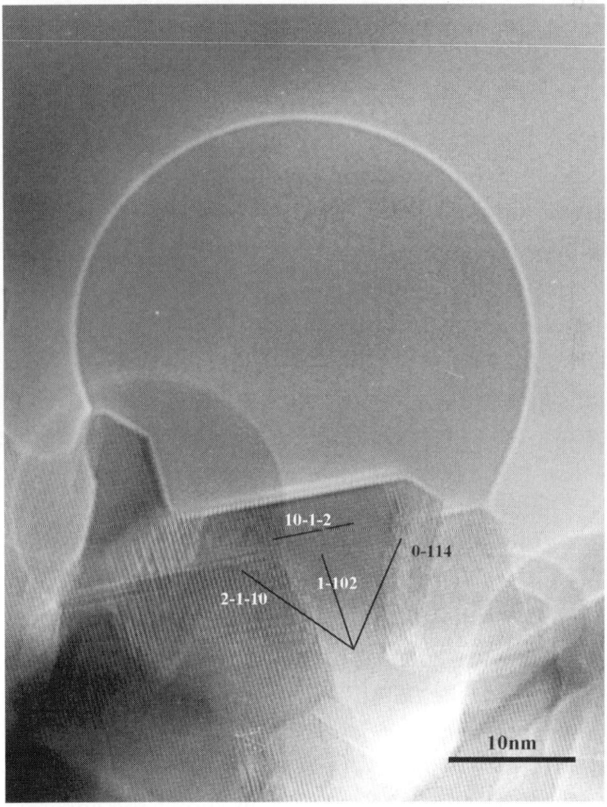

Fig. 33. HREM of liquid droplets sitting on whiskers of Al_2O_3.

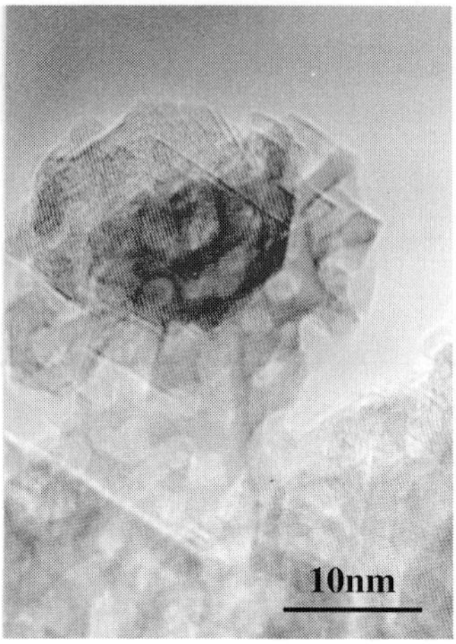

Fig. 34. Same as Fig. 33 but after cooled to room temperature.

When the temperature was decreased to room temperature, lattice fringes appeared in what had been a droplet, as shown in Fig. 34. The lattice fringes were coincident to those of alumina. Thus, the droplet was confirmed to be liquid alumina.

The S-L interface of alumina is very straight and anisotropic, being facetted along crystallographic orientation (or more exactly, projections of crystallographic planes) such as $(10\bar{1}\bar{2})$, $(1\bar{1}02)$ $(2\bar{1}\bar{1}0)$ and $(0\bar{1}14)$. The S-L interface changed its morphology significantly during its motion. Figure 35 shows the complete transformation of the S-L interface from $(2\bar{2}\bar{2}0)$ to $(10\bar{1}\bar{2})$.[12] The S-L interface parallel to $(2\bar{1}\bar{1}0)$ (hereafter denoted by $(2\bar{1}\bar{1}0)$S-L) encountered a small grain indicated by an arrow. The motion of the $(2\bar{1}\bar{1}0)$ was hindered by the grain, and during surmounting the grain, a new facet, $(10\bar{1}\bar{2})$ S-L, was formed at the left-hand edge of the grain. The growth of the $(10\bar{1}\bar{2})$ S-L was slower than that of the $(2\bar{1}\bar{1}0)$ S-L, and the latter was replaced by the former and the overall S-L interface transformed from the $(2\bar{1}\bar{1}0)$ facet to the $(10\bar{1}\bar{1})$ facet.

The S-L interface had a tendency to be parallel to $(10\bar{1}\bar{2})$. However, at the triple point of the vacuum and the S-L interface, the $(0\bar{1}14)$ and $(10\bar{1}\bar{2})$ facets competed with each other. Figure 36 shows the appearance and disappearance of $(0\bar{1}14)$ facet at the triple point at the right hand side.[12]

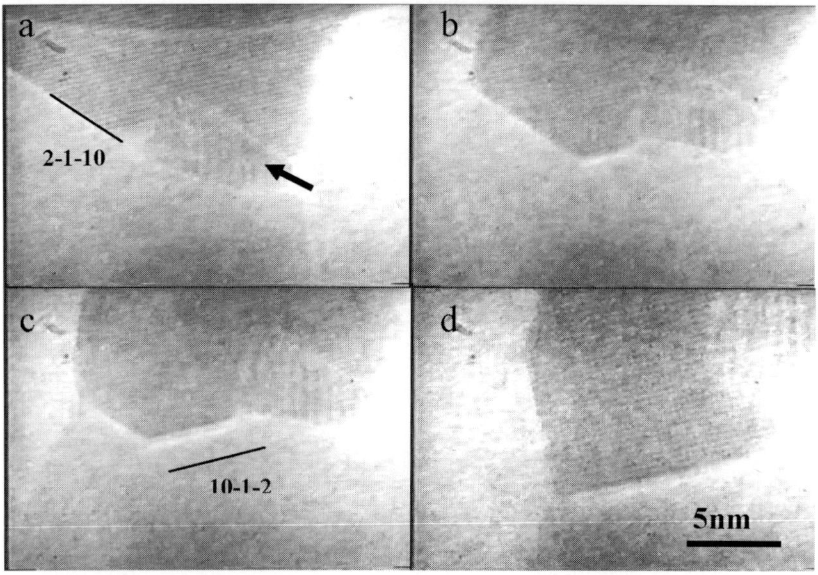

Fig. 35. Transformation of S-L interface from (2-1-10) to (10-1-2).

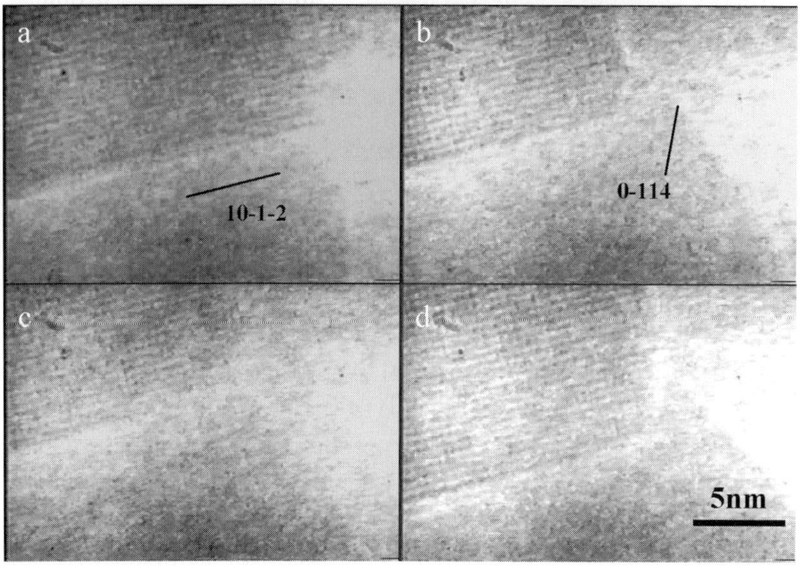

Fig. 36. The appearance and disappearance of $(0\bar{1}14)$ facet at the triple point between the S-L interface and vacuum.

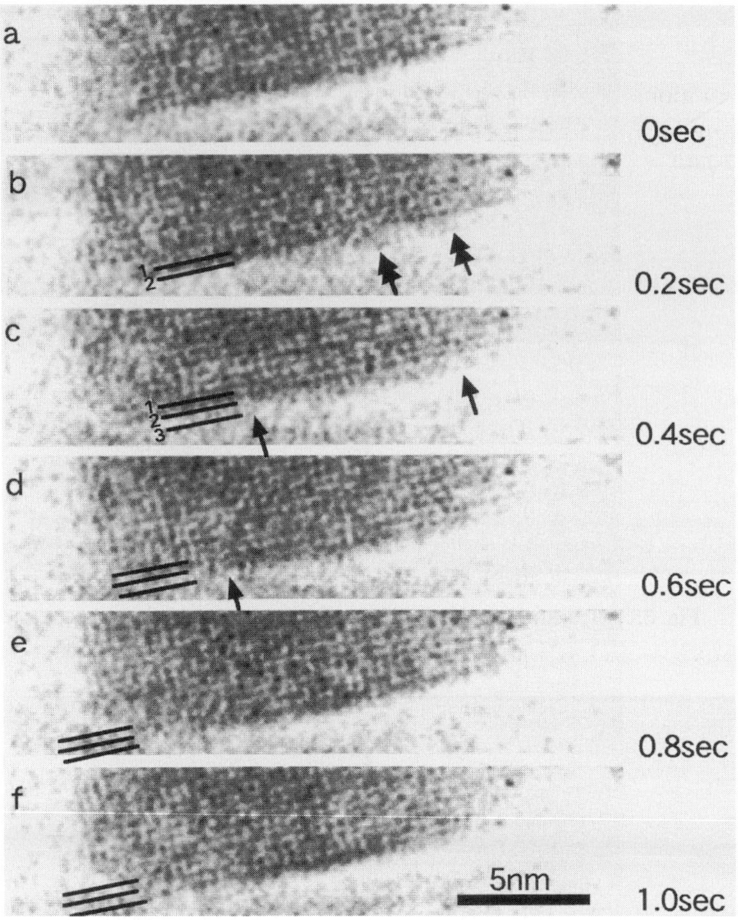

Fig. 37. Sequence of nucleation of crystal at the S-L interface of Al_2O_3.

Repeated transformation between $(0\bar{1}14)$ and $(10\bar{1}\bar{2})$ facets results in a rather zigzagged feature of the surface of the grown crystal.

Figure 37 shows a series of video-recorded images which reveal the process of formation of one monolayer. In Fig. 37(a) a molecularly flat $(10\bar{1}\bar{2})$ S-L interface is observed. Formation of the monolayer was initiated by nucleation of a cloud-like contrast with a thickness of about 2 monolayers at the central part of the overall solid-liquid interface (Fig. 37(b)).

The cloud-like contrast was elongated along the S-L interface and reduced the thickness. When the cloud extended to a width ranging from 5 to10 nm along the S-L interface, lattice fringes perpendicular to the interface were formed inside the cloud. The fringe contrasts inside the cloud correspond to those in the stem (Fig. 37(c)). These processes are believed to be formation of an island (or a terrace) of a monolayer with molecular steps on both sides. The island or the terrace expanded continuously, adding new lattices on both sides until the whole interface was completely covered by the new layer (Fig. 37(d)). The interface covered by the new layer existed stably until the next formation of a new terrace (Figs 37(e) and (f)).

Nucleation of the terrace took place always at the central part of the overall S-L interface. This suggests that the central part of the overall S-L interface is the preferential nucleation site of the terrace. In the theories of lateral growth of crystal, stochastic nucleation of terrace is assumed. The preferential nucleation of the initial terrace at the central part of the overall S-L interface strongly suggests that nucleation of terrace is not controlled stochastically. The obvious explanation is that the S-L interface is convex toward the liquid side. Figure 38 shows evidence for the convex

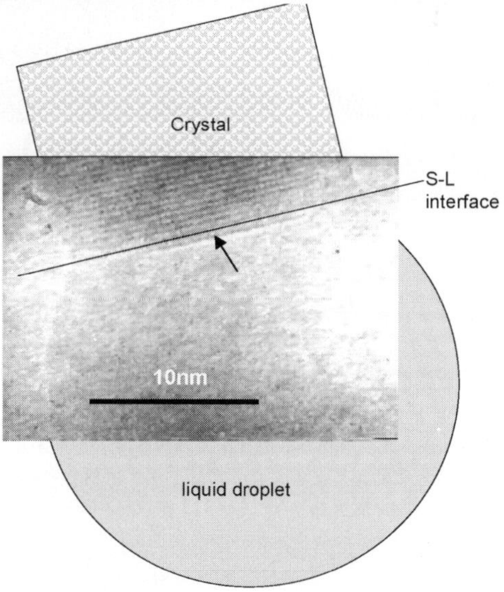

Fig. 38. Nucleation of a terrace at the central part of the overall S-L interface.

nature of the S-L interface. Under such conditions, it is not surprising that nucleation of a terrace should take place at the far front of the S-L interface, i.e., at the central part of the overall S-L interface.

4.2.3. *Al-Si alloy*

The S-L interfaces in Al-Si alloys were observed by heating a mixture of Al and Si powders.[43] Al-Si alloys were formed in the first run of heating. In the subsequent runs of heating, S-L interfaces were successfully observed at near atomic resolution. Figures 39(a) and (b) show low- and high-magnification micrographs of a typical S-L interface in an Al-Si alloy.

The S-L interface is parallel to Si (111) plane and flat over a considerable area (Fig. 39(a)). At a higher magnification (Fig. 39(b)) it is evident

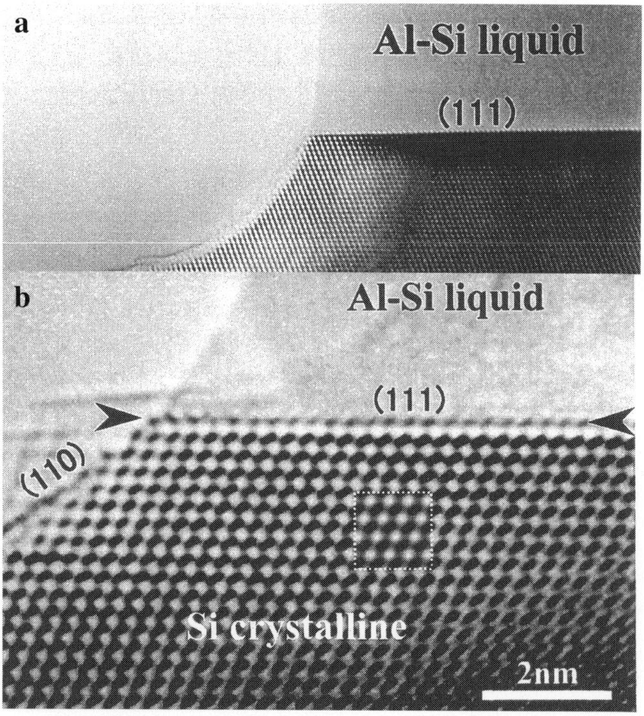

Fig. 39. A typical S-L interface in an Al-Si alloy.

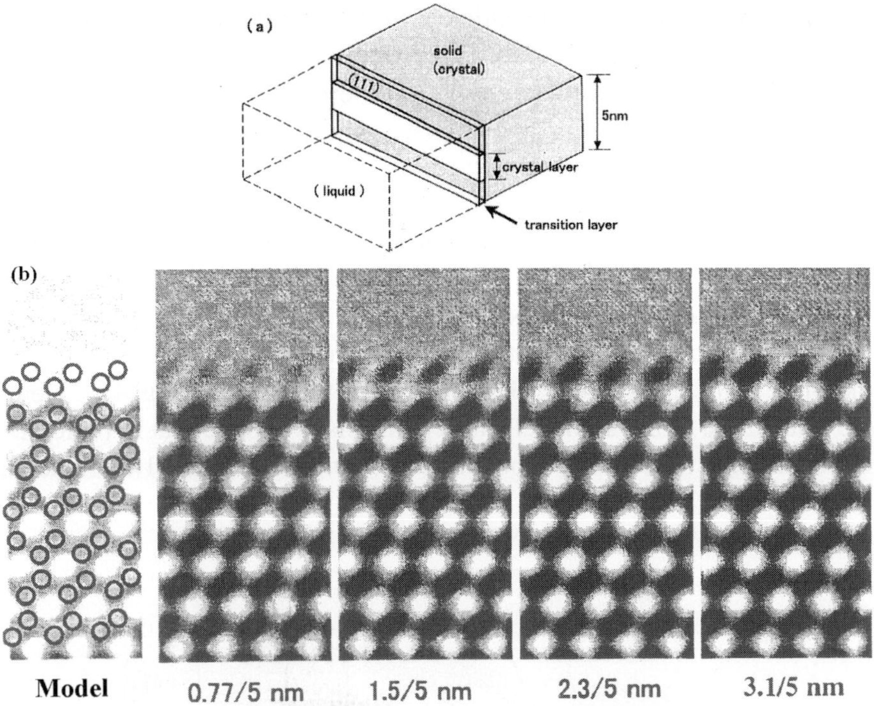

Fig. 40. Computer image simulation of the S-L interface. The total thickness of the S-L interface is assumed to be 5 nm.

that, between solid and liquid, a layer the intensity of which is just half of the perfect crystal exists. One possible explanation for this contrast is that it is an artifact such as the Fresnel fringes and/or oozing-out effect of wave function at the specimen cliff edge. Computer image simulation was carried out. An example is given in Fig. 40. Here, the first transition layer in the image is a mixture of Al-Si liquid and partially solid Si. This layer was modelled by adding an additional atomic layer on top of the Si (111) surface partway though the thickness of the specimen, as schematically shown in Fig. 40(a). The proportion of solid to liquid phase was varied in the simulations for a constant specimen thickness of 5nm, as indicated below each simulated image in Fig. 40(b), with the top number indicating the thickness of the crystalline region.

The closest match in appearance with the experimental image was obtained when the thickness of the crystalline region in the first layer at the S-L interface was between 1.5–2.3 nm. This result indicates that not only is there ordering in the first several liquid layers parallel to the interface, but there is also strong two-dimensional ordering within the first layer of liquid, since the positions of the dark dumbbells in this region are highly regular in the HREM image and accompanying simulations.

That the transition layer is reality and is not an artifact can be seen more straightforwardly by noticing the orientations of the dumbbells at the triple point between the S-L interface and the vacuum, Fig. 41.[45] Here, the orientation of the dumbbells in region A is reverse to that in region B, indicating that the region A contains a stacking fault (Fig. 41(b)). Such a

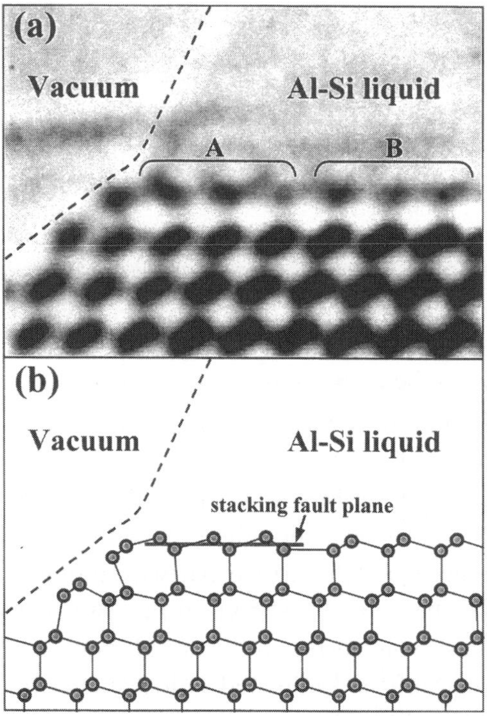

Fig. 41. Nano stacking fault at the triple point between the S-L interface and the vacuum.

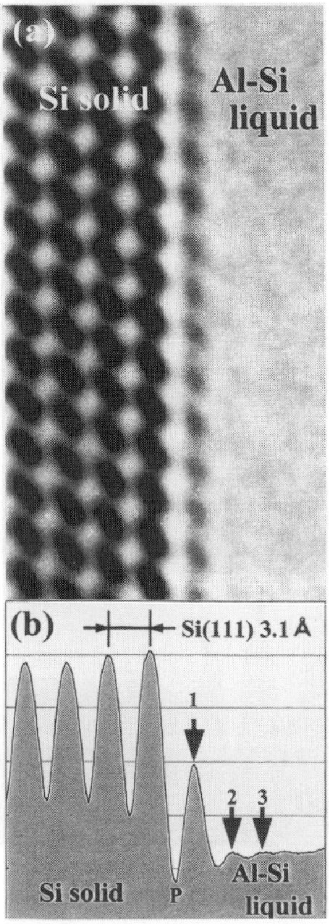

Fig. 42. Intensity profile across S-L interface.

stacking fault cannot be explained by any artifacts of the lattice fringes at the cliff of a crystal.

The intensity profile across the S-L interface is shown in Fig. 42. The transition layer actually extends over a few layers. Figure 43 are theoretical models based on dense random packing of hard spheres on a close-packed crystal surface.[46,47] The agreement between theory and experiment is excellent.

Close examination of the transition layer reveals that the intensity of the first transition layer fluctuates along the S-L interface. This is shown in Fig. 44.[48] Intensity profiles were taken along two rows, A_1 and A_2. A_1 is in the perfect crystal, while A_2 is along the first transition layer in the S-L

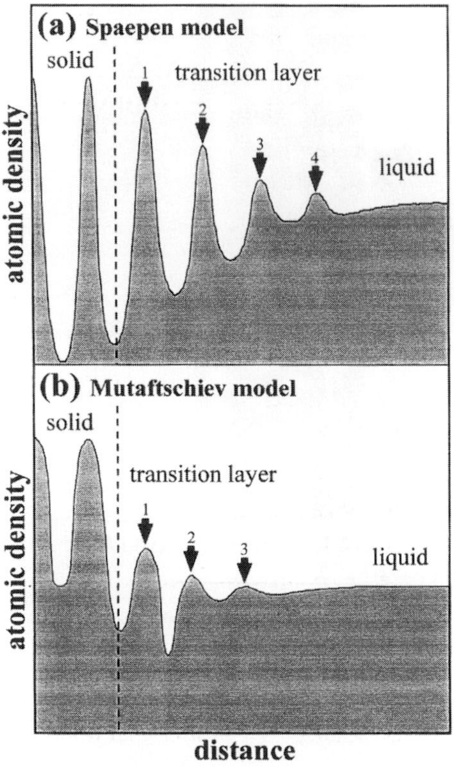

Fig. 43. Theoretical models. (a) Spaepen model46 (Reproduced by courtesy of Elsevier). (b) Mutaftschiev model47 (Reproduced by courtesy of Taylor & Francis).

interface. The intensity profile in layer A_1 is uniform, while that in layer A_2 varies along the S-L interface.

Figure 45(a–f) shows dynamical behavior of a solid Si-Al(-Si) alloy liquid interface during crystal growth. The S-L interface is moving from right to left. Fig. 45(g) shows the intensity profiles across the S-L interface shown in Fig. 45(a–f). In Fig. 45(a) the interface is at position **1**. In (b), 1/30 sec later, it advanced to the position **2**. The contrast of the atomic columns in the region between **1** and **2** is lower that that of the solid matrix but higher that that of the liquid. This suggests that this region between **1** and **2** is a mixture of the solid and the liquid. In other words, atoms in this region are in a half-molten state. In (c) the S-L interface has advanced to position **3**. Again, the lattice fringes between positions **1** and **3**

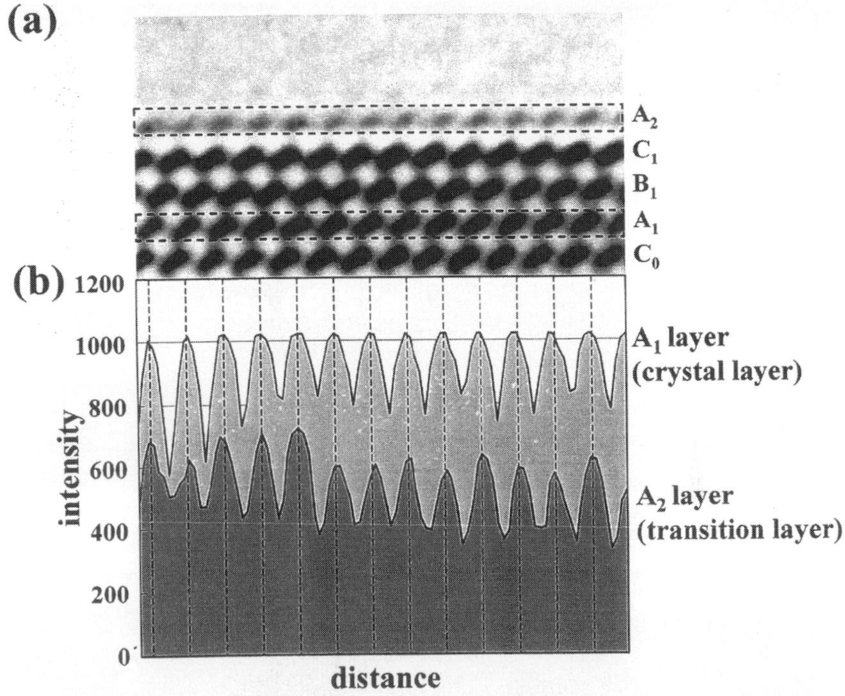

Fig. 44. Variation in intensity along the first transition layer of the S-L interface in an Al-Si alloy.

are fainter than that of the solid. It is only 5/30 sec later that the contrast of this region becomes comparable to that of the solid matrix. The velocity of the S-L interface in this particular event is estimated to be approximately 20 nm/sec.

The S-L interfaces with orientations other than {111} are smoothly curved. However, in some cases, faceting of the solid-liquid interface was observed. Figure 46 shows a solid-liquid interface along {773}.[12] The surface of solid Si is orientated along {773} and is reconstructed to a 2×5 structure due to wetting by molten Al, as will be described in detail in Sec. 4.3.2. At the very vicinity of the triple point between the S-L interface and the vacuum, the solid-liquid interface is parallel to {773} and at least three blocks of the 2×5 structure are observed at the S-L interface.

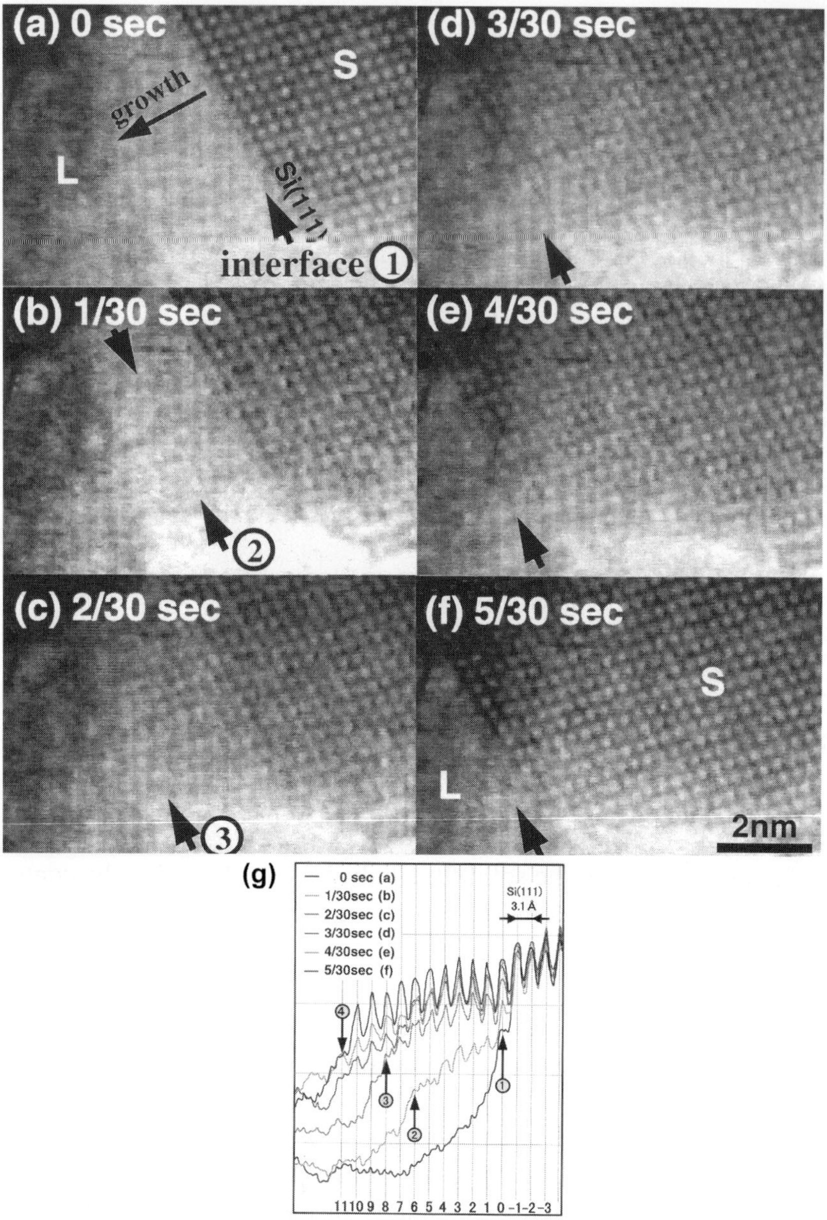

Fig. 45. (a–f) Series of video frames (1/30 s apart), showing motion of a Si(111) S-L interface. (g) Corresponding intensity profiles taken across the moving interface, showing the development of crystallinity with time.

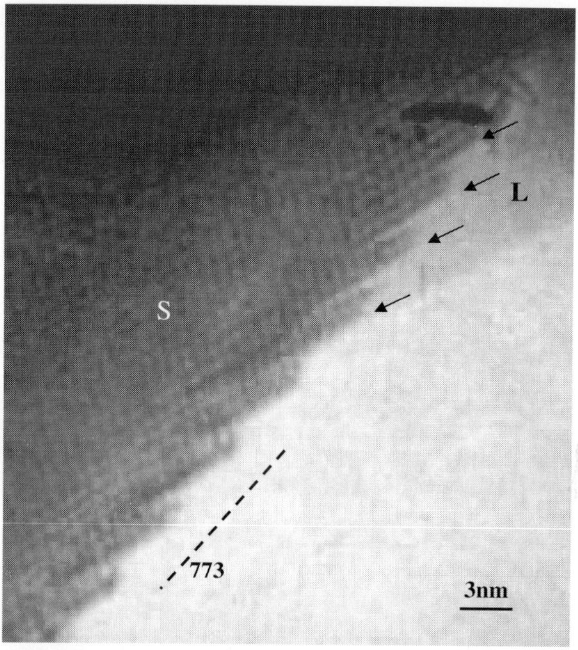

Fig. 46. Reconstructed S-L interface along Si(773) plane.

In Figs. 39 and 41, it was shown that stacking faults or twins can be nucleated at the triple point of the S-L interface and the vacuum. When more than one twins which were nucleated at different sites encounter, complicated arrangement of atomic columns is formed. Figure 47 shows an example. In Fig. 47(a) two twins, i.e., twin **1** and twin **2** were nucleated and propagated downward and upward, respectively. The mirror plane of twins **1** and **2** was (111) plane, but separated slightly. When such twins encountered a third twin **3** was formed to accommodate the misfit, as can be seen in Fig. 47(b). Furthermore, atomic arrangement in a region near the intersection of twins **1**, **2** and **3**, especially just beneath twin **2**, was perturbed considerably. Figures 47(c) and (d) show the diffractograms obtained by Fourier transformation from the areas denoted by A and B in Fig. 47(b).

In the region where the two twins with different step height encounter, the atomic arrangement is very complicated. This is to be attributed to a

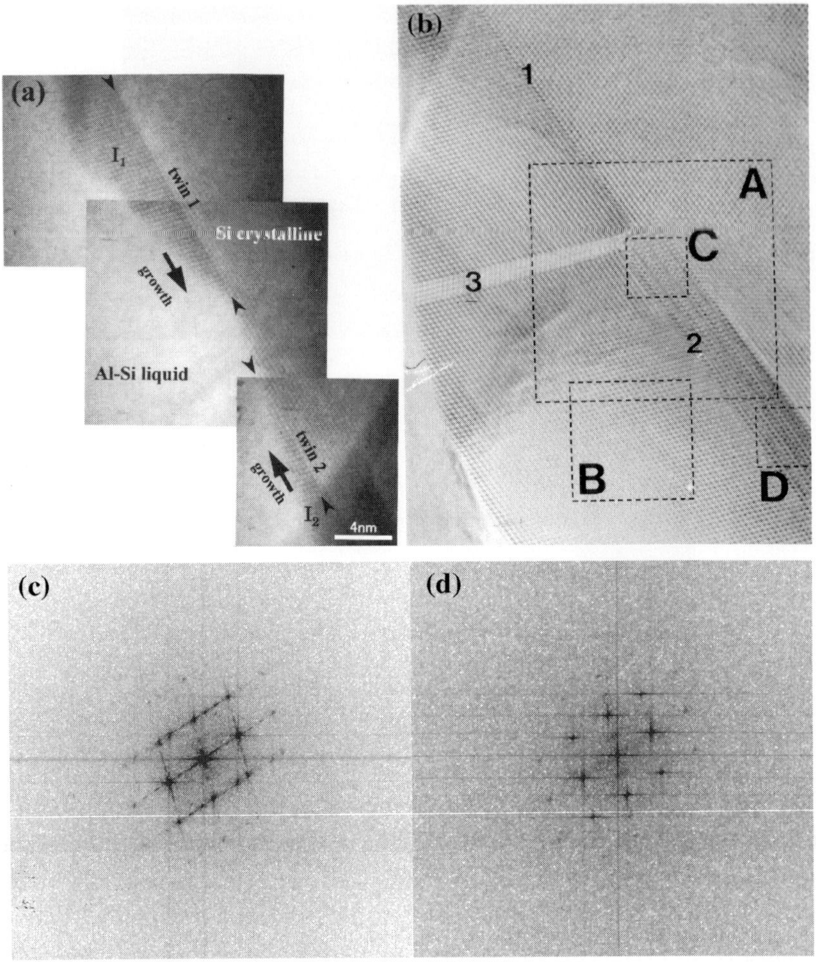

Fig. 47. (a) Propagation of two twins 1 and 2. (b) After twins 1 and 2 meet each other, twin 3 is formed. (c) Diffractogram taken from region A. (d) Diffractogram taken from region B.

large strain induced to accommodate misfit among crystallites with different orientations. Comparison of the diffratograms shown in Figs. 47(c) and (d) reveals that the diffraction pattern from area containing the perturbed area has extra spots of 1/3 1/3 $\overline{1/3}$ and 2/3, 2/3 $\overline{2/3}$. These spots can not be explained by twins, because these diffractograms were taken along

B = 011. A 180° inversion of the diffraction pattern along **B** = 011, which has an apparent twofold symmetry, should result in no apparent change in configuration. In the region C shown in Fig. 47(b), the three-fold structure is evident. Therefore, a novel structure, albeit thin, is formed at the perturbed region. In Si, phase transformation occurs under high pressure.[49] The perturbed region near the crossing of twins observed in the present study may be related with these high-pressure phases.

4.3. *Wetting of liquid metals on non-metallic substrates*

4.3.1. *Au liquid on Si substrate*

The Kamino holder with two-wire type facilitates observation of wetting behaviour of liquid metals on non-metallic substrates. Powder of Au was mounted on the upper heating **1** and Si on the lower heating element **2**, of the Kamino holder of two-wire type, shown in Fig. 1(b).[50] First, heating element **1** was heated to evaporate Au and deposit onto the Si particles mounted on heating element **2**, as shown in Fig. 48. The Si particle (or substrate) was kept at room temperature, so that Au deposited onto the Si substrate was solid.

Then, the heating element **2** was heated to melt the deposited Au particles. Figure 49 shows change in morphology of Si surfaces during thermal cycles between room temperature ((a) and (c)) and above the melting temperature ((b) and (d)). When Au particles melted, Au spread over the surface of the Si substrate. In this process, the followings are evident:

(1) On heating, the crystalline Au clusters melt.
(2) The surface of Si is covered with an amorphous layer initially.
(3) When Au clusters deposited onto a Si surface melt, the molten Au spreads over the Si surface, removing the surface amorphous layer.
(4) In doing so, an initially atomically rough surface of Si transforms into a well faceted atomically smooth surface.
(5) On {111},{001},{211} and {311} surfaces a reconstruction takes place.
(6) These change in morphology is quite reversible.

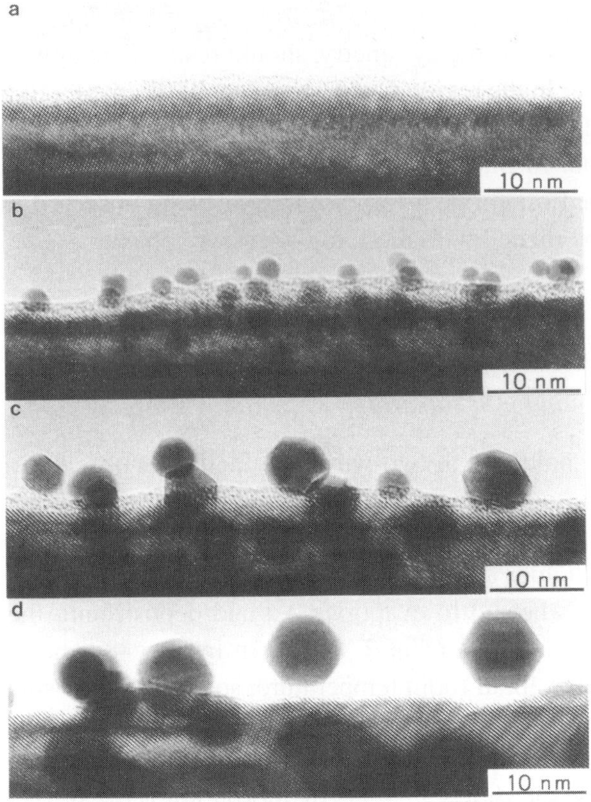

Fig. 48. Deposition of Au particles onto a Si substrate at room temperature.

The reconstruction of a Si surface takes place at an interface between a Si surface and the molten Au. Figure 50 shows the transformation of the Si surface. In Fig. 50(a), a large cluster of Au lay on the Si surface. At this stage the Au cluster was already molten. Also, both the surface of Si and the interface between Si and Au were still atomically rough. However, then, black dotty contrasts, which are atomic columns of Au, appeared at the left part of the interface between molten Au and Si, creeping to the left on the surface of Si. The S-L interface between Si and molten Au and the surface of Si, both of which were now covered with one layer of Au, became atomically flat. As the Au cluster became smaller and smaller, atomic columns of Au spread over the Si surface more and more, transforming

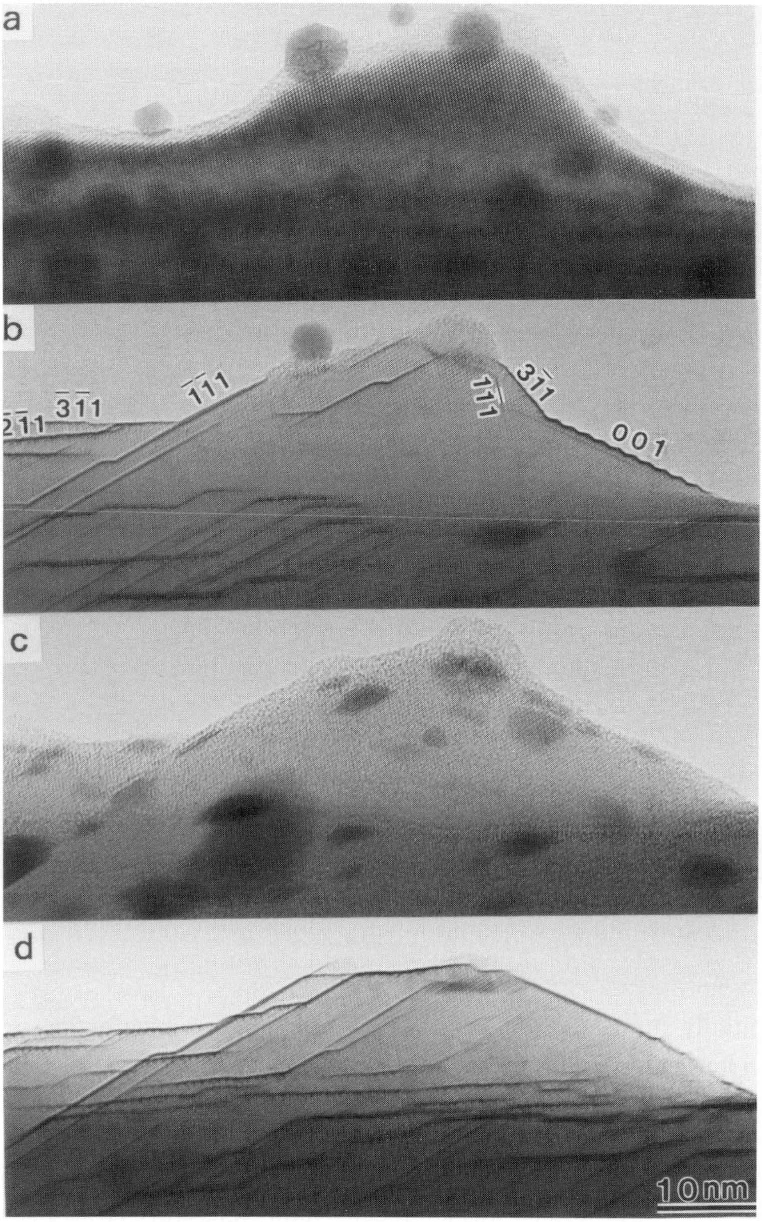

Fig. 49. Change in the morphology of Si surfaces during thermal cycles. (a) and (c) were taken at room temperature, while (b) and (d) were taken above the melting temperature of Au particles.

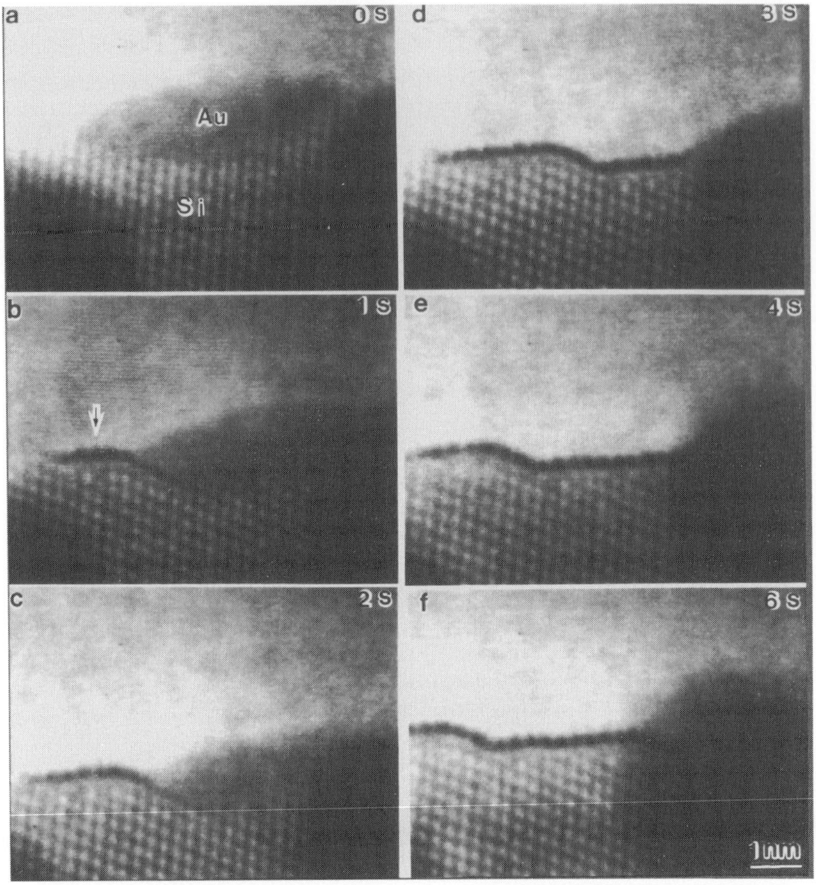

Fig. 50. Dynamic observation of the transformation of a Si surface.

the initially atomically rough surface of Si into the atomically flat surface, as can be seen in Figs. 50(c) and (d), and eventually the whole surface of Si was covered with atomic columns of Au (Figs. 50(e) and (f)).

Figure 51 shows a sequence of HREM profile-view images and shows the process of reconstruction of a (001) Si surface covered with Au atoms. Initially the (001) surface was atomically flat (Fig. 51(a)). After 1 second, the left corner of the (001) surface rose by about 0.2 nm (position A in Fig. 51(b)), and the surface approximately 0.7–0.8 nm way from A to the right sank by about 0.2 nm (B in Fig. 51(b)), leading to the formation of

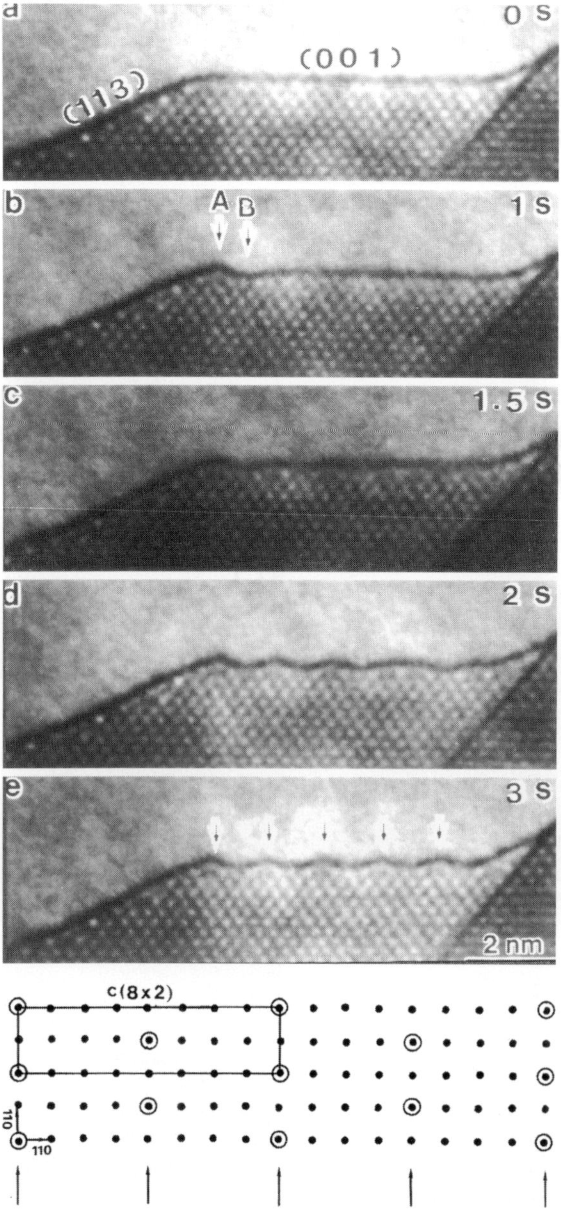

c(8×2) structure as viewed along the ⟨110⟩ direction.

Fig. 51. Sequence of HREM profile-view images showing the processes of reconstruction of a Si (001).

a characteristic chevron-shaped surface profile. The measured distance between the two neighbouring tops of the chevron was 1.54 nm; this is exactly four times the periodicity of the lattice spacing of (110) plane of Si. This reconstructed surface most probably corresponds to the c(8 × 2) structure observed in an UHV. It is noted that the present experiment was carried out in a non-UHV between 4×10^{-6} and 6×10^{-6}Pa. Thus, the clean Au/Si(001) surface can be obtained in a non-UHV.

4.3.2. *Al on Si*

The reconstruction of a Si surface due to wetting of liquid Al was also observed.[51] Figure 52(a–d) shows a sequence of HREM images showing a change in the morphology of a Si surface before and during heating

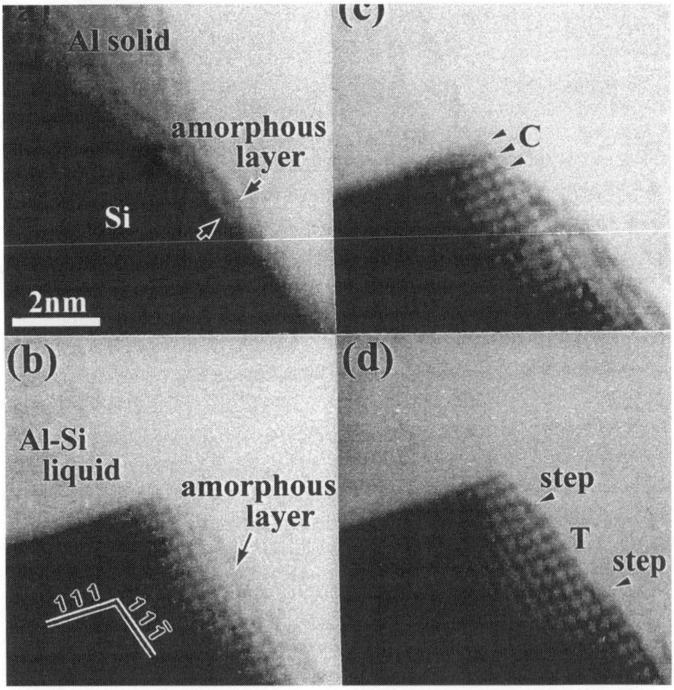

Fig. 52. HREM showing change in morphology of a Si surface (a) before and (b-d) during heating above the melting point of Al.

above the melting temperature of Al particles. The size of Al particles is a few hundreds of nanometers. Thus, the melting point of the Al particles should be same as that of a bulk sample, i.e., 660°C. Before heating (Fig. 52(a)), both Al (upper) and Si (lower) were crystalline and the surface of Si was atomically rough and covered with a native oxide amorphous layer. When the specimen was heated above the melting temperature of Al, the Al crystal became molten and a liquid Al-solid Si (S-L) interface was formed. For a few seconds just after the melting of Al, the surface amorphous layer persisted on the Si surface, as shown in Fig. 52(b). The amorphous layer, however, gradually disappeared from the vicinity of the Al-Si interface as show by C in Fig. 52(c), and eventually a clean surface appeared over a considerably wide area of the Si surface (Fig. 52(d)). At the same time, the surface of Si was transformed from atomically rough to atomically flat. The surface of Si now constituted a terrace-step structure, as indicated by *T* in Fig. 52(d).

Once the surface of Si became clean and atomically flat, the individual atomic columns near the triple points among solid Si, liquid Al-Si alloy and the vacuum became very active. Figure 53 shows a typical example. At the S-L interface between solid Si and liquid Al-Si alloy, a transition layer is present, as already described in detail in Sec. 4.2.3. In Fig. 53(a) a step of one monolayer thickness is observed near the triple point. The height of the step became two in Fig. 53(b). The height of the step fluctuated between one and two repeatedly, as can be seen from Fig. 53(b–h). Furthermore, the configuration of the atomic columns near the triple point is different from that in the matrix. Fluctuation of the position of a step with respect to the S-L interface is also seen in Fig. 53(e–g).

A long-range migration of a step is shown in Fig.54.[48] Step **1** was nucleated at the right-hand side of a Si surface (Fig. 54(a)), and migrated to the left (Fig. 54(b)). Then, another step **2** was nucleated where step **1** had been nucleated (Fig. 54(c)), and then migrated to the right as well (Fig. 54(d)). The two steps were blocked and piled up to form a stable configuration (Fig. 54(e)).

Figure 55 shows HREM images of those Si surfaces with different orientations which were made clean by molten Al atoms. The transformation from an atomically rough to an atomically flat surface takes place not only on low-index surfaces such as (110) and (111) but also on high-index

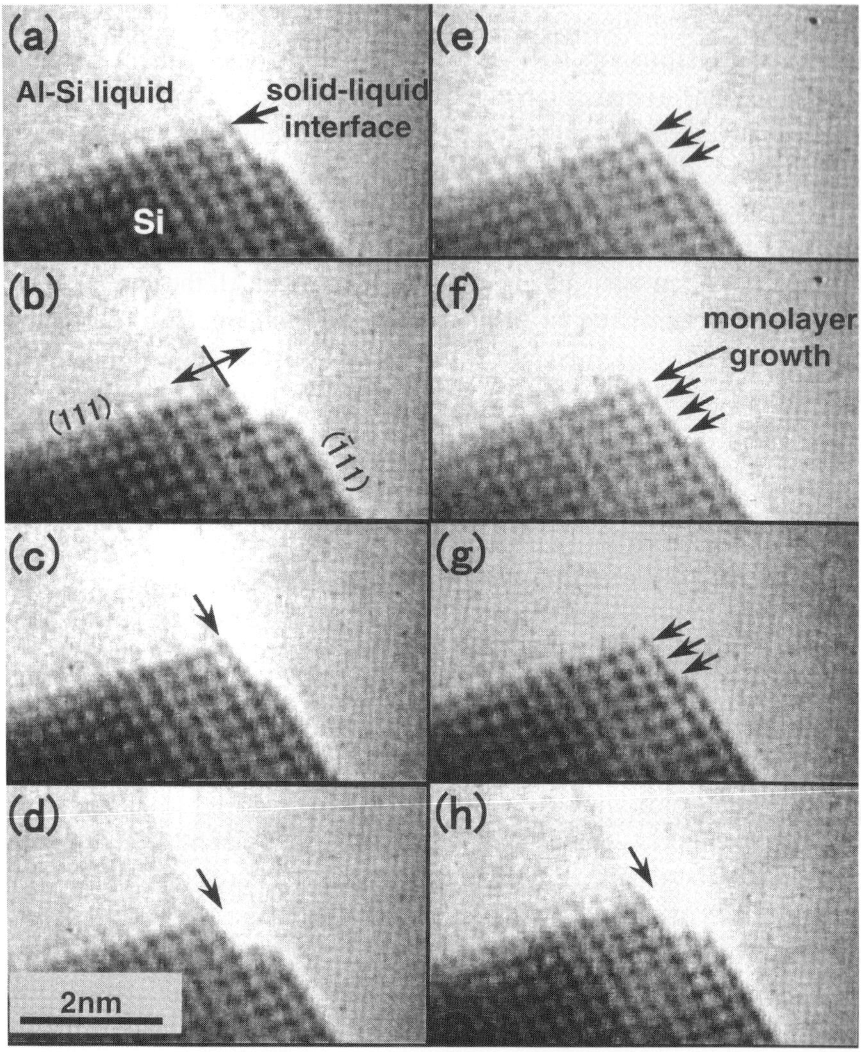

Fig. 53. Motions of individual atomic columns near the triple point between the vacuum and solid Si-liquid Al (S-L) interface.

surfaces such as (112), (115) and (773). Among these, the (112), (115) and (773) surfaces have periodic features as shown by arrows in Figs. 55(c), (d) and (e), respectively. The (112) surface consists of (111)-oriented terraces and (001)-oriented steps. The (773) surface consists of (111)-oriented

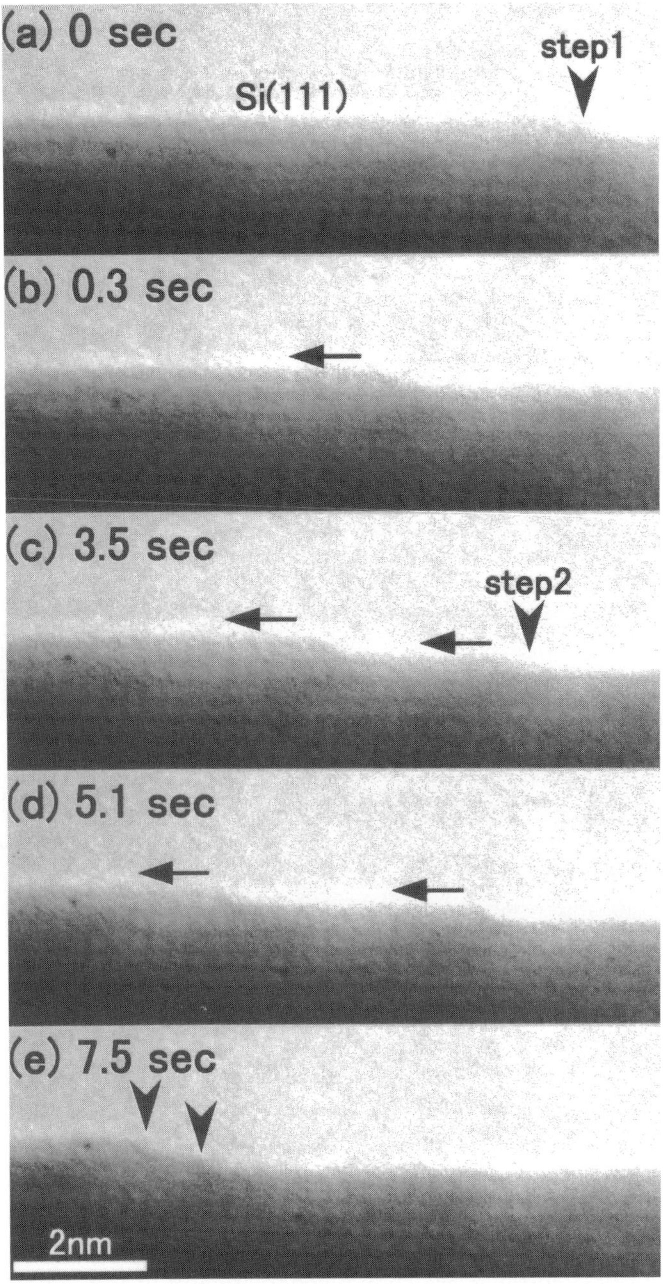

Fig. 54. Migration of steps on a Si surface.

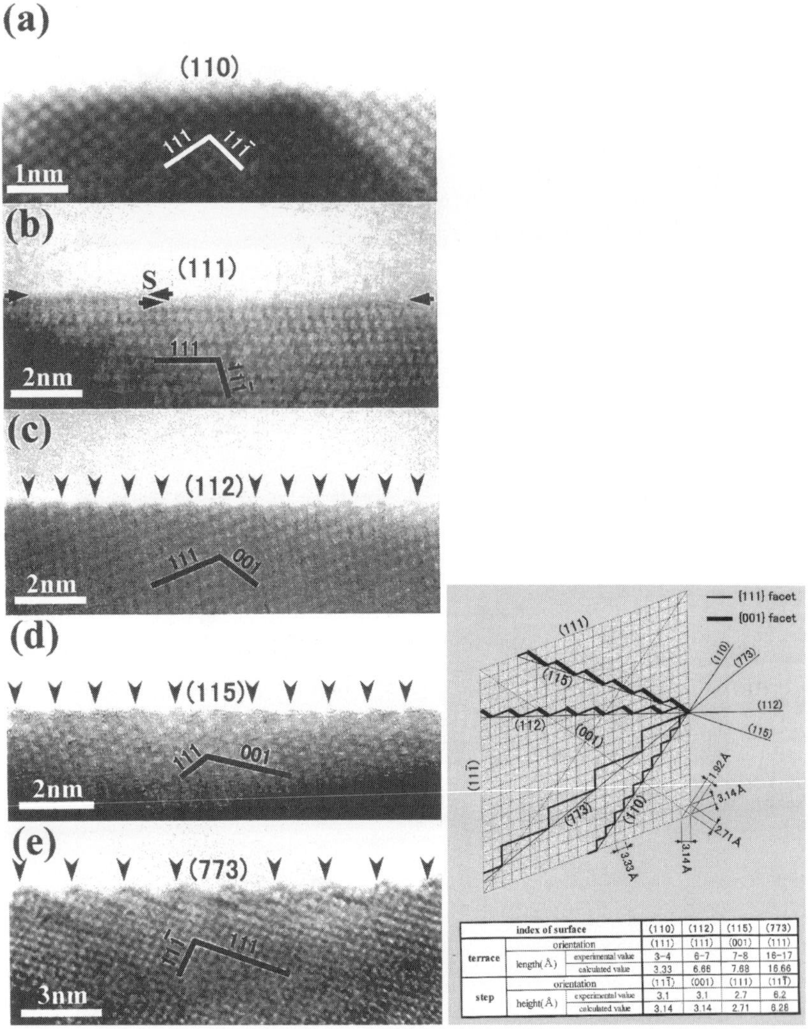

index of surface			(110)	(112)	(115)	(773)
terrace	orientation		(111)	(111)	(001)	(111)
	length(Å)	experimental value	3–4	6–7	7–8	16–17
		calculated value	3.33	6.66	7.68	16.66
step	orientation		(11$\bar{1}$)	(001)	(111)	(11$\bar{1}$)
	height(Å)	experimental value	3.1	3.1	2.7	6.2
		calculated value	3.14	3.14	2.71	6.28

Fig. 55. HREM images of Si surfaces with different orientations which were made clean by molten Al atoms.

terraces and (11$\bar{1}$)-oriented steps. The (115) surface consists of (001)-oriented terraces and (111)-oriented step. The structures of these (112), (115) and (773) surfaces are shown schematically in Fig. 55(f), together with that of the (110) surface.

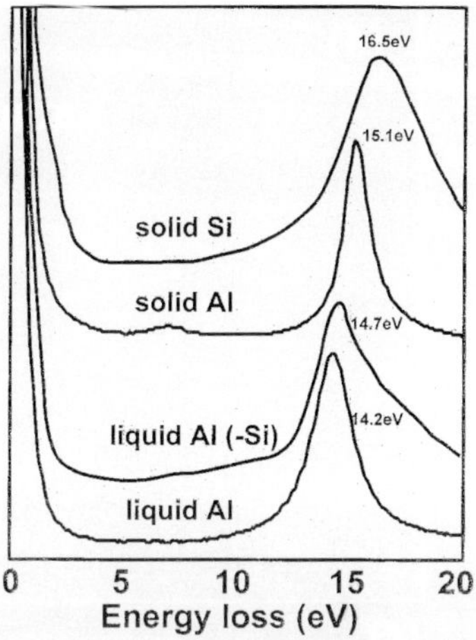

Fig. 56. Low-energy EELS spectra of pure liquid Al, liquid Al-Si alloy, solid Al and Si.

Similar change in morphology of a Si surface by wetting of a liquid Au was observed, as already described in Sec. 4.3.1. In the case of Au-Si, the outermost layer of the modified surface has a very strong contrast (Figs. 49, 50 and 51). However, in the case of Al-Si, the outermost layer of the modified surface will not show such a strong contrast, if exists, since Al and Si have similar atomic weights. In order to confirm the existence of the outermost layers containing Al, mapping by EELS was carried out.[52] Figure 56 shows low loss spectra from liquid Al, solid Al, liquid Al-Si alloy and solid Si. The liquid Al has a plasmon loss peak at 14.2 eV and the liquid Al-Si alloy has a plasmon loss peak at 14.7 eV, while the solid Al and Si have plasmon loss peaks at 15.1 and 16.5 eV, respectively. Thus, it was possible to map the liquid Al-Si and the solid Si using 1.5 eV windows centred at 14.2 eV and 16.5 eV.

Figure 57 shows examples of elemental mapping. Figure 57(a) is a conventional bright-field image and Figs. 57(b) and (c) are maps of the solid Si and the liquid Al-Si, respectively. In Fig. 57(c) the liquid Al-Si becomes bright as expected. In addition, it is noted in Fig. 57(c) that on the surface of Si there is a thin bright layer. This indicates that the Si surface is wetted by Al. Fig. 57(d) shows another example of mapping but obtained using the Al-$L_{2,3}$ edge. It is clear that the surface of Si is covered with a layer

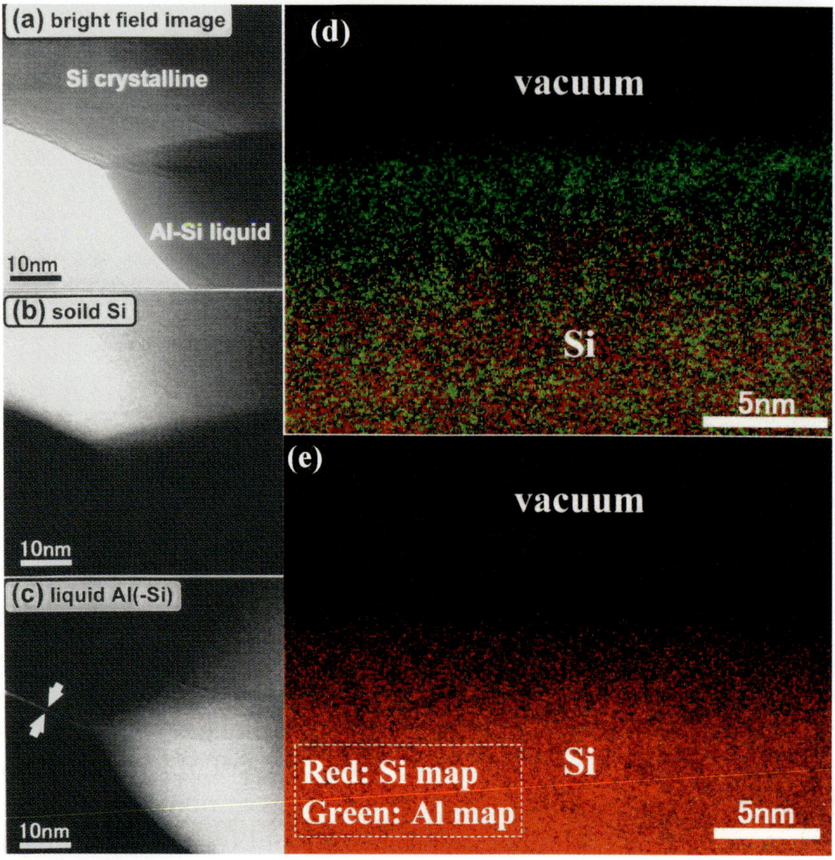

Fig. 57. (a) BF image near a triple point of solid Si-liquid Al-vacuum. (b) Plasmon loss mapping by solid Si and (c) by liquid Al-Si. (d) Elemental mapping on Si surface near the triple point. (e) Elemental mapping far away from the liquid droplet. Mapping was performed just above the melting point of Al(~943 K).

containing Al. By contrast, the surface of Si that is far away from the molten Al shows no evidence of segregation of Al, as shown in Fig. 57(e).

4.3.3. *Size dependence of the wetting angle of liquid metals on non-metallic substrates*

Apart from the reconstruction of the surface by wetting of liquid metals such as Au and Al, wetting of liquid metals on non-metallic substrates is

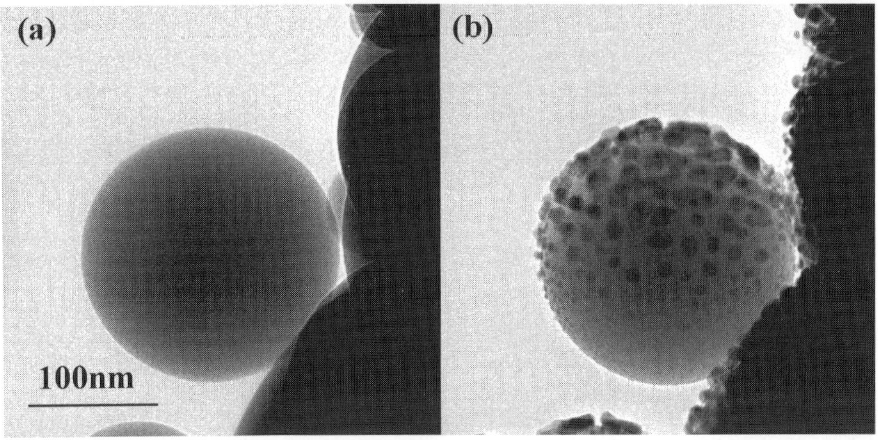

Fig. 58. SiO_2 particle (a) before and (b) after deposition of Bi.

a very important issue in metal processing. Now, the wetting angle can be measured accurately with *in-situ* heating experiments.[53] For this purpose, again the Kamino holder of two-wire type is useful. On the lower heating element **2**, particles of amorphous SiO_2 with perfect spherical shape (donated by Admatechs, Corp.) were mounted. On the upper heating element **1** were mounted Bi particles, and heated to evaporate Bi, which was deposited onto the surface of SiO_2 sitting on the heating element **2**. Figure 58 shows a SiO_2 particle before and after deposition of Bi. The Bi particles are crystalline and facetted at room temperature (Fig. 59(a)), but on heating the heating element above the melting point of Bi particles, the Bi particles became molten and spherical, as shown in Fig. 59(b).

Since the SiO_2 substrate is perfectly spherical, it is very convenient to measure the wetting angle on those Bi particles for which the interfaces between the substrate and the particles under consideration were end-on, some examples of which being indicated by arrows.

Figure 60 shows an example of the size dependence of the contact angle of Bi liquid cluster supported on a SiO_2 substrate. It is apparent that the contact angle of the smaller Bi liquid cluster B is smaller than that of the larger one A. It is to be noted that both of clusters A and B sat nearby on the same SiO_2 particle, so that their histories should have been very similar except their sizes.

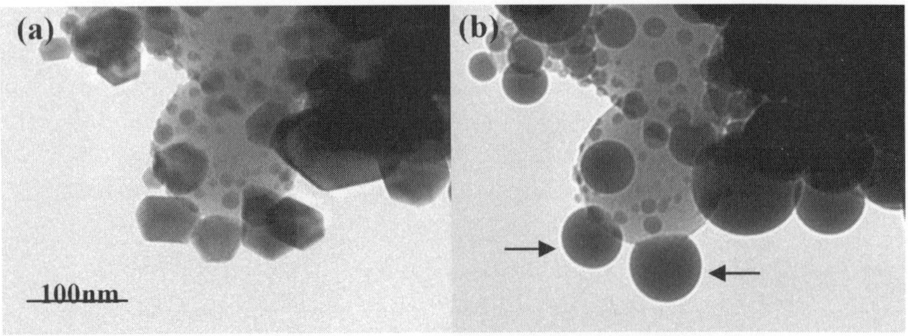

Fig. 59. Bi particles on SiO_2 substrate (a) before and (b) after melting of Bi particles.

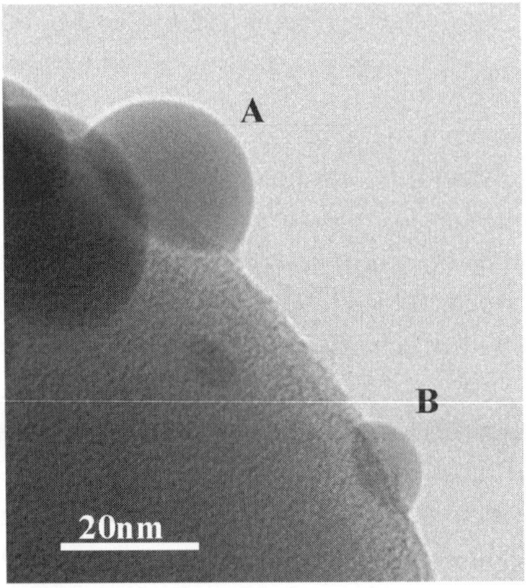

Fig. 60. Example of size dependence of the wetting angle of Bi liquid on SiO_2 substrates.

Figure 61 shows HREM images of Bi particles before and after melting. Before melting (Fig. 61(a)), the cluster shows lattice fringes. At the edge of the cluster existed the so-called clouds, as indicated by arrows. After melting (Fig. 61(b)), the cluster became perfectly spherical and the lattice fringes disappeared. It is also clear that there is no evidence of the intermediate layer between the Bi particle and the SiO_2 substrate.

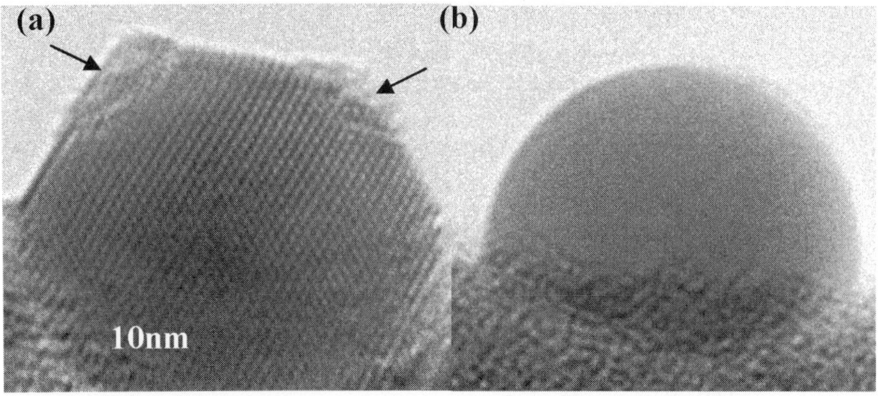

Fig. 61. HREM images of Bi particle (a) before and (b) after melting.

Similar measurements were carried out on a variety of substrate for liquid droplets of Bi and Sn, and the results are summarized in Figs. 62(a) and (b) for Bi and Sn, respectively. The wetting angle of Bi and Sn liquid clusters supported on non-metallic substrates shows essentially similar behaviour, that is, the wetting angle of liquid clusters of Bi and Sn remains unchanged or decreases very slowly with decreasing the size of cluster till around 20–40 nm in diameter, then decreasing precipitously with further decreasing the diameter. The wetting angles of Bi and Sn particles with a diameter larger than, say, 50 nm coincide with those obtained by more macroscopic methods. Thus, this technique should be very powerful in measuring the wetting angles of various materials on various substrates.

5. Solid-Gas Reactions

5.1. *Oxidation of Si*

In-situ experiments on solid-gas reactions have gathered much attention because of increasing interests in catalysis.[54,55] Now with a dedicated environmental transmission electron microscope (ETEM), it is possible to carry out *in-situ* HREM observation under a pressure of a few hundreds Pa. However, this type of ETEM usually requires massive modifications to the specimen chamber and the column of the TEM. The Kamino holder

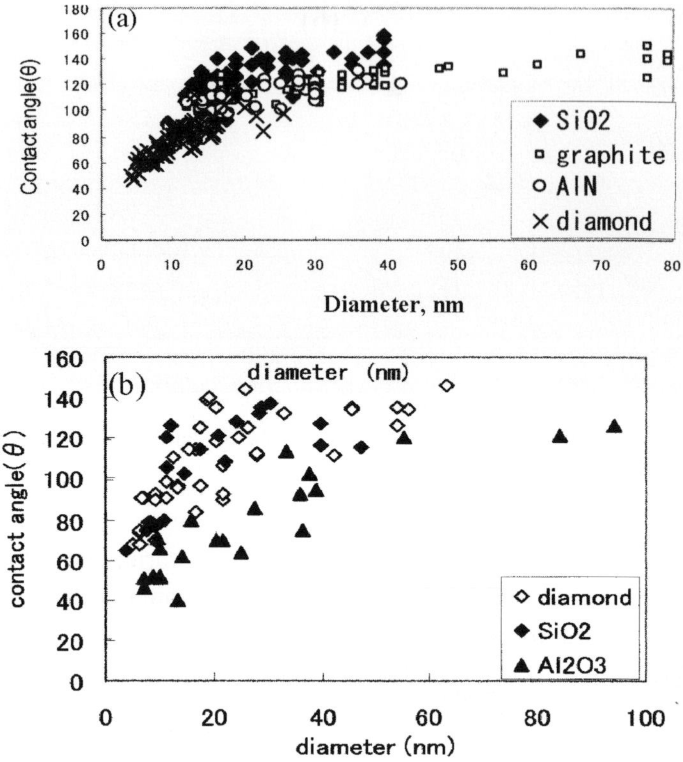

Fig. 62. Wetting angles of (a) Bi and (b) Sn liquid droplets on a variety of substrates.

with a gas-injecting nozzle, shown in Fig. 1(c), can be attached to a conventional TEM, and allows HREM observation under a pressure of up to 10^{-2}Pa.[16] This holder was employed to observe reduction of a native SiO_2 layer which covered a surface of Si.

Figure 63 shows a typical example of *in-situ* observations of reduction of SiO_2 and re-oxidation of a fresh surface of Si which appeared as a result of reduction, together with relevant EELS spectra. Figure 63(a) shows a HREM image of Si at room temperature, the surface of which was covered with a 3.0 nm thick amorphous layer. EELS spectrum taken from the amorphous layer indicates that the layer is SiO_2 (Fig. 63(d) and (e)). The specimen was heated at 973 K in the vacuum

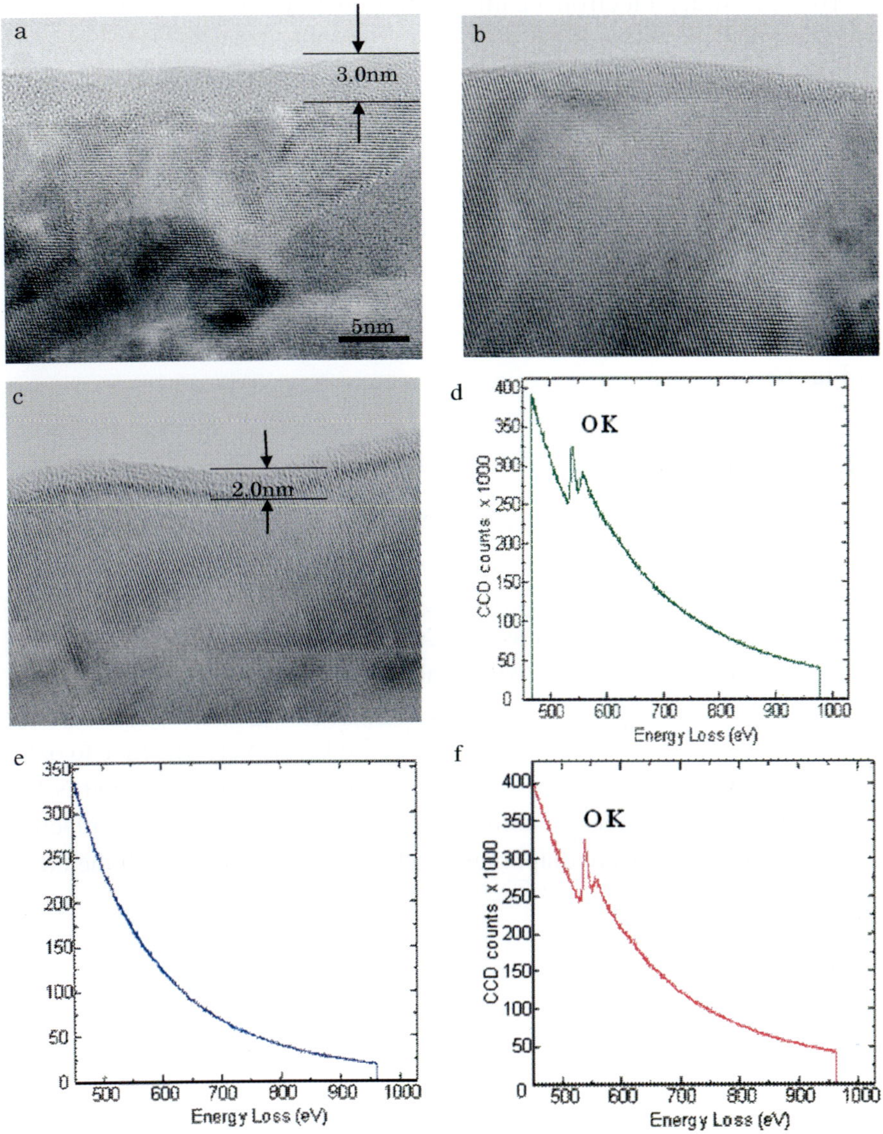

Fig. 63. HREM images of a Si particle (a) before and (b) after reduction under a high vacuum. (c) is HREM image of the Si re-oxidized. (d) and (e) are EELS spectra from SiO_2 layer on Si shown in (a). (f) is EELS spectrum from reduced Si in (b).

of 10^{-5} Pa under electron beam irradiation with the electron density of 20 A cm^{-2}.

This resulted in decrease in the thickness of SiO_2 layer down to as small as 0.3 nm (Fig. 63(b)). EELS spectrum obtained from such a reduced particle shows no evidence of O line (Fig. 63(f)). Then, oxygen gas with 5N purity was injected to the specimen in such a way that the pressure of the specimen chamber was increased gradually from 3.0×10^{-5} to 8.0×10^{-3} Pa in 1 h, while keeping the specimen temperature at 973 K. The surface of the once-reduced Si particle was oxidized again and SiO_2 layer with the thickness of 20 nm was formed again as shown in Fig. 63(c).

5.2. *Three-way catalyst*

CeO_2-based three-way catalysts (TWCs) have attracted much attention as a candidate for controlling the air pollution from automotive emissions. A ceria-zirconia solid solution ($Ce_2Ze_2O_{7+x}$; $0 \leqq x \leqq 1$) having the pyrochlore structure is now widely used as TWCs in automobiles because of a particularly excellent ability for oxygen absorption/release.

Oxidation and reduction of $Ce_2Ze_2O_{7+x}$ ($0 \leqq x \leqq 1$) were studied by *in-situ* experiments.[56–59] It was found that $Ce_2Ze_2O_{7+x}$ is very sensitive to oxygen. Thus, TEM observation was carried out in a Hitachi H-9000 NAR operated at 300 kV under a controlled vacuum of 4×10^{-4}Pa by replacing the selected area aperture diaphragm with an air leak valve, with the electron flux of $\sim 2 \times 10^{21}$e/m^2 s[59].

Figure 64(a) shows HREM image of as-prepared $Ce_2Ze_2O_7$ viewed along [100] direction, together with the corresponding diffraction pattern. The dark area at the left side is a Pt particle. In the HREM image, lattice fringes with 0.38 nm spacing are observed, which correspond to 220 spot in the diffraction pattern. Figure 64(b) was taken 20 m after (a). By this time $Ce_2Ze_2O_7$ had absorbed oxygen to be partially oxidized into $Ce_2Ze_2O_{7.5}$, as can be seen from HREM and the diffraction pattern. In the diffraction pattern, 200 spots, which are forbidden in pyrochlore structure, appeared and in the HREM lattice fringes with a spacing of 0.53 nm were observed. After 2 hr oxidation into $Ce_2Ze_2O_8$ was completed, as can be seen in the HREM and the diffraction pattern in (c); 100 spots now

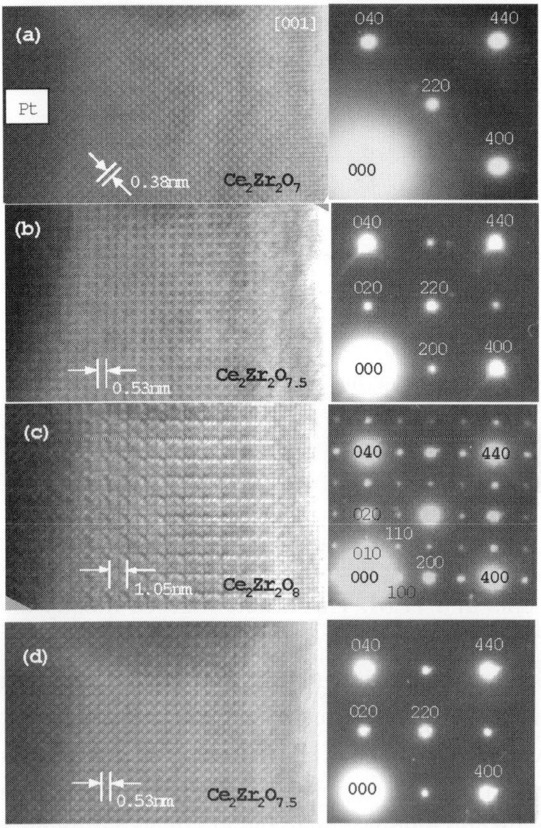

Fig. 64. HREM images and the corresponding diffraction patterns of $Ce_2Ze_2O_{7+x}$ ($0 \leqq x \leqq 1$). (a) initial, (b) after 20m, (c) after 2h, and (d) irradiated with a high-density electron beam after (c).

appeared and the corresponding lattice fringes with a spacing 1.05 nm appeared in the HREM image. When this completely oxidized $Ce_2Ze_2O_8$ was irradiated with a high density flux of electrons, $Ce_2Ze_2O_8$ was partially reduced back to $Ce_2Ze_2O_{7.5}$ (d).

Figures 65(a), (b) and (c) show the Ce-$M_{4,5}$ white-line peaks corresponding to Fig.64(a),(b) and (c), respectively. The ratio of intensities of Ce-M_5 and Ce-M_4 peaks $I(M_4)/I(M_5)$ changed from 0.95 for $Ce_2Ze_2O_7$ to 1.25 for $Ce_2Ze_2O_8$. This indicates that the valence state of Ce changed from Ce^{3+} to Ce^{4+} by the oxidation from $Ce_2Ze_2O_7$ to $Ce_2Ze_2O_8$

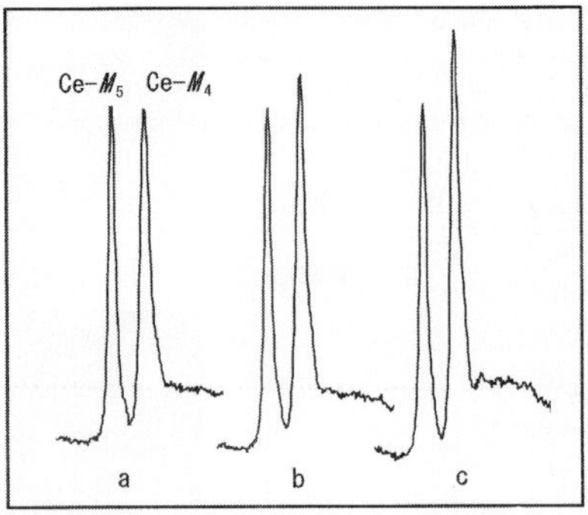

Fig. 65. (a),(b),(c) Ce-$M_{4,5}$ white-line peaks of $Ce_2Ze_2O_{7+x}$; $0 \leqq x \leqq 1$, corresponding to (a), (b) and (c) in Fig. 64, that is, $Ce_2Ze_2O_7$ (a), $Ce_2Ze_2O_{7.5}$ (b) $Ce_2Ze_2O_8$ (c).

6. Conclusions and Outlook

The author hopes that this article has shown a variety and usefulness of the *in-situ* experiments in a TEM. This article focused mostly on *in-situ* experiments with high-resolution electron microscopy (HREM) mode. However, it should be emphasized that there are many problems of materials science and engineering that can be studied effectively by applying the *in-situ* experiments even in the conventional (non-HREM) mode. Another point is that the application of combined use of a variety of technique such as elemental analysis including mapping and holography to the *in-situ* experiments will open a further new way to the characterization of real materials and real processing.

Acknowledgments

The works described in this review are a compilation of the researches carried out by the author's group. The author thanks his colleagues, especially Dr. T. Kamino, Professor K. Sasaki, Dr. S. Arai, Dr. S. Tsukimoto, Professor K. Kuroda and many students.

References

1. P. B. Hirsch, R. W. Horne, and M. J. Whelan, *Phil. Mag.*, **1**, 677 (1956).
2. E. P. Butler and K. F. Hale, *Dynamic Experiments in the Electron Microscope*, (North-Holland, New York, 1981).

3. T. Imura and H. Saka, *Memoirs Faculty Engg Nagoya University*, **28**, 54 (1976).
4. D. J. H. Cockayne, I. L. F. Ray, and M. J. Whelan, *Phil. Mag.***20**, 1265 (1969).
5. H. Saka, Y. Sueki, and T. Imura, *Phil. Mag.*, A, **37**, 273 (1978).
6. H. Saka, T. Iwata, and T. Imura, *Phil. Mag.*, A, **37**, 291 (1978).
7. H. Saka, T. Kondo, and T. Imura, *Phil. Mag.*, A, **47**, 859 (1983).
8. H. Saka, T. Kondo, and N. Kiba, *Phil. Mag.*, A, **44**, 1213 (1981).
9. H. Saka and T. Kamino, *In situ Microscopy in Materials Research*, (ed.) P.L. Gai (Dordrecht: Kluwar, 1997) p. 173.
10. T. Kamino, K. Sasaki, and H. Saka, *Microsc. Microanal.*, **3**, 393 (1997).
11. J. M. Howe and H. Saka, *MRS Bulletin*, **29**, 951 (2004).
12. H. Saka, K. Sasaki, S. Tsukimoto, and S. Arai, *J. Mat. Res.*, **20,** 1629 (2005).
13. H. Saka, S. Tsukimoto, and K. Sasaki, *Korean J. Electron Microsco.*, **36**, (Special Issue, 1), 9 (2006).
14. T. Kaimino and H. Saka, *Microsco. Microanal. Microstruct.*, **4**, 127 (1993).
15. H. Mori, H. Yasuda, and T. Kamino, *Phil. Mag. Lett.*, **68**, 279 (1994).
16. T. Kamino, T. Yaguchi, M. Konno, A. Watabe, T. Marukawa, T. Mima, K. Kuroda, H. Saka, S. Arai, H. Makino, Y. Suzuki, and K. Kishita, *J. Electron Microsco.*, **54**, 497 (2005).
17. T. Kamino, T. Yaguchi, and H. Saka, *J. Electron Microsco.*, **43**, 10 (1994).
18. T. Kamino, T. Yaguchi, M. Ukiana, Y. Yasutomi, and H. Saka, *Mater. Trans. JIM*, **36**, 73 (1995).
19. K. Sasaki, T. Murase, and H. Saka, *Ultramicroscopy*, **56**, 184 (1994).
20. K. Sasaki, H. Saka, and T. Arii, *Mater. Trans. JIM.*, **37**, 1037 (1996).
21. S. Tsukimoto, K. Sasaki, T. Hirayama, and H. Saka, *Phil. Mag. Lett.*, **76**, 173 (1997).
22. K. Sasaki and H. Saka, (unpublished data).
23. S. Frabboni, G. Matteucci, G. Pozzi, and M. Vani, *Phy. Rev. Lett.*, **55**, 2196 (1985).
24. E. P. ErNisse, *Appl. Phys. Lett.*, **35**, 8 (1979).
25. K. Sasaki, S. Tsukimoto, M. Konno, T. Kamino, and H. Saka, *J. Microsco.*, **203**, 12 (2001).
26. M. Takagi, *J.Phys.Soc.Jpn.*, **9**, 359 (1954).
27. G. L. Allen, R. A. Bayles, W. W. Gile, and W. A. Jesser, *Thin Solid Films*, **144**, 297 (1986).
28. H. Saka,Y. Nishikawa, and T. Imura, *Phil. Mag.*, A, **57**, 895 (1988).
29. T. Ohashi and K. Kuroda, and H. Saka, *Phil. Mag.*, B, **65**, 1041 (1992).
30. K. Sasaki and H. Saka, *Phil. Mag.*, A, **63**, 1207 (1991).
31. K. Sasaki and H. Saka, *Microsco. Microanal. Microstruct.*, **4**, 287 (1993).
32. E. J. Siem and E. Johnson, *J. Mater. Sci.*, **41**, 2703–2710 (2006).
33. Y. Senda, K. Sasaki, and H. Saka, *Phil. Mag.*, **84**, 2635 (2004).
34. N. Kato, N. Miura, and N. Tsutsui, *J. Vac. Sci. Technol.*,A, **16**, 1127 (1998).
35. H. Matsuki, H. Ibuka, and H. Saka, *Sci & Technol. Advanced Materi.*, **3**, 261 (2002).
36. T. Kato, K. Nunome, K. Y. Morimoto, K. Nishikawa and H. Saka, *Phil. Mag. Lett.*, **80**, 187 (2000).
37. J. Chang, T. Sakai, and H. Saka, *Phil. Mag. Lett.*, **85**, 247 (2005).
38. J. M. Howe, *Interfaces in Materials* (John Wiley, New York, 1997).
39. K. A. Jackson, *Liquid Metals and Solidification* (ASM, Cleveland, OH, 1958) p. 174.

40. D. E. Temkin, *Molecular Roughness of Crystal-Melt Boundary in Crystallization Process* (Consultant Burea, New York, NY, 1966) p. 151.
41. S. Arai, S. Tsukiomoto, and H. Saka, *Microsco. Microanal.*, **4**, 264 (1998).
42. K. Sasaki and H. Saka, *MRS Symp. Proc.*, **466**, 185 (1997).
43. S. Arai, S. Tsukimoto, H. Miyai, and H. Saka, *J. Electron Microsco.*, **48**, 317 (1999).
44. S. Arai, S. Tsukimoto, S. Muto, and H. Saka, *Microsc. Microanal.*, **6**, 358 (2000).
45. S. Arai, S. Tsukimoto, and H. Saka, *J. Electron Microsco.*, **52**, 79 (2003).
46. F. Spaepen, *Solid State Physics*, **47**, 1 (1994).
47. A. Bonissent and B. Mutaftschiev, *Phil. Mag.*, **35**, 65 (1997).
48. S. Tsukimoto, *Doctoral Dissertation*, (Nagoya University, 1999).
49. J. Z. Hu, L. D. Merkle, C. S. Menoni, and I. L. Spain, *Phys.Rev.*, B**34**, 4679 (1986).
50. T. Kamino,T. Yaguchi, M. Tomita, and H. Saka, *Phil. Mag.*, A**,** **75**, 105 (1997).
51. S. Tsukimoto, S. Arai, and H. Saka, *Phil. Mag. Lett.*, **79**, 913 (1999).
52. S. Tsukimoto, S. Arai, M. Konno, T. Kamino, K. Sasaki, and H. Saka, *J. Microsc.*, **203**, 17 (2007).
53. J. Murai, T. Marukawa, T. Mima, S. Arai, K. Sasaki, and H. Saka, *J. Mater. Sci.*, **41**, 2723 (2006).
54. P. L. Gai and E. D. Boyes, *Electron Microscopy in Heterogeneous Catalysis*, IOP Publishing (Bristol and Philadelphia, 2003).
55. R. Sharma, Chapter 2 in this book.
56. T. Sasaki, Y. Ukyo, A. Suda, M. Sugimoto, K. Kuroda, S. Arai, and H. Saka, *J. Ceram. Soc. Jpn.*, **111**, 382 (2003).
57. T. Sasaki, K. Ukyo, K. Kuroda, S. Arai, and H. Saka, *J. Electron Microsco.*, **52**, 309 (2003).
58. S. Arai, S. Muto, J. Murai, T. Sasaki, Y. Ukyo, K. Kuroda, and H. Saka, *Mater. Trans.*, **45**, 2951 (2004).
59. S. Arai, S. Muto, T. Sasaki, Y. Ukyo, K. Kuroda, and H. Saka, *Electrochemical & Solid-State Lett.*, **9**, E1 (2006).

CHAPTER 4

IN-SITU TRANSMISSION ELECTRON MICROSCOPY: NANOINDENTATION AND STRAINING EXPERIMENTS

Wouter A. Soer and Jeff T. De Hosson*

Department of Applied Physics, Netherlands Institute for Metals Research and Materials Science Centre, University of Groningen Nijenborgh 4, 9747 AG Groningen, the Netherlands
**j.t.m.de.hosson@rug.nl*

In the field of transmission electron microscopy there are still fundamental and practical reasons which hamper a straightforward correlation between microscopic structural information with the properties of materials. In this chapter it is argued that one should focus more on the generic features of defects, using a mesoscopic approach including various length scale transitions, and in particular on *in-situ* rather than on postmortem observations of solely atomic structures. This viewpoint has been exemplified with *in-situ* TEM indentation and *in-situ* straining studies at elevated temperatures of Al and Al-alloys with grain sizes ranging between submicrometers to submillimeters. It is concluded that *in-situ* Transmission Electron Microscopy has provided new insights into the interaction between dislocations and grain boundaries on various length scales, in which specifically the effect of Mg in Al-Mg alloys on these interaction mechanisms has been clarified.

1. Introduction

Microscopy in the field of materials science is generally devoted to linking *microstructural* observations to *properties*. The microstructural features in turn are determined by chemical composition and processing, and consequently, advanced microstructural investigations, certainly in the field of nano-science and technology, require a microscope with a resolving power in the order of sub-nanometer. However, the actual linkage between the microstructure studied by microscopy on one hand and the physical property of a material is almost elusive. The reason is that various physical

At these grain sizes, stress-induced movement of grain boundaries is an important deformation mechanism in pure Al. In Al-Mg however, the grain boundaries are found to be effectively pinned by solute Mg atoms. Such pinning effects may significantly enhance the mechanical properties of ultrafine-grained or nanocrystalline alloys. Subsequently, Sec. 6 deals with the mechanical behavior of Al-Mg alloys with a substantially larger grain size. Under specific conditions of strain rate and elevated temperature, coarse-grained Al-Mg alloys exhibit superplastic properties, i.e. they show very high elongations prior to failure, typically in excess of a few hundred percent. This makes them attractive candidates for the production of components with a large freedom of design. The physical mechanisms by which coarse-grained superplastic alloys deform are markedly different from those involved in conventional superplasticity of fine-grained materials. In particular, they allow for much higher forming rates, which is a considerable advantage from the perspective of commercial viability. This section shows *in-situ* straining experiments so as to unravel the deformation mechanisms responsible for the superplastic properties of coarse-grained Al-Mg alloys. To this end, the microstructure and dislocation substructure of the alloys are analyzed as a function of the deformation parameters. The observations are discussed in relation to dynamic reconstruction mechanisms and their influence on the ductility of the alloys.

2. *In-Situ* Nanoindentation in a TEM

The observation of the plastic deformation introduced by conventional nanoindentation has been restricted for a long time to post mortem studies of the deformed material, mostly by atomic force microscopy or scanning or transmission electron microscopy. This post mortem approach entails some significant limitations to the analysis of the deformation mechanisms. Most importantly, it does not allow direct observation of the microstructure during indentation and thus lacks the possibility to monitor deformation events and the evolution of dislocation structures as the indentation proceeds. Moreover, the deformed microstructure observed after indentation is generally different from that of the material under load, due to recovery during and after unloading. In the case of post mortem analysis by transmission electron microscopy, the preparation of the indented surface in the form of a thin foil often leads to mechanical damage to the specimen or relaxation of the stored deformation due to the proximity of free surfaces, thereby further obscuring the indentation-induced deformation.

The recently developed technique of *in-situ* nanoindentation in a transmission electron microscope[6–11] does not suffer from these limitations and allows for direct observation of indentation phenomena. Furthermore, as the indenter can be positioned on the specimen accurately by guidance of the TEM, regions of interest such as particular crystal orientations or grain boundaries can be specifically selected for indentation. *In-situ* nanoindentation measurements by Minor *et al.*[11] on polycrystalline aluminum films have provided experimental evidence that grain boundary motion is an important deformation mechanism when indenting thin films with a grain size of several hundreds of nanometers. This is a remarkable observation, since stress-induced grain boundary motion is not commonly observed at room temperature in this range of grain sizes.

Grain boundary motion in metals typically occurs at elevated temperatures driven by a free energy gradient across the boundary, which may be presented by the curvature of the boundary or stored deformation energy on either side of the boundary.[12] In the presence of an externally applied shear stress, Winning *et al.*[13] found that migration of both low-angle and high-angle grain boundaries in pure Al occurs at temperatures above 200°C. This type of stress-induced grain boundary motion (known as dynamic grain growth) is considered by many researchers to be the mechanism responsible for the extended elongations obtained in superplastic deformation of fine-grained materials (see Sec. 6). The occurrence of grain boundary motion in room temperature deformation of nanocrystalline fcc metals was anticipated recently by molecular dynamics simulations[14] and a simple bubble raft model.[15] Experimental observations of such grain boundary motion have subsequently been provided by *in-situ* straining experiments of nanocrystalline Ni thin films[16] and *in-situ* nanoindentation of nanocrystalline Al thin films.[17] In both the simulations and the experiments, grain boundary motion was observed for grain sizes below 20 nm. The dislocation mobility is greatly restricted at such grain sizes and other deformation mechanisms become more relevant. In contrast, the grain size for which grain boundary motion was found by *in-situ* nanoindentation[11] was of the order of 200 nm. In simple deformation modes such as uniform tension or compression, dislocation-based plasticity is still predominant in this regime and grain boundary motion generally does not occur. In the case of nanoindentation however, the stress field is highly inhomogeneous and consequently involves large stress gradients.[18] These stress gradients are thought to be the primary factor responsible for the observed grain boundary motion at room temperature. Another aspect that may contribute to the occurrence of this phenomenon is the specific geometry of the *in-situ* indentation specimens, which will be discussed in Sec. 2.2.

Since the properties of high purity metals such as pure Al are less relevant for the design of advanced materials, we have focused on the indentation behavior of Al-Mg films and the effect of Mg on the deformation mechanisms described above. To this end, *in-situ* nanoindentation experiments have been conducted on ultrafine-grained Al and Al-Mg films with varying Mg contents.[19–21] The classification "ultrafine-grained" in this respect is used for materials having a grain size of the order of several hundreds of nanometers. In this chapter, the TEM observations are interpreted and related to quantitative load-displacement data, obtained both directly from the *in-situ* indentation experiments and indirectly through conventional *ex-situ* nanoindentation on the same specimens.

2.1. *Stage design*

In-situ nanoindentation inside a TEM requires a special specimen stage designed to move an indenter towards an electron-transparent specimen on the optic axis of the microscope. The first indentation holder was developed in the late 1990s by Wall and Dahmen[6,7] for a high-voltage microscope at the National Center for Electron Microscopy (NCEM) in Berkeley, California. In the following years, several other stages were constructed at NCEM with improvements made to the control of the indenter movement and the ability to measure load and displacement. In the work described in this chapter, two of these stages were used: a homemade holder for a JEOL 200CX microscope,[8] and a prototype holder for a JEOL 3010 microscope with dedicated load and

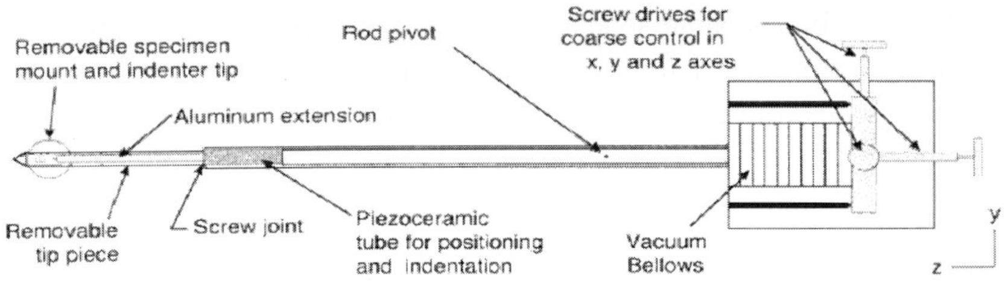

Fig. 1. Schematics of *in-situ* nanoindentation holder for a JEOL 200CX microscope.[8]

displacement sensors, developed in collaboration with Hysitron (Hysitron Inc., Minneapolis, MN).

The principal design of both holders is roughly the same. The indenter tip is mounted on a piezoceramic tube as illustrated in Fig. 1. This type of actuator allows high-precision movement of the tip in three dimensions, the indentation direction being perpendicular to the electron beam. Coarse positioning is provided by manual screw drives that move the indenter assembly against the vacuum bellows. The indenter itself is a Berkovich-type diamond tip, which is boron-doped in order to be electrically conductive in the TEM. The goniometer of the TEM provides a single tilt axis, so that suitable diffraction conditions can be set up prior to indentation.

The motion of the indenter into the specimen during indentation is controlled by the piezoceramic tube. In the holder for the JEOL 200CX, the voltage applied to the tube is controlled manually and recorded together with the TEM image. Since the compliance of the load frame is relatively high, the actual displacement of the indenter into the material depends not only on the applied voltage, but also to a certain extent on the response of the material. Consequently, this indentation mode is neither load- nor displacement-controlled. In the prototype holder for the JEOL 3010, a capacitive sensor monitors the load and displacement during indentation. The displacement signal is used as input for a feedback system that controls the voltage on the piezoceramic tube based on a proportional-integral-derivative (PID) algorithm.[22] The indentation is therefore displacement-controlled and can be programmed to follow a predefined displacement profile as a function of time.

The need for a separate load and displacement sensor as implemented in the prototype holder is mainly due to the complex response of the piezo tube. If the response were fully known, the load could be calculated at any time during indentation from the displacement (which can be determined directly from the TEM image) and the characteristics of the load frame.[23] Ideally, the correlation between the applied voltage and the displacement of the piezo element is linear. However, hysteresis and saturation effects lead to significant nonlinearities. Moreover, as lateral motion is achieved by bending the tube, the state of deflection strongly affects the response in the indentation direction as well. Calibration measurements of the piezo response in vacuum at 12 points across the lateral range showed an average proportionality constant of 0.12 μm/V with a standard deviation as large as 0.04 μm/V. Although during indentation, the deflection of the tube is approximately constant and the response becomes more reproducible, the abovementioned hysteresis and saturation effects still complicate the measurement of the load. The implementation of a dedicated load sensor, as in the new prototype holder, is therefore essential for obtaining reliable quantitative indentation data. The *in-situ* indentation load-displacement curves presented in this chapter have all been produced with this displacement-controlled holder.

2.2. *Specimen geometry*

The geometry of the specimens used for *in-situ* nanoindentation has to comply with two basic requirements: (i) an electron-transparent area of the specimen must be accessible to the indenter in a direction perpendicular to the electron beam, and (ii) this area of the specimen must be rigid enough to support indentation without bending or breaking. A geometry that fulfills both these requirements is a wedge that is truncated to a cap width large enough to provide the necessary rigidity while still allowing the electron beam to pass through. For the present investigation, we used wedge specimens prepared by bulk silicon micro-machining. Using this technique, wedge-shaped protrusions are routinely prepared on Si (001) substrates with a resolution of the order of 1 μm. The side planes of the ridge are aligned with {111} planes of the silicon crystal, so that repeated annealing and oxide removal subsequently leads to the sharpening of the wedge driven by a reduction of the surface energy. In this way, a cap width

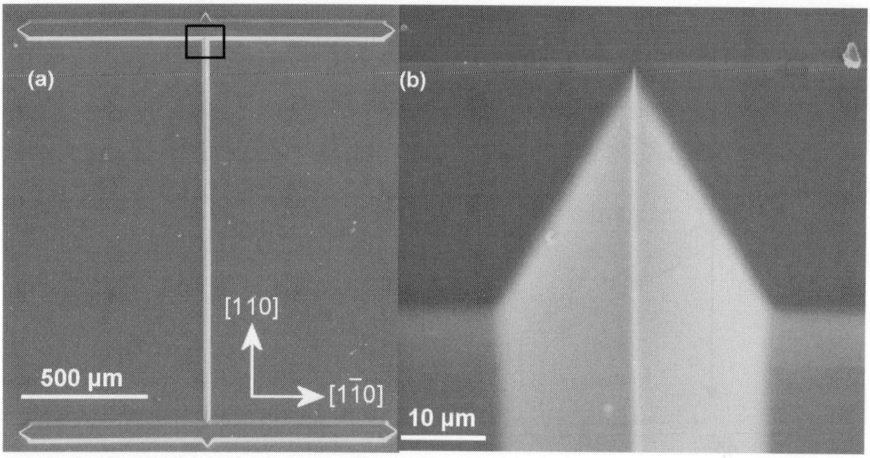

Fig. 2. (a) Low-magnification scanning electron micrograph of the microfabricated Si specimen geometry, consisting of an H-shaped structure in which the crossbar of the H is a sharp ridge. (b) Magnified image of one of the ends of the ridge as indicated by the rectangle in image (a).

of the order of 100 nm can be achieved. Figure 2 shows the micrographs of the specimen design used. The ridge has a length of 1.5 mm and a height of 23 μm above the substrate. The included angle between the {111} side planes is 54.7°.

The silicon ridge specimen geometry provides a means to investigate any material that can be deposited as a thin film onto the silicon substrate. Metals with a low atomic number such as aluminum are particularly suitable for this purpose, since films of these metals can be made to several hundreds of nanometers thickness and still be transparent at the cap of the wedge to electrons with typical energies of 200–300 keV, as schematically depicted in Fig. 3(a). An example of a resulting TEM image is shown in Fig. 3(b).

3. Experimental Procedure

3.1. *Specimen preparation and microstructure*

The Al and Al-Mg films that will be discussed in this chapter were deposited by thermal evaporation. The substrate was kept at 300°C to

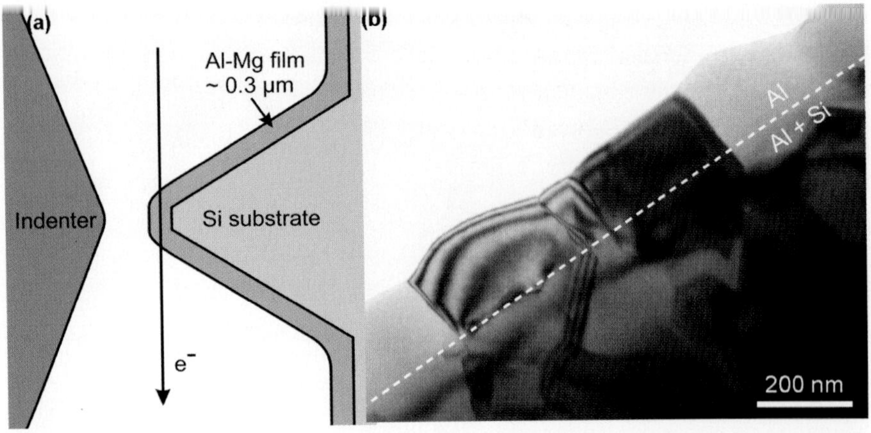

Fig. 3. (a) Schematic of *in-situ* indentation setup. The deposited Al-Mg film is electron-transparent and accessible to the indenter at the tip of the Si wedge. (b) Typical bright-field image of a deposited film. The dashed line shows the top of the Si ridge.

establish a grain size of the order of the layer thickness, which was 200 to 300 nm for all specimens. After evaporation, the substrate heating was switched off, allowing the specimen to cool down to room temperature in approximately one hour. One pure Al film was prepared by evaporating a high purity (5N) aluminum source. Deposition of the Al-Mg alloy films was achieved by evaporating alloys with varying Mg contents. Since Al and Mg have different melting temperatures and vapor pressures, the Mg content of the deposited film is not necessarily equal to that of the evaporated material. Moreover, the actual evaporation rates depend on the quality of the vacuum and the time profile of the crucible temperature. The composition of the deposited alloy films was therefore determined by energy dispersive spectrometry (EDS) in a scanning electron microscope at 5 kV. The measured Mg concentrations of the four Al-Mg films prepared were 1.1, 1.8, 2.6 and 5.0 wt%.

Since the solubility level of Mg in Al is 1.9 wt% at room temperature,[24] β' and β precipitates were formed in the 2.6 and 5.0 wt% Mg specimens due to the relatively long cooling time. The attainable image resolution in the indentation setup was not high enough to resolve these precipitates, being compromised by the thickness of the specimen and possibly by the fact that the electron beam travels very closely to the substrate over a large distance. Nevertheless, the presence of precipitates both in the matrix and at the grain boundaries could be confirmed by strain contrast and distorted grain boundary fringes, respectively, which were not observed in the 1.1 and 1.8 wt% Mg specimens (Fig. 4). Furthermore, the presence of the brittle β phase on the grain boundaries leads to the appearance of intergranular cracks in the 2.6 and 5.0 wt% Mg specimens, as shown in the scanning electron micrographs in Fig. 5. While Al deposited on a clean Si (001) surface may give rise to a characteristic

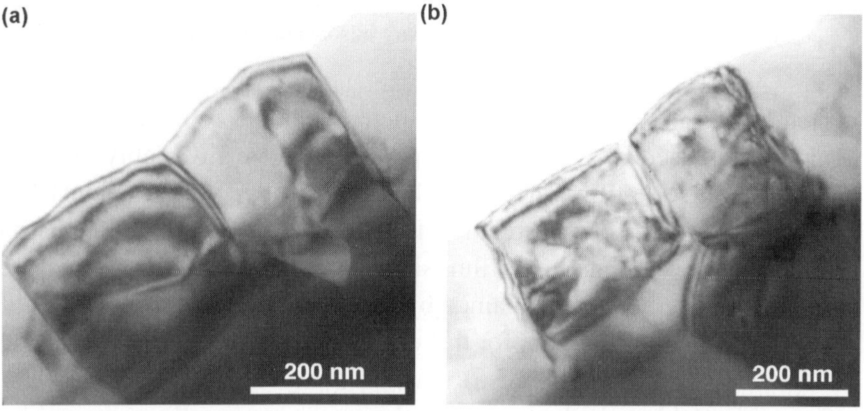

Fig. 4. Bright-field images of evaporated Al-Mg layers with (a) 1.1 and (b) 5.0 wt% Mg. The presence of Al-Mg precipitates in (b) is revealed by strain contrast.

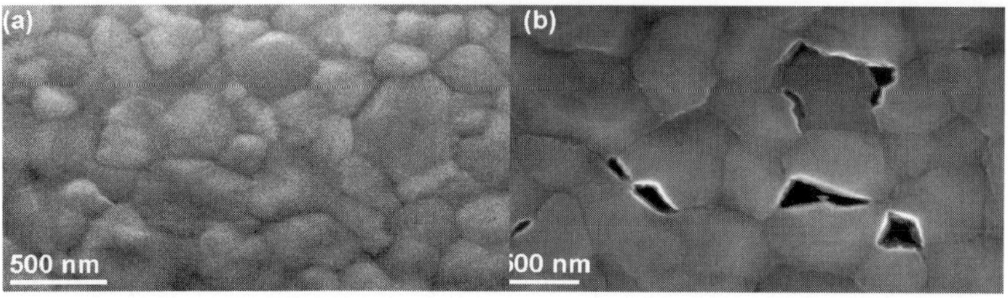

Fig. 5. Scanning electron micrographs of (a) pure Al film and (b) Al-5.0%Mg film away from the ridge. Cusped grain boundaries give rise to considerable surface roughness in both films. Grain boundary embrittlement by β precipitates leads to the appearance of intergranular cracks in the Al-5.0%Mg film.

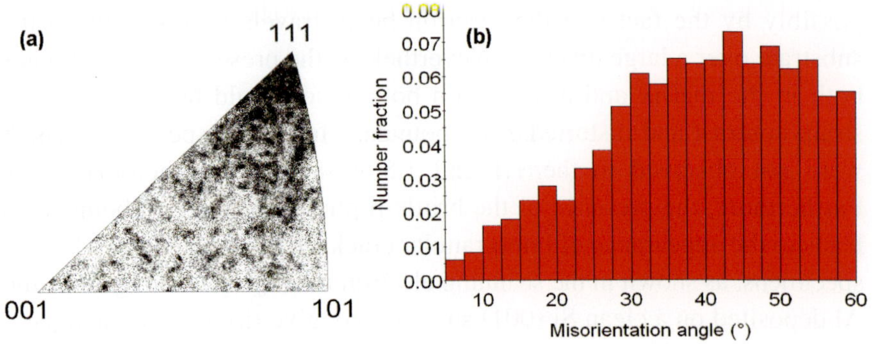

Fig. 6. EBSD scan Al film: (a) discrete pole figure showing the $\langle 111 \rangle$ texture of the evaporated film; (b) distribution of the grain boundary misorientation angle.

mazed bicrystal structure due to two heteroepitaxial relationships,[25] the Si substrates used in the present experiments were invariably covered with a native oxide film. Therefore, the orientations of the Al and Al-Mg grains of the film show no relation to that of the Si surface. An EBSD scan on the evaporated Al film showed a significant $\langle 111 \rangle$ texture (Fig. 6(a)), which can be explained by the fact that the surface energy of fcc materials has a minimum for this orientation. Furthermore, the EBSD measurements provided the distribution of the grain boundary misorientations (Fig. 6(b)), which shows that the grains are mostly separated by random high-angle grain boundaries with no significant preference for particular CSL orientations.

3.2. *In-situ and exsitu nanoindentation experiments*

On each of the evaporated films, three to four *in-situ* experiments were carried out with maximum depths ranging from 50 to 150 nm, using the indentation stage for the JEOL 200CX. The indentation rate, being controlled manually through the piezo voltage, was of the order of 5 nm/s. In addition, several quantitative *in-situ* indentation experiments were conducted with the prototype holder for the JEOL 3010 microscope on the Al and Al-2.6%Mg films. These displacement-controlled indentations were made to a depth of approximately 150 nm with a loading time of 20 s. In order to be able to resolve grain boundary phenomena during each *in-situ* indentation, the specimen was tilted to such an orientation that two adjacent grains were both in (different) two-beam conditions.

Conventional nanoindentation measurements were carried out *ex-situ* on the same films away from the wedge. As in the *in-situ* experiments, a pyramidal Berkovich tip was used. Load-controlled indentations were executed to maximum depths of 50, 100 and 150 nm at a targeted strain rate of 0.05 s^{-1}, defined as loading rate divided by load. At this strain rate the indenter velocity during loading was of the order of 2 nm/s, which is comparable to the *in-situ* measurements.

4. Dislocation Dynamics in Al and Al-Mg Thin Films

4.1. *In-situ observations of dislocation propagation*

The effect of Mg on the propagation of dislocations is particularly visible during the early stages of loading. While in pure Al the dislocations instantly spread across the entire grain (i.e. faster than the 30 frames per second video sampling rate), they advance more slowly and in a jerky type fashion in all observed Al-Mg alloys. Figure 7 shows a sequence of images from an indentation in Al-2.6%Mg. The arrows mark the consecutive positions where the leading dislocation line is pinned by solutes. From these images, the mean jump distance between obstacles is estimated to be of the order of 50 nm. Due to the single-tilt axis limitation of the indentation stage, the orientation of the slip plane relative to the electron beam is unknown; therefore, the measured jump distance is a projection and a lower bound of the actual jump distance.

At the low strains for which jerky-type dislocation motion is observed, solute atoms are the predominant barriers to mobile dislocations, as has been shown in earlier *in-situ* pulsed nuclear magnetic resonance (NMR) experiments.[5,26–28] Consequently, the mean jump distance can be predicted by Mott-Nabarro's model of weakly interacting diffuse forces between Mg solutes and dislocations in Al.[27] A calculation of the effective obstacle spacing, assuming that the maximum internal stress around a solute atom has a logarithmic concentration dependence, yields a value of 30 nm in Al-2.6%Mg.

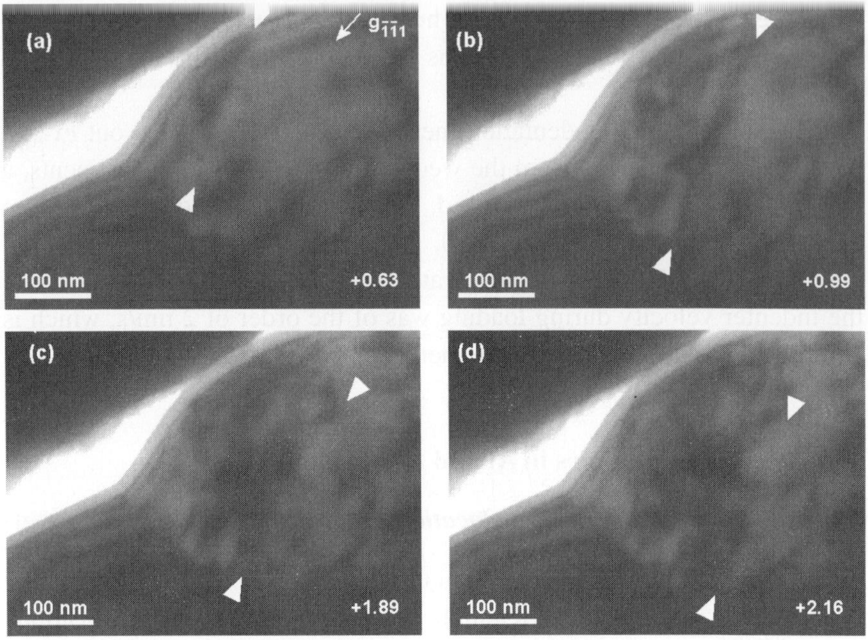

Fig. 7. Series of bright-field images showing jerky motion of dislocations during indentation of Al-2.6wt%Mg. The time from the start of the indentation is given in seconds. Note the presence of a native oxide layer on the surface.

This is in fair agreement with our experimental observation of a mean jump distance of the order of 50 nm. Besides solute atoms, (semi-) coherent β′/β precipitates in Al-Mg alloys can also provide significant barriers to dislocation motion. As aforementioned, the mean spacing of these precipitates could not be measured very accurately due to the limited resolution of the microscope combined with the specific indentation stage. However, we can make an estimate based on the solid solubility of magnesium in Al at room temperature of 1.9 wt%. The calculated volume fraction f_V is 2.4% for the β phase at 300 K. The mean planar separation, which is a relevant measure for the interaction of a gliding dislocation with a random array of obstacles in its slip plane, is given by[30,31]

$$\lambda \cong \frac{2\sqrt{2}\pi r}{3f_V}, \tag{1}$$

provided that the size of the particles r is negligible in comparison with their center-to-center separation, i.e. if $\lambda \gg r$. It is reasonable to assume that the minimum size of the semicoherent precipitates is at least 10 nm to produce sufficient strain contrast in Fig. 4(b). As a result, the mean planar separation of the precipitates is calculated to be at least 92 nm, i.e. larger than the mean separation between the solutes. In this approach, the obstacles are assumed to be spherical and consequently, we ignore the effect that the precipitation in Al may become discontinuous or continuous depending on the temperature. However, even in the case of a Widmanstätten structure, the effective separation between the needle-shaped precipitates is larger than the effective solute obstacle spacing.[32] Therefore, based on the experimental observations in the alloys below and above the solid solubility of magnesium, the strain contrast depicted in Fig. 4(b) and the abovementioned theoretical considerations, solute atoms are assigned as the main obstacles to dislocation motion.

4.2. *Serrated yielding in Al-Mg alloys*

A considerable part of the research effort on Al-Mg alloys has been devoted to understanding the pronounced, repeated yielding that occurs during plastic deformation of these alloys. The physical basis for this phenomenon, known as the Portevin–Le Châtelier (PL) effect or serrated yielding,[33] is a negative strain rate sensitivity of the flow stress, caused by interaction between dislocations and mobile solute atoms.[34] This self-repeating process consists of pinning of the dislocations by the solutes, the breakaway of the dislocations from the solutes, and diffusion of the solute atoms to the dislocations, which are consequently pinned again. In uniaxial deformation, the most characteristic features of the PL effect are serrations, i.e. stress drops or steps, in the stress-strain curve. The PL effect in Al-Mg has been investigated in several deformation modes, including depth-sensing indentation.[35,36] The associated dislocation dynamics have been characterized by *in-situ* straining in a high-voltage electron microscope[37,38] and pulsed NMR experiments.[26]

The repeated yielding due to the PL effect occurs within specific limits of temperature, strain, strain rate and impurity concentration. Based on the theoretical model by Kubin and Estrin,[39] Chinh *et al.*[36] calculated a minimum concentration of 0.62 wt% Mg for instabilities to occur in binary Al-Mg at room temperature. The strain required for serrated yielding to start during indentation depends on the Mg concentration; an estimate for the equivalent indentation depth can be obtained from the following empirical relation for Vickers indentation of bulk Al-Mg[36]:

$$h_c = A(C - C_0)^n, \tag{2}$$

where $A = 2.91$ μm, $C_0 = 0.86$ wt% and $n = -0.23$ are the fitting parameters experimentally determined at a loading rate of 14 mN/s. Given that the critical load P_c at which the instabilities start is proportional to the loading rate ζ [35], we have

$$h_c \propto \sqrt{P_c} \propto \sqrt{\zeta}. \tag{3}$$

Assuming an average loading rate in the present experiments of 0.03 mN/s, we find a critical depth ranging from 0.10 μm for Al-5.0%Mg to 0.19 μm for Al-1.1%Mg. This is consistent with the results from the ex situ quantitative indentations, as will be shown in the next section.

In-situ straining studies in a TEM have related the PL effect to sudden activation, multiplication and coordinated motion of dislocations.[37,38] Such behavior was not observed in our *in-situ* experiments. Moreover, the indentation depths at which dislocation motion was studied were considerably lower than the estimated critical depths as obtained above. Therefore, it is concluded that the jerky motion observed *in-situ* is due to solute drag without appreciable diffusion of solute Mg.

4.3. *Effect of solute drag on load-controlled indentation curves*

The extraction of mechanical properties from the quantitative indentation measurements on the evaporated thin films was compromised by the surface roughness and the grain size at shallow depths and by the film thickness at deeper depths. Recent numerical studies[40,41] suggest that for a soft film on a hard substrate, the influence of the substrate may not be appreciable until the depth exceeds one half of the film thickness. Still at these relatively high indentation depths, the probed volume was not sufficiently large to give reliable hardness and modulus data. As illustrated in Fig. 5, most of the films show considerable surface roughness due to cusps at the grain boundaries. This leads to an ill-defined contact area during initial loading. Furthermore, the size of the indents was of the order of the grain size, causing scatter in the indentation results due to microstructural variations. For these reasons, our analysis of the quantitative data focuses on characteristic features of the load-displacement curves and their relation to the *in-situ* observations, rather than on the calculation of hardness and elastic modulus. The *ex-situ* load-controlled indentation measurements on the pure Al film showed abrupt displacement bursts during loading up to a depth of around 70 nm, as illustrated in Fig. 8(a).

Between the bursts, the slope of the loading curve increases continuously. No such discontinuities were observed in indentations of any of the Al-Mg films, as illustrated in Fig. 8(b) showing loading curves of the Al-2.6%Mg film. As would be expected from the critical indentation depths for the PL effect obtained in the previous section, no pronounced serrated yielding was observed in the Al-Mg films during indentation to 150 nm depth, except for the Al-5.0%Mg film (Fig. 8(c)). Indeed in this case, the serrations start between 80 and 100 nm depth as predicted by the calculations. The initially "soft" response of the Al-Mg films during the first tens of nanometers can be attributed to their surface roughness.

Analysis of the curvature of the loading portions prior to the first excursion and between subsequent excursions in the pure Al film shows that these are well described by elastic loading by a sharp Berkovich indenter. The yield behavior is therefore classified as staircase yielding due to sudden dislocation nucleation and propagation. Staircase yielding has been reported for indentation of both single crystal and polycrystalline Al thin films.[42] The absence of these yield events during indentation of Al-Mg films, both below and above the solubility limit, shows that initial plasticity is significantly affected by solute Mg. Presumably, solute drag prevents dislocation bursts from propagating through the crystal, i.e. the stored elastic energy is insufficient to push a series of dislocations through the solute atmosphere at constant indentation load. As the load increases further, some of the available dislocations are able to overcome the force associated with solute pinning, thereby allowing plastic relaxation to

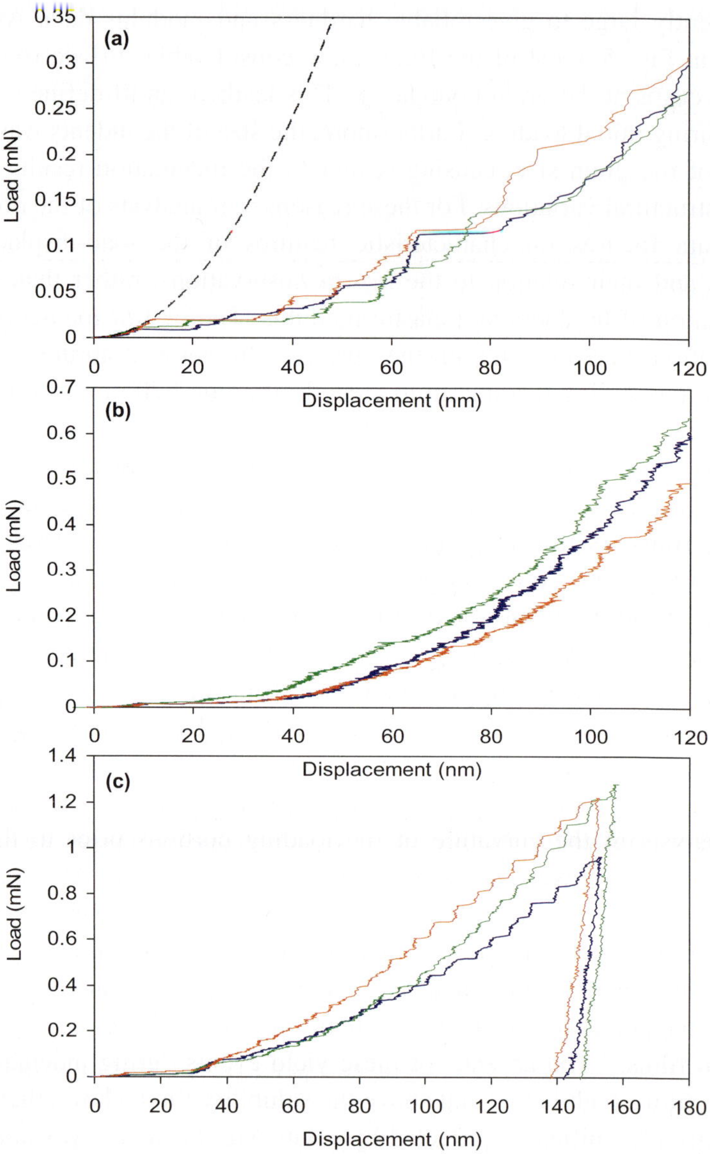

Fig. 8. (a) Loading curves for load-controlled indentations on pure Al film, showing discrete displacement jumps. The dashed line represents Berkovich indentation of an elastic material with $E_r = 74$ GPa. (b) Loading curves without pronounced yield excursions for indentations in Al-2.6%Mg film. (c) Serrated yielding during indentation of Al-5.0%Mg film.

proceed smoothly. Since there is no collective motion of dislocations as in pure Al, the measured loading response is essentially continuous. This perception is supported by the extensive solute drag observed *in-situ*.

4.4. *Effect of solute drag on displacement-controlled indentation*

Interestingly, the difference in initial yield behavior between the pure Al and Al-Mg films was not observed in the quantitative displacement-controlled indentations performed *in-situ*.

The stage was equipped with a Berkovich indenter with an end radius of approximately 150 nm as measured by direct imaging in the TEM. The displacement rate during indentation was 7.5 nm/s. Given the significant rounding of the indenter the initial loading is well described by spherical contact up to a depth of the order of 10 nm. In Tabor's approximation, the elastoplastic strain due to spherical loading is proportional to $\sqrt{\delta/R}$, where δ is the indentation depth and R the indenter radius; the equivalent strain rate is therefore proportional to $1/\sqrt{(4\delta R)}\ d\delta/dt$. Using the above-mentioned values it is easily seen that at a depth of 10 nm the initial strain rates in both types of experiments compare reasonably well to one another, with values of 0.14 s^{-1} and 0.10 s^{-1} for the *ex-situ* and *in-situ* experiments, respectively.

Figure 9(a) shows the data recorded during an indentation on pure Al. The loading curve shows pronounced load drops, which have the same physical origin as the displacement excursions in load-controlled indentation, i.e. stress relaxation by bursts of dislocation activity. Also in this case, the loading behavior up to the first load drop appears to follow closely the elastic Berkovich response, although this comparison may not be entirely valid because of irregularities on the tip surface as observed in TEM. In contrast with the *ex-situ* load-controlled indentations, the measured response of Al-Mg follows roughly the same behavior (Fig. 9(b)): load drops occur with approximately the same size and frequency as in pure Al. These observations illustrate that while the physical mechanism underlying the instabilities in load-controlled and displacement-controlled indentation are the same, the criteria for them to occur may depend on the indentation mode used. One rationale for this difference may be as follows. When the critical shear stress for a dislocation source under the

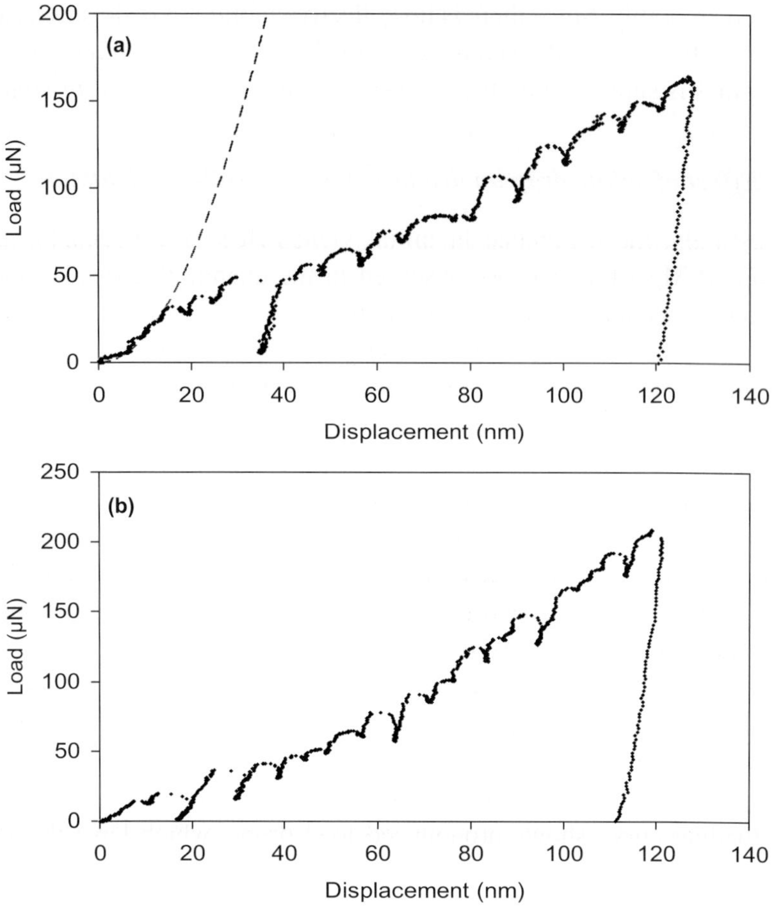

Fig. 9. Displacement-controlled indentation behavior measured *in-situ* on (a) pure Al and (b) Al-2.6%Mg.

indenter is reached under load control, a significant strain burst results only if the source is able to generate many dislocations at constant load. This again is possible only if the newly nucleated dislocations can freely propagate through the lattice, as in pure Al. Under displacement control however, the feedback system reduces the load during a yield event so as to keep the error in the constant displacement rate minimal. The observed instabilities in particular lead to large and rapid changes in contact stiffness, which are very challenging from the perspective of feedback control. If the feedback bandwidth is sufficiently high, the system may respond to the decrease in contact stiffness when only a few dislocations are nucleated; in this case, the occurrence of a detectable load drop does not require collective propagation of many dislocations and as such may be observed under solute drag conditions as well. Warren *et al.*[22] reported that the density of load drops in displacement-controlled indentation of an Al (100) surface is significantly higher than that of displacement bursts in load-controlled indentation of the same surface, demonstrating that the former is indeed a more sensitive technique to detect discontinuous yielding than the latter. Besides the indentation control mode, also the loading rate may affect the initial yield phenomena to some extent, especially in the case of Al-Mg alloys where strain-induced diffusion of Mg is appreciable. The quantitative *in-situ* indentations show a considerable amount of dislocation activity prior to the first macroscopic yield point. This is illustrated in Fig. 10. While the indented grain is free of dislocations at the onset of loading (Fig. 10(a)), the first dislocations are already nucleated within the first few nanometers of the indentation (Fig. 10(b)), i.e. well before the apparent initial yield point. At the inception of the first macroscopic yield event, dislocations are present throughout the entire grain (Fig. 10(c)). The yield event itself is associated with collective motion of these dislocations, which significantly changes the appearance of the dislocation structure (Fig. 10(d)).The *in-situ* observations of Al-Mg furthermore provide a self-consistency check for the dynamics of a yield event. With solute drag preventing full load relaxation, the size of a forward surge Δh is essentially determined by the dislocation velocity v and the mechanical bandwidth of the transducer f. Therefore, ignoring the drag exerted by the feedback system, the dislocation velocity may to a first approximation be estimated as $v \sim \Delta h f$, which, using $\Delta h = 7$ nm and $f = 125$ Hz, yields a velocity of the order of 1 μm/s. This is of the same order as observed *in-situ* for the initial dislocations in Fig. 10(b), which traversed the 300 nm film thickness in about 130 ms (four video frames at a frame rate of 30 frames per second). These observations provide strong evidence in support of the claim that dislocations are nucleated prior to the first detectable yield point.[44-46] In the present *in-situ* experiments, the geometry of the indenter tip is not so accurately defined as to conclusively validate the

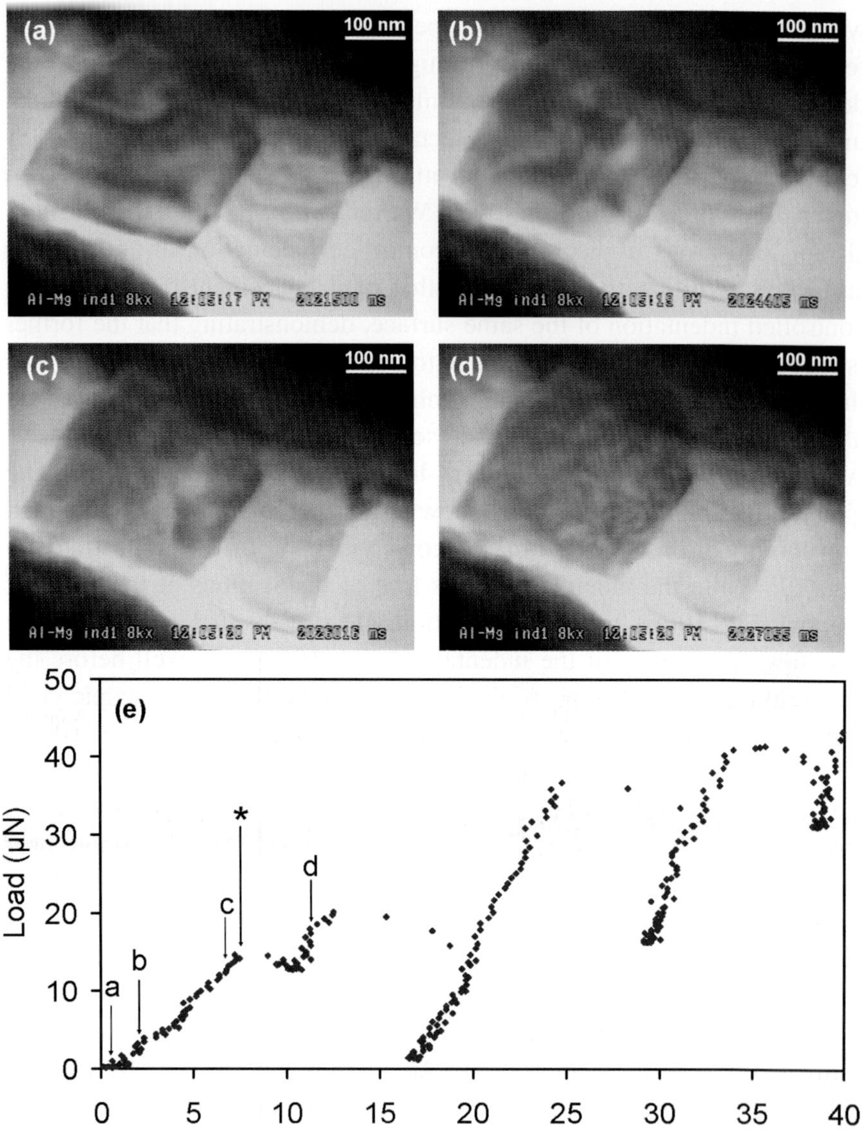

Fig. 10. TEM bright- field Image sequence (a–d) from the initial loading portion (e) of the indentation on Al-2.6%Mg depicted in Fig. 9(b). The first dislocations are nucleated between (a) and (b), i.e. prior to the apparent yield point. The nucleation is evidenced by an abrupt change in image contrast: before nucleation, only thickness fringes can be seen, whereas more complex contrast features become visible at the instant of nucleation. See http://www.dehosson.fmns.rug.nl/.[43]

correspondence of the loading curve to purely elastic loading. Furthermore, the geometry and the microstructure of the specimens may affect the nucleation behavior through the presence of nearby grain boundaries and free surfaces. To further clarify the dislocation dynamics at this initial stage of nanoindentation, *in-situ* experiments on more carefully defined systems have recently been conducted.[46]

5. Grain Boundary Dynamics in Al and Al-Mg Thin Films

To confirm the occurrence of grain boundary movement in aluminum as had been reported earlier,[11] several *in-situ* indentations were performed near grain boundaries in the pure Al film. Indeed, significant grain boundary movement was observed for both low and high-angle boundaries. This is illustrated in Fig. 11 by image frames of subsequent stages of the loading part of an indentation near a high-angle boundary. After initial contact (Fig. 11(a)) and plastic deformation of grain B

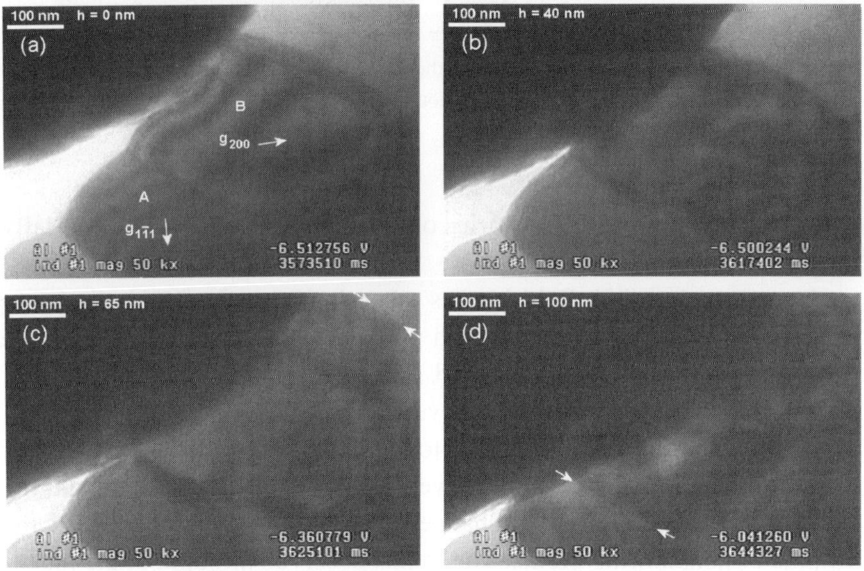

Fig. 11. Series of bright-field images from an indentation on Al, which is accommodated by movement of the grain boundaries (marked with arrows). The approximate indentation depth *h* is given in each image.

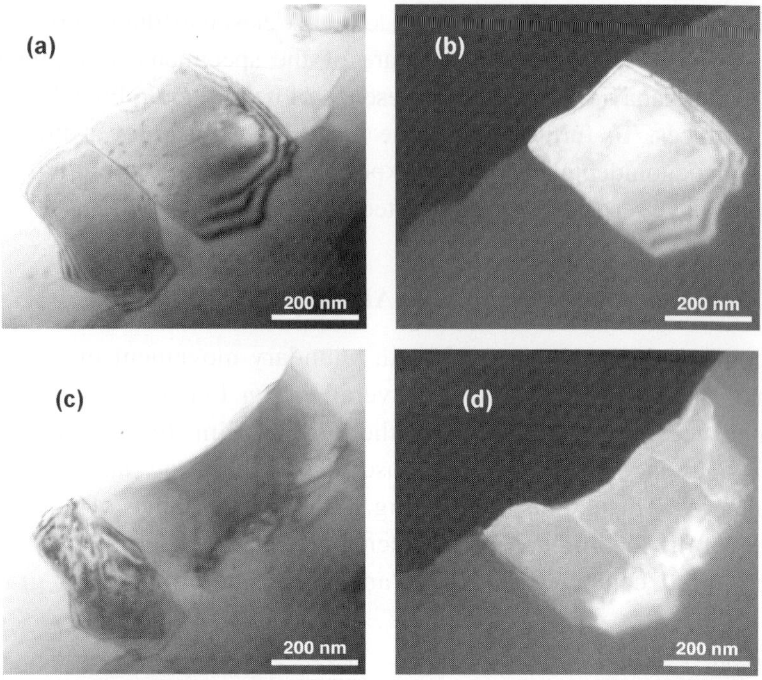

Fig. 12. Bright- and dark-field images of the indented grain (a, b) before and (c, d) after the indentation depicted in Fig. 11. Grain boundary motion leads to a significant volume increase of the indented grain.

(Fig. 11(b)), both grain boundaries outlining grain B move substantially (Figs. 11(c), (d)).

By comparing dark-field images taken before and after the indentation, as shown in Fig. 12, the grain boundary shifts are measured to be 0.04 μm for the left boundary and 0.22 μm for the right boundary. It should be emphasized that the observed grain boundary motion is not simply a displacement of the boundary together with the indented material as a whole; the boundary actually moves through the crystal lattice and the volume of the indented grain changes accordingly at the expense of the volume of neighboring grains. The trends observed throughout the indentations suggest that grain boundary motion becomes more pronounced with decreasing grain size and decreasing distance from the indenter to the boundary. Moreover, grain boundary motion occurs less frequently as the end radius of the indenter increases due to tip blunting or contamination. Both these observations are consistent with the view that the motion of grain boundaries is promoted by high local stress gradients as put forward in the introduction of this chapter. The direction of grain boundary movement can be both away from and towards the indenter, and small grains may even completely disappear under indentation.[17] Presumably, the grain boundary parameters play an important role in the mobility of an individual boundary, since the coupling of the indenter-induced stress with the grain boundary strain field depends strongly on the particular structure of the boundary.

The quantitative *in-situ* indentation technique offers the possibility to directly relate the observed grain boundary motion to features in the load-displacement curve. While this relation has not been thoroughly studied in the present investigation, preliminary results suggest that the grain boundary motion is associated with softening in the loading response. Softening can physically be accounted for by the stress relaxation that occurs upon grain boundary motion. However, the quantification of overall mechanical behavior is complicated by the frequent load drops at this stage of indentation, and further *in-situ* indentation experiments are needed to investigate this phenomenon more systematically and quantitatively. The movement of grain boundaries as observed in Al was never found for high-angle boundaries in any of the Al-Mg specimens, even when indented to a depth greater than half of the film thickness. Figure 13 shows a sequence of images from an indentation on an Al-1.8%Mg layer.

At an indentation depth of approx. 85 nm into grain B (Fig. 13(c)), plastic deformation is initiated in grain A by transmission across the grain boundary. However, no substantial grain boundary movement occurs; small grain boundary shifts (~10 nm) that were measured occasionally can be attributed to displacement of the material under the indenter as a whole, with conservation of grain volume, rather than to actual grain boundary motion (Fig. 14). Our observations as such indicate a significant pinning effect of Mg on high-angle grain boundaries in these alloys. In contrast to high-angle grain boundaries, the mobility of low-angle boundaries in Al-Mg was found to be less affected by the presence of Mg. This is illustrated by the rapid disintegration of a low-angle tilt boundary in Al-5.0%Mg as shown in Fig. 15. At a relatively low indentation depth of

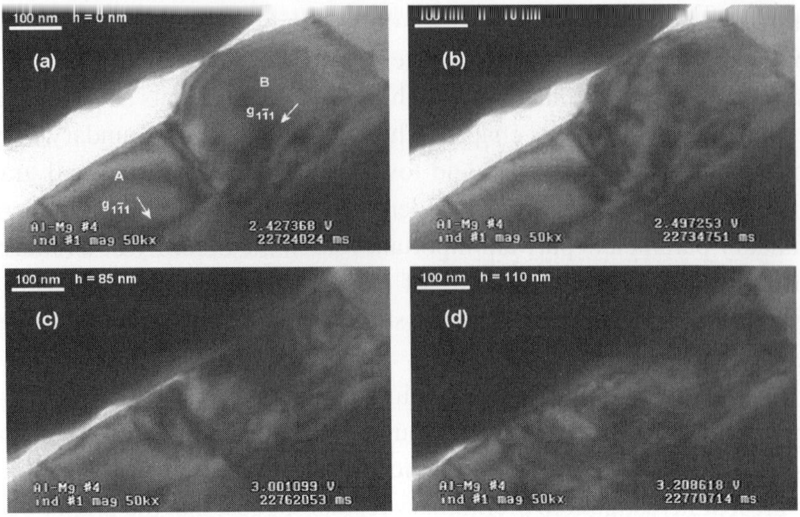

Fig. 13. Series of bright-field images from an indentation on Al-1.8%Mg. No movement of the high-angle grain boundaries is observed.

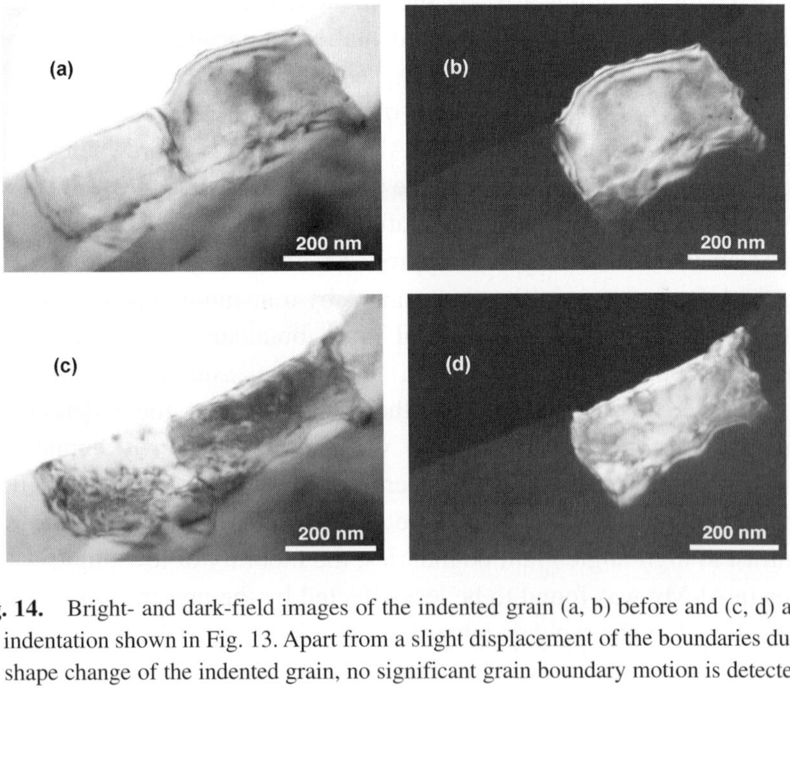

Fig. 14. Bright- and dark-field images of the indented grain (a, b) before and (c, d) after the indentation shown in Fig. 13. Apart from a slight displacement of the boundaries due to the shape change of the indented grain, no significant grain boundary motion is detected.

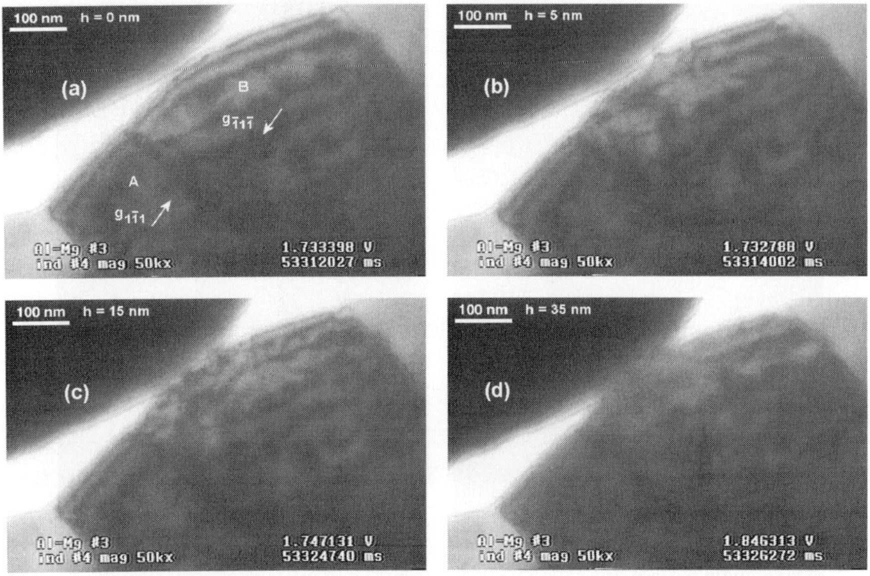

Fig. 15. Series of bright-field images from an indentation on Al-5.0%Mg, showing the disintegration of a low-angle $\langle 110 \rangle$ tilt boundary between (c) and (d).

about 20 nm, the dislocations that were initially confined to the indented grain spread across both grains without being visibly obstructed by the tilt boundary. The boundary effectively disappears at this point with the end result of the two grains becoming one. Figures 16(a)–(c) show the orientation of the two grains before indentation. The grains share the same $\langle 112 \rangle$ zone axis, but are in different two-beam conditions due to their slight misorientation (~ 0.7°). Figure 16(d) shows the grains after the indentation to be both in the same diffracting condition as the grain in Fig. 16(a).

Ideally, in order to compare the observed grain boundary behavior between different measurements, the indenter-induced stress at the boundary should be known. However, due to surface roughness, tip imperfections and the complicated specimen geometry, it is difficult to accurately measure or calculate the local stress fields. Comparisons between different measurements are therefore mainly based on indentation depth. Our observation of grain boundary pinning in Al-Mg in this context means that no motion of high-angle boundaries was observed in Al-Mg in more than

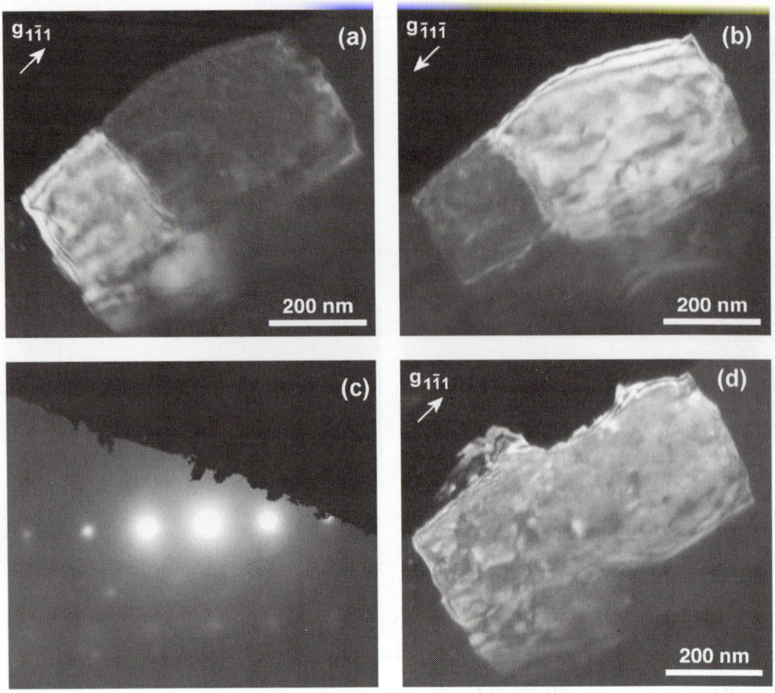

Fig. 16. (a,b) Dark-field images of the two Al-5.0%Mg grains shown in Fig. 15 before indentation. (c) Diffraction pattern showing the $\langle 112 \rangle$ orientation of both grains; the cut-off is due to the *in-situ* specimen geometry. (d) Dark-field image after indentation.

fifteen indentations to a depth of the order of 100 nm, while in pure Al, grain boundary motion was frequently observed at indentation depths of 50 nm or less.

The Al-Mg films used in this chapter include compositions both below and above the solubility limit of Mg in Al. However, no differences in indentation behavior between the solid solution and the precipitated microstructures were observed. Consequently, the observed pinning of high-angle boundaries in Al-Mg is attributed to solute Mg. The pinning is presumably due to a change in grain boundary structure or strain fields caused by solute Mg atoms on the grain boundaries.

Relatively few direct experimental observations have been reported of this type of interaction. Sass and co-workers observed that the addition of Au and Sb impurities to bcc Fe changes the dislocation structure of $\langle 100 \rangle$

twist boundaries of both low-angle[48] and high-angle[49] misorientation. Rittner and Seidman[50] calculated solute distributions at $\langle 110 \rangle$ symmetric tilt boundaries with different boundary structures in an fcc binary alloy using atomistic simulations. However, the influence of solutes on the structure of such boundaries has not been experimentally identified.

Possible changes in atomic boundary structure due to solute atoms may be observed by high-resolution TEM (HRTEM). Atomic-scale observation of grain boundaries using this technique requires that the crystals on both sides share a close-packed direction so that both lattices can be atomically resolved at the same time. The mazed bicrystal structure that forms when an Al film is deposited epitaxially onto a Si (001) surface meets this condition. The epitaxial relationships Al (110) // Si (001), Al [001] // Si [110] and Al (110) // Si (001), Al [001] // Si lead to two possible orientations that are separated exclusively by 90° $\langle 110 \rangle$ tilt boundaries.[25,51] The structure of such boundaries has been successfully studied in HRTEM studies of Al films on Si substrates [25,52,53] and Au films on Ge substrates,[54–57] which exhibit the same epitaxial relationships. Moreover, the effect of alloying elements in Al has been explored by evaporating alloys such as Al-Cu and Al-Ag.[58]

In order to study the effect of Mg on these tilt boundaries, we deposited Al and Al-Mg films onto Si (001) substrates that had been stripped from their native oxide film. Indeed, we found that in epitaxial films evaporated from pure Al, the 90° $\langle 110 \rangle$ tilt grain boundaries are facetted on {100}A//{110}B and {557}A//{557}B planes, which can be atomically resolved (Fig. 17(a)). The addition of Mg however drastically changes the microstructure of the deposited film: evaporation of Al-Mg on a Si substrate heated to 300°C (which is necessary to reduce the lattice mismatch between Al and Si) leads to the formation of the intermetallic compound Mg2Si, which prohibits any further epitaxial growth (Fig. 17(b)). Even in a two-step evaporation consisting of a pure Al deposition to provide a basis for the bicrystal structure and a subsequent Al-Mg deposition to introduce the Mg, the Mg diffuses to the substrate driven by the reaction with the Si substrate. This method could therefore not be used to study the effect of Mg on the atomic structure of the grain boundaries.

Another effect that may contribute to the pinning of special boundaries is solute drag on extrinsic grain boundary dislocations (EGBDs) as

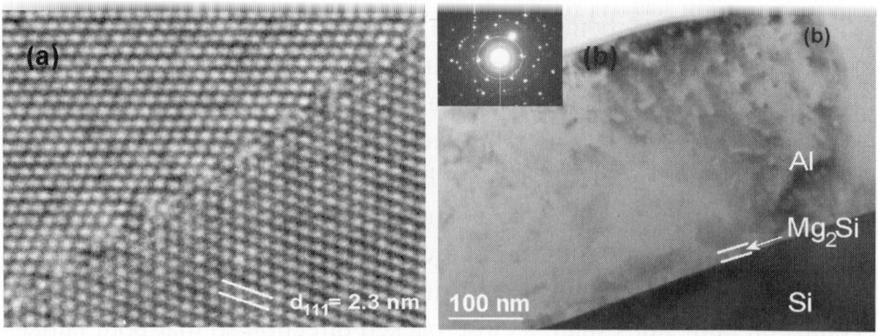

Fig. 17. (a) High-resolution micrograph of a 90° ⟨110⟩ asymmetrical tilt boundary in an epitaxial Al thin film, showing a periodic structure along the boundary plane. The orientation of the boundary plane is $\{100\}_A // \{110\}_B$. (b) Cross section of a film deposited from an Al-2.2wt%Mg source onto a Si (001) substrate; the intermetallic compound Mg_2Si, identified by its diffraction ring pattern (inset), forms a 15 nm thick layer on the interface.

reported by Song *et al.*,[59] who showed that the dissociation rate of EGBDs in Al alloys is reduced by the addition of Mg. This implies that the indenter-induced deformation is accommodated more easily by these boundaries in pure Al than by those in Al-Mg.

The fact that low-angle grain boundaries were found to be mobile regardless of the Mg content can be explained by their different boundary structure. Up to a misorientation of 10–15°, low-angle boundaries can be described as a periodic array of edge and screw dislocations by Frank's rule.[60] In such an arrangement, the strain fields of the dislocations are approximated well by individual isolated dislocations and their interaction with an external stress field can be calculated accordingly. Since there is no significant interaction between the individual grain boundary dislocations, the stress required to move a low-angle boundary is much lower than for a high-angle boundary. Low-angle pure tilt boundaries consisting entirely of parallel edge dislocations are fully glissile and therefore particularly mobile. In general, a combination of glide and climb is required to move a low-angle boundary.[61] As a corollary, the structural difference between low and high-angle boundaries also affects the extent of solute segregation. Because solutes generally segregate more strongly to high-angle boundaries,[62] the observed difference in mobility may partly be a compositional effect.

6. Superplastic Behavior of Coarse-Grained Al-Mg Alloys

Superplasticity is the ability of a polycrystalline material to undergo very large uniform tensile deformation prior to failure, at a temperature well below its melting point. Typical values of the elongation to failure in uniaxial tension under superplastic conditions are of the order of a few hundred percent, and in some alloys can even exceed 1000%. Although initial experimental observations of superplasticity in metals date back to the 1920s, for a long time the phenomenon was mainly regarded as a laboratory curiosity.[63] However, research interests in superplasticity greatly increased in the 1960s,[64,65] when it was demonstrated that in this regime metal sheets could easily be formed to complex shapes. Superplastic forming (SPF) presents a potentially attractive alternative to other forming techniques. Due to the low flow stress characteristic of superplastic deformation, the tooling costs are minimal; blow forming is commonly used to form metal sheets under superplastic conditions. Furthermore, the exceptionally high ductility allows for a large freedom of design. At present, the main limitation towards mass application of SPF is the relatively low strain rate that is associated with conventional fine-structure superplasticity. Forming of a typical component can take up to one hour at these strain rates. For this reason, SPF has mostly been restricted to the production of prototypes or small series of metallic components so far. However, research efforts are increasingly being directed towards new classes of superplastic materials, some of which exhibit superplastic behavior at considerably higher forming rates.[66,67] This so-called high-strain-rate superplasticity is expected to receive broad industrial interest and may replace existing forming techniques if such materials can efficiently be produced on a large scale. The present section is concerned with the high-strain-rate superplastic behavior of coarse-grained Al-Mg alloys, which are a promising candidate in this category.[68]

The hallmark of superplastic deformation is a low flow stress σ that shows a high strain-rate sensitivity m as defined by

$$\sigma = k\,\dot{\varepsilon}^{m}, \tag{4}$$

in which k is a constant and $\dot{\varepsilon}$ is the strain rate. A high strain-rate sensitivity is necessary to stabilize the plastic flow so as to avoid necking during tensile deformation. The incipient formation of a neck leads to a local increase of the strain rate, which, in the case of a positive strain-rate sensitivity, leads to an increase of the flow stress in the necked region. If the strain-rate sensitivity is sufficiently high, the local flow stress increases to such an extent that further development of the neck is inhibited. Most common metals show a strain-rate sensitivity exponent lower than 0.2, whereas values around 0.3 or higher are needed to delay necking long enough to produce the strains characteristic of superplasticity. Besides a high strain-rate sensitivity, a low rate of damage accumulation (e.g. cavitation) is required to allow large plastic strains to be reached.

The possibility to conduct high-strain-rate forming operations by deforming solid solution alloys in the viscous glide regime has received little attention compared to the vast amount of work on superplasticity based on grain boundary sliding. This is likely due to the fact that the tensile elongations obtained under solute-drag creep conditions are generally lower than in fine-structure superplasticity, owing to the difference in strain-rate sensitivity ($m \approx 0.3$ vs. $m \approx 0.5$, respectively). Nevertheless, for coarse-grained Al-Mg alloys deformed in the viscous glide regime, values for the maximum strain in excess of 300% can be obtained.[69–71] Such elongations are close to those found in conventional superplasticity of fine-grained Al-Mg alloys and are sufficient for many practical applications. Moreover, forming by viscous-glide controlled creep has two important advantages over conventional superplastic forming: (i) the rate of viscous glide is not restricted by dislocation climb and consequently higher strain rates can be achieved, and (ii) since viscous glide is independent of grain size, the preparation of the materials is less complex. It should be noted that since the deformation under viscous-glide control does not follow the original definition of superplasticity in the strictest sense, the deformation behavior has also been referred to as "enhanced ductility" or "quasi-superplasticity" by some researchers.[69,70,72] In this chapter, we will use the designation "coarse-grain superplasticity" because of the low flow stress and relatively high strain-rate sensitivity associated with this regime, both of which are characteristic of superplastic deformation.

Viscous-glide creep, or solute-drag creep, results from the impediment of gliding dislocations by their interaction with solute atoms. There are two competing mechanisms in this regime, dislocation glide and climb;

the slower of the two is rate-controlling. A physical interpretation of the empirically found three-power-law relation for viscous-glide creep is readily based on the Orowan equation, relating the macroscopic strain rate $\dot{\varepsilon}$ to the mobile dislocation density ρ_m and the average dislocation velocity $\overline{v}_d$:

$$\dot{\varepsilon} = \chi b \rho_m \overline{v}_d \,. \tag{5}$$

Although no direct measurements of the relation between applied stress and dislocation velocity under solute-drag conditions are available, most models suggest that $\overline{v} \propto \sigma$ in this regime, i.e. the stress exponent $n_v = 1$.[73] Furthermore, experimental observations have shown that the stress exponent n_d of the mobile dislocation density $\rho_m \propto \sigma^{n_d}$ lies between 1.6 and 1.8 for Al-Mg alloys.[73,75] This is in reasonable agreement with theoretical predictions[76] suggesting that $\rho_m \propto \sigma^2$. The discrepancy between the theoretical and experimental values of n_d has been attributed to the fact that dislocations are increasingly located in subgrain boundaries at lower stresses, or alternatively to the incomplete relaxation of dislocation loops during unloading of specimens at room temperature.[77] Within the formulation of Eq. (5), the strain-rate sensitivity index depends critically on the stress dependence of the product $\rho(\sigma)\overline{v}_d \propto (\sigma)$. Assuming $n_v \approx 1$ and $n_d \approx 2$, it follows from Eq. (5) that $\dot{\varepsilon} \propto \sigma^3$, or as formulated in the original model by Weertman[76]:

$$\dot{\varepsilon} \approx \frac{0.35}{\xi} \mu \left(\frac{\sigma}{\mu} \right)^3 , \tag{6}$$

where ξ is a parameter that characterizes the interaction between the solutes and the dislocations. From Eq. (6) it follows that the stress exponent $n \approx 3$ and hence the strain-rate sensitivity $m = 1/n \approx 0.33$. From the abovementioned considerations it is clear that the three-power law (m = 0.33) is no more than an approximate relationship arising from the stress dependence of the dislocation density and the drag stress.

6.1. *In-situ TEM straining experiments*

The alloys used in this chapter are two coarse-grained Al-4.4%Mg and Al-4.4%Mg-0.4%Cu alloys with minor amounts of Ti, Mn and Cr

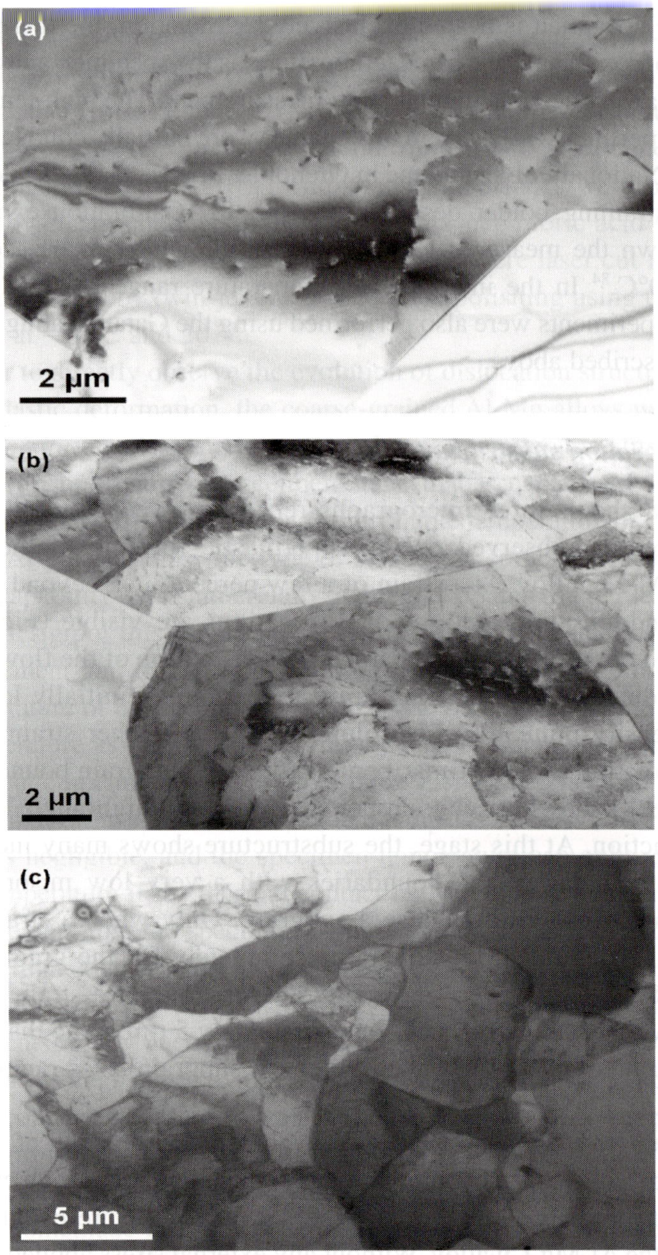

Fig. 18. Dislocation substructure in Al-Mg deformed at 440°C and $5 \cdot 10^{-3}$ s^{-1} to (a) 4%, (b) 20% and (c) 170%.

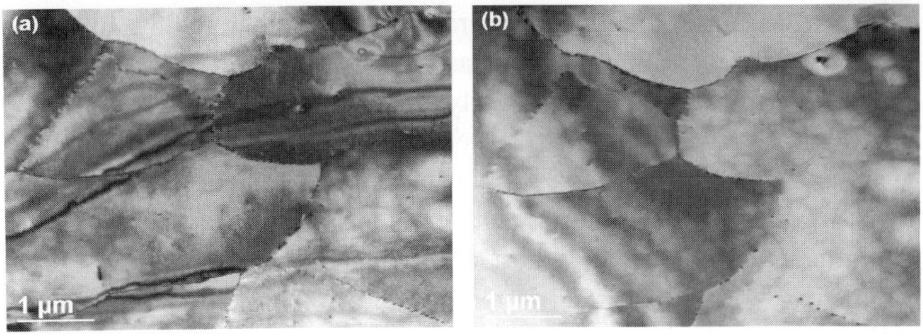

Fig. 19. Subgrains in Al-Mg-Cu (a) at room temperature and (b) after *in-situ* annealing at 450°C for 10 minutes.

At this temperature, most of the lattice dislocations are absorbed into the subgrain boundaries, and some rearrangement of the dislocations in the subgrain boundaries is observed. However, the subgrain structure as a whole remains intact for at least 10 minutes, even in a thin TEM foil, where dislocations can easily escape to the free surfaces of the specimen. Given the strain rate around 10^{-2} s^{-1}, this is long enough a time to ensure that the observed recovery mechanisms are dynamic rather than static. Our observations of subgrain formation are similar to those of binary Al-Mg alloys in torsion[78] and compression[79] in the solute-drag regime. However, since the maximum achievable strain in tensile mode is considerably lower, the grains do not thin to such an extent that so-called geometric dynamic recrystallization[88] becomes relevant. In all deformation modes, the substructure formation in the Al-Mg alloys is more sluggish than in pure Al, presumably due to a lowering of the stacking fault energy by the solute Mg.

The effect of the Mg content on the tensile ductility is twofold. On the one hand, a higher Mg content increases the extent of solute drag, thereby stabilizing the plastic flow. However, beyond a few percent Mg, the effect on the strain-rate sensitivity becomes fairly marginal; Taleff *et al.*[70] reported an increase of $m = 0.29$ to $m = 0.32$ in going from 2.8% to 5.5% Mg. On the other hand, the presence of Mg significantly reduces dynamic recovery as evidenced by the slow formation of subgrains. As a result, Mg concentrations above 5% can easily give rise to dynamic recrystallization within a certain domain of temperature and strain rate, which in the absence of grain refining second phase particles leads to rapid coarsening of the microstructure. The currently used composition with 4.4% Mg appears to be a good balance between solute drag and dynamic recovery, leading to enhanced tensile ductility in excess of 300%. In torsional deformation, where a high strain-rate sensitivity to avoid necking is less important, the ductility benefits most from dynamic recovery (leading to geometric dynamic recrystallization at high strains) and is consequently higher for pure Al than for Al-Mg alloys.[89]

The initially inhomogeneous formation of subgrains gives rise to a "core and mantle" microstructure, in which most deformation is concentrated along the grain boundaries. In fine-grained superplasticity, this type of microstructure has been associated with grain mantle deformation processes as an accommodating mechanism for grain boundary sliding.[86,87,90] In the present case, dynamic recovery is initially confined to the mantle region, but extends throughout the microstructure at higher strains.

6.3. *In-situ observations of substructure evolution*

The evolution of the dislocation substructure during superplastic deformation can be directly observed by *in-situ* tensile experiments in a TEM. A difficulty inherently associated with this technique is presented by the image forces resulting from the proximity of free surfaces, which may significantly influence the dislocation dynamics compared to bulk behavior (e.g. Ref. 82). In the present investigation we have attempted to minimize such effects by preparing tensile TEM specimens from macroscopically prestrained alloys and studying only the initial motion of dislocations from their starting configuration. However, for the present case of Al-Mg alloys, it turns out that surface diffusion of Mg severely limits the temperature range at which the *in-situ* experiments can be conducted.

At temperatures in excess of 200°C, the tensile TEM specimens were consistently found to fracture intergranularly at very low loads (typically ~30 gf). This is evidently not representative of the bulk behavior at high temperature showing very high tensile ductility. Below 200°C, the specimens showed ductile transgranular failure at loads of the order of 300 gf. By EDS and electron diffraction analysis it was found that the intergranular fracture areas of the TEM specimens deformed at high temperature contained large amounts of Mg and MgO. Presumably, surface diffusion of Mg becomes appreciable at high temperature and leads to segregation of Mg to the grain boundaries and consequently to grain boundary embrittlement.

In the temperature range below 200°C, dislocation climb is not activated and therefore extensive dynamic recovery is not to be expected. However, even at low temperature, some rearrangements of the substructure were observed that may be illustrative of those occurring during dynamic recovery. Figure 20 shows dislocation motion leading to the formation of a subgrain boundary, which is the initial stage of grain refinement in the coarse-grained alloys. The absorption of dislocations by a subgrain boundary is shown in Fig. 21; this process contributes to the

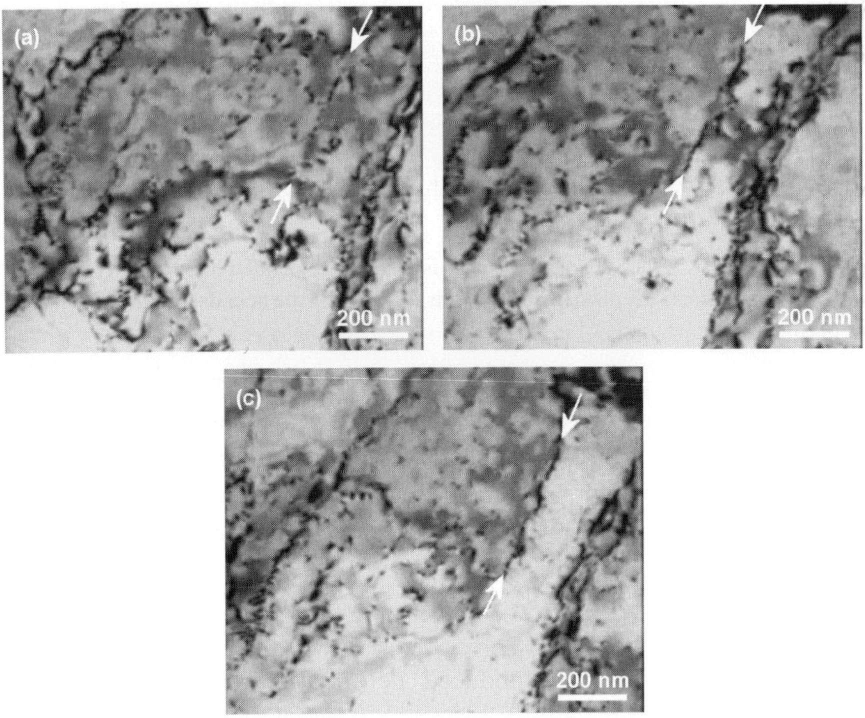

Fig. 20. Rearrangement of dislocations leading to the formation of a subgrain boundary (marked with arrows) in Al-Mg during *in-situ* straining at ~150°C.

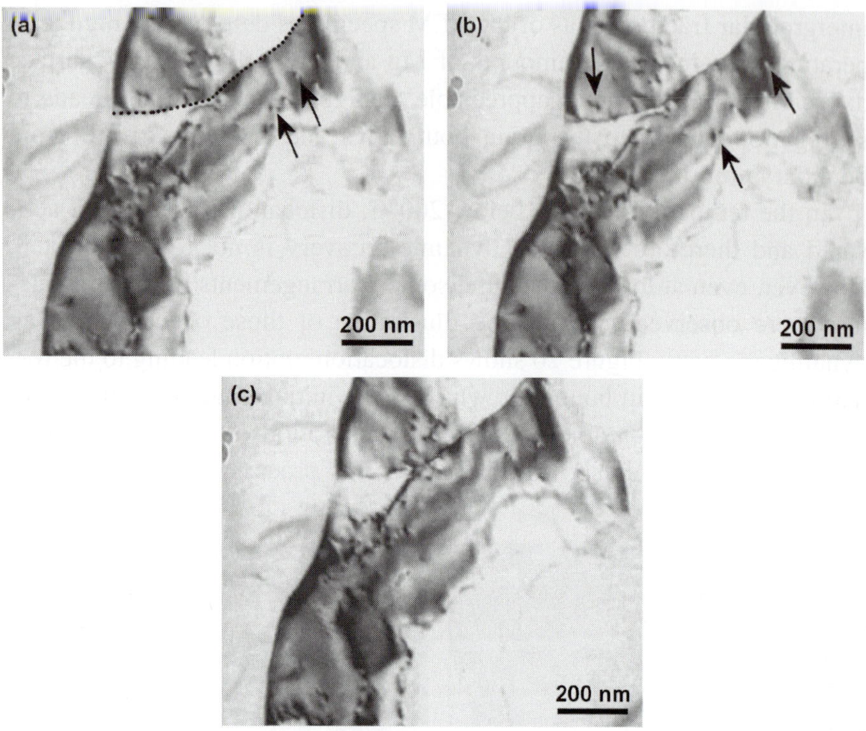

Fig. 21. Absorption of dislocations into a subgrain boundary in Al-Mg during *in-situ* straining at ~150°C. The subgrain boundary is marked by a dotted line in (a). The arrows in (a) and (b) indicate the dislocations that are absorbed by the boundary in the next image respectively.

increase in grain boundary misorientation that is associated with dynamic recovery. In other words: although dislocation motion is solely due to glide at low temperatures, the observed rearrangements resemble the processes that contribute to dynamic recovery at high temperature.

7. Conclusions

Observation of indentation-induced deformation is commonly performed post mortem by atomic force microscopy (AFM) or transmission electron microscopy (TEM). This indirect approach entails several disadvantages: it is not possible to monitor time-dependent deformation processes during deformation; furthermore, the microstructure observed after indentation may differ substantially from that during indentation due to relaxation processes. The recently developed technique of *in-situ* nanoindentation in a TEM allows for direct observation of indentation-induced dynamical processes and consequently does not suffer from the abovementioned limitations. Using this technique the deformation behavior of Al and Al-Mg thin films with a grain size of several hundreds of nanometers can be studied.

The movement of dislocations through the Al-Mg films proceeds in a jerky fashion due to their interaction with solute Mg atoms (solute drag). The observed jump distance in an Al-2.6%Mg film is of the order of 50 nm. This value compares well to the theoretical average distance between obstacles in the presence of diffuse interaction. For load-controlled indentation, the solute drag significantly influences the load-displacement curve. In pure Al, several plateaus are observed which are attributed to dislocation nucleation and propagation. These plateaus are significantly smaller or even absent in the Al-Mg alloy films, where dislocation propagation is hindered by the interaction with solutes.

During indentation of the ultrafine-grained Al film, extensive movement of both low- and high-angle grain boundaries is observed. The occurrence of this deformation mechanism, which under uniform stress conditions is restricted to nanocrystalline materials, is attributed to the high stress gradients involved in sharp indentation. In contrast to the observations in pure Al, no such movement of high-angle grain boundaries is found in any of the Al-Mg alloy films. Since this apparent pinning effect is observed for Mg concentrations both below and above the solubility limit in Al at room temperature, it is considered to be due to solute Mg. Presumably the presence of Mg atoms alters the atomic structure of the grain boundary, which in turn changes its local stress field. Low-angle grain boundaries are not susceptible to this pinning effect, since they can be regarded as a periodic arrangement of dislocations with negligible mutual interaction. The stress required to move a low-angle boundary is therefore much lower than for a high-angle boundary.

With grain sizes decreasing below several hundreds of nanometers, the deformation of metals is increasingly accommodated by grain boundaries rather than conventional dislocation mechanisms. The observation that grain boundaries may be effectively pinned by solute Mg is therefore very interesting from the perspective of the development of ultrafine-grained and nanocrystalline aluminum alloys.

The final section of this chapter concentrates on deformation of Al-Mg alloys with a substantially larger grain size of about 70 μm. Under specific conditions of strain rate and high temperature, these alloys show superplastic behavior. Superplastic materials can be deformed to very high elongations prior to failure, usually in excess of a few hundred percent, at a relatively low flow stress. This characteristic property makes them very attractive for the production of components with a large freedom of design. The hallmark of superplastic deformation is a high strain-rate sensitivity of the flow stress, which suppresses necking during deformation and consequently allows high elongations to be reached.

In coarse-grained Al-Mg alloys, superplastic deformation is accomplished by viscous glide of dislocations. In this mechanism, the interaction between dislocations and solute Mg atoms accounts for the high strain-rate sensitivity. The viscous glide is accompanied by dynamic reconstruction of the microstructure, the appearance of which depends on the deformation parameters and can be both detrimental and beneficial to the ductility. These reconstruction mechanisms have been investigated by analyzing the microstructure of the alloys as a function of tensile strain at different strain rates and temperatures.

Dynamic recrystallization is dominant at strain rates in excess of 10^{-1} s^{-1} and results in rapid coarsening of the microstructure and premature failure. The optimum strain rate for superplasticity of the coarse-grained Al-Mg alloys lies around 10^{-2} s^{-1}; in this regime, dynamic recovery prevails, leading to values of the maximum elongation in excess of 300%. During dynamic recovery, grain refinement occurs by the formation of subgrain boundaries and low-angle grain boundaries. TEM observations show that subgrain formation proceeds slowly, presumably due to the low stacking fault energy in Al-Mg compared to pure Al. During initial straining, subgrains are formed primarily along the original grain boundaries. A uniform substructure is established at a strain of the order of 1.

With a high optimum strain rate of approximately 10^{-2} s^{-1}, the coarse-grained Al-Mg alloys are an attractive alternative to conventional fine-grained superplastic Al-Mg alloys. Superplastic deformation of these fine-grained alloys is attributed to grain boundary sliding, which generally requires low strain rates of 10^{-4} to 10^{-3} s^{-1} in order to be accommodated by diffusion processes. An additional advantage of superplasticity based on viscous glide is that this mechanism has virtually no grain size dependence and therefore the preparation of such materials is less complex.

In summary, *in-situ* Transmission Electron Microscopy has provided new insights into the interaction between dislocations and grain boundaries on various length scales, in which specifically the effect of Mg in Al-Mg alloys on these interaction mechanisms has been clarified.

Acknowledgments

The work was part of the research program of the Netherlands Institute for Metals Research, project MC4.01104. The authors are grateful to the support and collaboration with Andy M. Minor, Steven Shan, Eric A. Stach, all from LBL Berkeley USA, and with S.A. Syed Asif (Hysitron, USA) Oden L. Warren (Hysitron, USA), and with Béatrice Doisneau-Cottignies (Grenoble-France).

References

1. F. R. N. Nabarro, *Theory of Crystal Dislocations*, Oxford University Press, Oxford, 1967.
2. J. P. Hirth and J. Lothe, *Theory of Dislocations*, (McGraw-Hill, New York, 1968).
3. E. O. Hall, *Proc. Phys. Soc. London* **B 64**, 747 (1951).
4. N. J. Petch, *J. Iron Steel Inst.* **174**, 25 (1953).
5. J. Th. M. De Hosson, O. Kanert, and A. W. Sleeswyk, *Dislocations in Solids*, Vol. 6, F. R. N. Nabarro (ed.) North-Holland, Amsterdam, 1983, pp. 441–534.
6. M. A. Wall, U. Dahmen, *Microsc. Microanal.* **3**, 593 (1997).
7. M. A. Wall, U. Dahmen, *Microsc. Res. Tech.* **42**, 248 (1998).
8. E. A. Stach, T. Freeman, A. M. Minor, D. K. Owen, J. Cumings, M. A. Wall, T. Chraska, R. Hull, J. W. Morris, Jr., A. Zettl, and U. Dahmen, *Microsc. Microanal.* **7**, 507 (2001).
9. A. M. Minor, J. W. Morris, Jr., and E. A. Stach, *Appl. Phys. Lett.* **79**, 1625 (2001).
10. A. M. Minor, E. T. Lilleodden, E. A. Stach, and J. W. Morris, Jr., *J. Electron. Mater.* **31**, 958 (2002).
11. A. M. Minor, E. T. Lilleodden, E. A. Stach, and J. W. Morris, Jr., *J. Mater. Res.* **19**, 176 (2004).
12. R. D. Doherty, D. A. Hughes, F. J. Humphreys, J. J. Jonas, D. Juul Jensen, M. E. Kassner, W. E. King, T. R. McNelley, H. J. McQueen, and A. D. Rollett, *Mater. Sci. Eng.* **A 238**, 219 (1997).

81. J. Pelissier and P. Debrenne, *Microsc. Microanal. Microstruct.* **4**, 111 (1993).
82. A. Couret, J. Crestou, S. Farenc, G. Molenat, N. Clement, A. Coujou, and D. Caillard, *Microsc. Microanal. Microstruct*. **4**, 153 (1993).
83. U. Messerschmidt and M. Bartsch, *Ultramicroscopy* **56**, 163 (1994).
84. B. Doisneau-Cottignies, private communication.
85. A. R. Chezan and J. Th. M. De Hosson, *Mater. Sci. Forum* **405–497**, 883 (2005).
86. A. R.Chezan and J. Th. M. De Hosson, *Mater. Sci. Eng.* **A410–411**, 120 (200).
87. W. A.Soer, A. R. Chezan, and J. Th. M. De Hosson, *Acta Mater*. **54**, 3827 (2006).
88. H. J. McQueen, O. Knustad, N. Ryum, and J. K. Solberg, *Scripta Metall.* **19**, 73 (1985).
89. S. Gourdet and F. Montheillet, *Mater. Sci. Eng.* **A 283**, 274 (2000).
90. R. C. Gifkins, *Metall. Trans.* **A 7**, 1225 (1976).

CHAPTER 5

IN-SITU HRTEM STUDIES OF INTERFACE DYNAMICS DURING SOLID-SOLID PHASE TRANSFORMATIONS IN METAL ALLOYS

James M. Howe

Department of Materials Science and Engineering
University of Virginia, Charlottesville, VA 22904-4745, USA
jh9s@virginia.edu

Understanding the details of atomic motion at interphase boundaries is critical to understanding the mechanisms and behavior of phase transformations in materials. This article presents results obtained from *in-situ* hot-stage high-resolution transmission electron microscope (HRTEM) experiments that were used to determine the atomic structure, dynamics and mechanisms of transformation at order-disorder, precipitation and massive transformation interfaces in metal alloys at the atomic level. The similarities and differences among these various types of transformations are emphasized, as are the strengths and limitations of *in-situ* HRTEM for obtaining such information.

1. Introduction

Interfaces which separate two crystals that differ in composition, Bravais lattice, or both composition and lattice, are referred to as interphase boundaries.[1,2] Understanding how atoms arrange at interphase boundaries and cross from one phase to form the other is essential to understanding the nature of phase transformations in materials.[3,4] Phase transformations are important, because they are one of the main techniques that materials scientists employ to develop material structures with sizes ranging from a few atoms to micrometers.[5] Since interphase boundaries are internal interfaces, *in-situ* hot-stage high-resolution transmission electron microscopy (HRTEM) is one of the few techniques capable of providing both atomic-level structural and dynamic information about such interfaces. This chapter describes how *in-situ* hot-stage HRTEM can be used to determine the dynamic behavior of interphase boundaries at the atomic level.

Interphase boundaries are often divided into three classes based on the degree of atomic matching or coherency across the interface as:

(i) coherent interfaces, where there is complete continuity of atomic planes across the interface between two phases,
(ii) partly coherent interfaces, where the mismatch between two crystal structures across the interface is accommodated by periodic misfit dislocations, i.e. strain localization, in the interface, and
(iii) incoherent interfaces, where atomic matching is sufficiently poor that there is no correspondence of atom planes across the interface or localization of misfit into dislocation defects.[6]

In addition to this classification, which is based on the degree of atom matching and/or misfit localization, an interphase boundary may be sharp or diffuse, depending on whether the composition or structural changes occur abruptly across a single plane, or smoothly over several or more planes.[1,2] The examples in this paper illustrate how HRTEM has been a valuable technique for determining the dynamic behavior and mechanisms of motion of both diffuse and sharp interphase boundaries displaying the three different degrees of coherency listed above. Before discussing the examples, the next section provides a brief description of HRTEM imaging and particular requirements for performing *in-situ* hot-stage HRTEM studies, used to induce phase transformations in materials.

2. *In-Situ* Hot-Stage HRTEM of Interphase Boundaries

HRTEM imaging is phase contrast imaging, where two or more beams are allowed to interfere to form an image.[7,8] Since the lens system of the TEM must preserve the coherence of the image-forming beams, the spacing that one wants to resolve in the image must be within the resolution limits of the microscope. This requirement generally limits HRTEM to observations of projected specimen structures along relatively low-index zone axes with wide interplanar spacings that are within the resolution limits of the microscope. Under optimum conditions, one can directly interpret a HRTEM image directly as a map of the projected crystal structure (crystal potential) along the electron-beam direction. Such interpretation is usually possible only for a narrow range of specimen and microscope conditions, but this situation and its variations are well understood. In any case, the resulting HRTEM image is a two-dimensional projection of the three-dimensional structure, with the columns of atoms being either dark or bright, depending on the exact specimen and imaging conditions. There are many factors that can potentially contribute to HRTEM image contrast and it is common practice to compare experimental images with calculated images under the same conditions. In the following examples, the atomic columns usually appear as bright spots in the images, contrast that is typical when imaging thin metal specimens with strong atomic potentials.

The specimen and microscope requirements for *in-situ* hot-stage HRTEM imaging are not different from those of usual HRTEM, except that one must have a heating holder and some method of recording and analyzing dynamic images.[9,10] At present, most HRTEM's are equipped with charge-coupled device (CCD) or TV-rate cameras that are fiber optically coupled to the HRTEM for recording high-resolution images. TV-rate cameras generally acquire 30 frames/s, so that the time resolution of the video recording is ~0.03 s. Faster video systems are available, but the signal to noise in the images then starts to become an issue. Specimen drift is usually not a significant problem when making small temperatures changes (a few degrees) and images are being acquired at TV rates, but substantial drift makes subsequent batch processing and computer comparison of digital images difficult. It is possible to install drift compensators on the microscope that alleviate this problem, but these are not readily available commercially. Drift can be a significant problem when making large temperature changes (tens of degrees for example) and this is a limiting factor when using current heating holders at high magnifications.

The most versatile specimen holder for materials science applications is a double-tilt hot-stage and several manufacturers offer such holders.[11] A typical water-cooled holder can achieve a temperature of 800–1000°C with a full range of tilt. As in static HRTEM, the area of interest ideally needs to be in a zone-axis orientation in order to interpret the atomic structure in the images. Knowledge of the temperature of the specimen in the area of interest as well as the heating/cooling rate when the current is changed are critical for quantitative studies and can vary from the hot-stage thermocouple readout depending on the sample conductivity, how well it is in contact with the holder, etc. In most cases, experience has shown that the sample temperature is usually within about ±15°C of the thermocouple readout for holders of this type.

Just about any type of phase transformation that can be induced by temperature changes within the limits of the holder can be examined by *in-situ* hot-stage HRTEM. (The same applies to *in-situ* cooling experiments using liquid nitrogen or liquid helium holders, although holder vibration due to bubbling can be a factor when using cooling liquids.) These include order-disorder reactions,[12] grain-boundary motion,[13] melting-freezing of materials,[14] precipitation-dissolution,[15] interfacial reaction,[16] crystallization,[17] twin and martensitic motion,[18] oxidation-reduction,[19] faceting-roughening transitions,[20] nanoparticle reactions,[21] coarsening,[22] quasimelting,[23] catalytic reaction,[10] etc.

3. Example Studies

3.1. *Diffuse coherent interface in Au-Cu alloy*

Figure 1(a) shows a HRTEM image of a diffuse interphase boundary between the long-range ordered AuCu-I phase and disordered α phase in a Au-41Cu (at.%) alloy sample taken as a frame from a videotape at approximately 305°C during *in-situ* heating in the HRTEM.[12] The position where a 25-pixel wide intensity profile was taken across the interphase boundary is indicated by the white rectangle in the figure. The corresponding intensity profile is shown in Fig. 1(b). In this profile, the ends of the diffuse interphase boundary on the ordered and disordered sides are indicated by the lines labeled O and D, respectively, and the same locations are also indicated in Fig. 1(a). Detailed HRTEM image simulations (e.g., inset in Fig. 1(a)) show that the Au-rich (001) planes in the ordered AuCu-I phase appear as bright lines in the image in Fig. 1(a) and corresponding high intensity peaks on the left side of the intensity profile in Fig. 1(b). The Cu-rich planes in between are barely visible because they

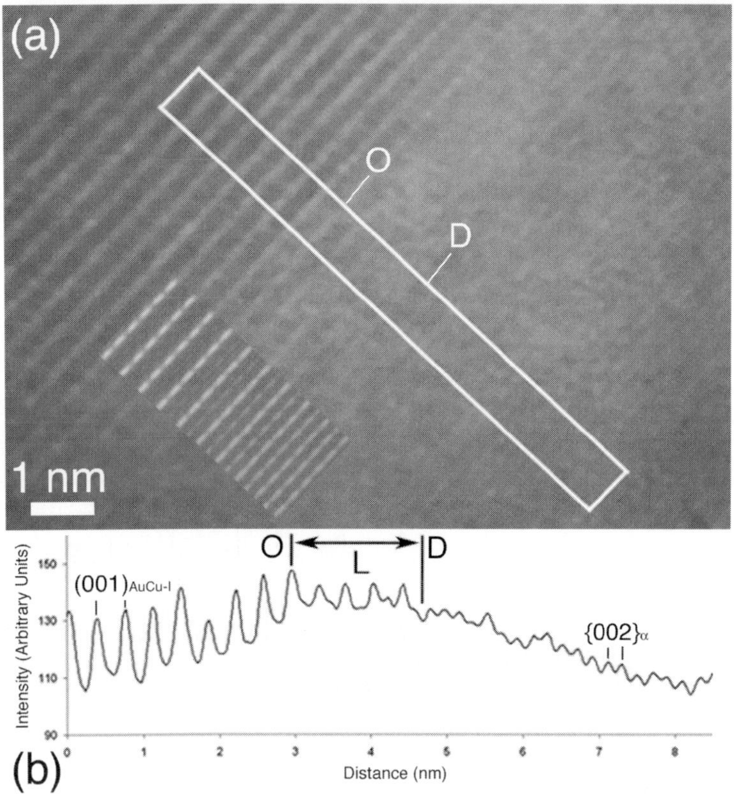

Fig. 1. (a) HRTEM image of an order-disorder interphase boundary at 305°C with a simulated image of the interphase boundary for the same sample and microscope conditions shown superimposed on the experimental image. The Au-rich (001) planes in the AuCu-I phase appear bright on the left side of the interphase boundary as do the {002} planes that are spaced approximately half as far apart in the α phase on the right side. (b) A 25-pixel wide intensity profile taken across the interphase boundary over the region indicated by a white box in (a). The positions O and D, which indicate the ordered and disordered sides of the diffuse interphase boundary, respectively, are indicated in both (a) and (b) (from Ref. 12).

appear dark in the HRTEM image in Fig. 1(a) under these experimental conditions. The lower intensity peaks spaced only half this distance apart to the right of position D correspond to the {002} planes in the disordered α phase. In the region between O and D, the intensities change from

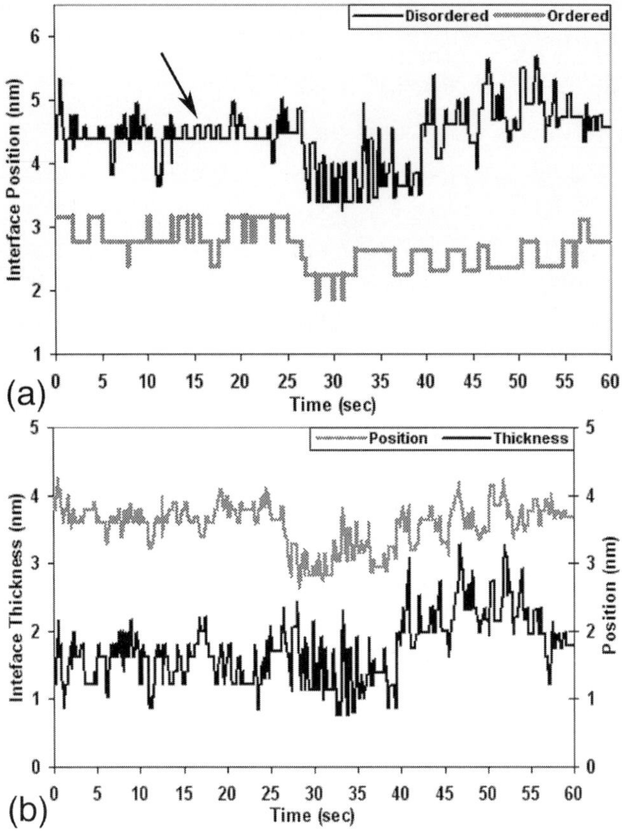

Fig. 2. (a) Graphs showing the positions of the disordered (top) and ordered (bottom) sides of the diffuse interphase boundary in Fig. 1(a) over a period of 60 s. (b) Corresponding graphs showing the interphase boundary mean position (top) and thickness (bottom) over the same period. The interphase boundary mean position in Fig. 2(b) fluctuates around a value of approximately 3.5 nm, which is roughly the midpoint between the two intensity profiles in Fig. 2(a) (from Ref. 12).

left to right, leading to the interphase boundary thickness labeled L in Fig. 1(b).

The positions O and D in Fig. 1 were determined for each frame (time interval of 0.03 s) over a period of 60 s and these are plotted in Fig. 2(a), which shows several important features. First, it is evident that the disordered side of the diffuse interphase boundary (top graph) generally moves more frequently and often over larger distances than the ordered side of the interphase boundary (bottom graph) with time. The smallest distance present in the top graph (arrow) is the spacing of one {002} plane in the disordered α phase. Following the disordered side of the interphase boundary with time, it is evident that the interphase boundary fluctuates over a distance of one to six {002} α planes. In contrast, using the {002} plane spacing as a reference, it is apparent that the ordered side of the interphase boundary in the lower graph typically moves twice this distance, which is the spacing between the Au-rich (001) planes in the AuCu-I phase. In addition, it fluctuates with a frequency that is several times less than that of the disordered side. Thus, the position of the disordered side of the interphase boundary rapidly fluctuates between a number of adjacent {002} α planes with time, while the ordered side fluctuates more slowly and only between Au-rich planes in the ordered AuCu-I phase.

The different interphase boundary behaviors observed in Fig. 2(a) are likely due to a combination of several factors occurring at the interphase boundary, namely: (i) the tetragonality of the AuCu-I structure, which leads to an energetically stable (001) plane that tends to keep the ordered side of the interphase boundary parallel to this plane, (ii) a lower average diffusivity in the AuCu-I phase as compared to the disordered phase at 305°C and related to this, (iii) a longer atomic jump-distance (0.37 nm) for diffusion perpendicular to the ordered (001) planes in the AuCu-I ordered phase, and (iv) the fact that the interphase boundary must move this same distance between Au-rich planes on the ordered side.[24,25] A combination of these effects leads to a significantly higher activation energy for movement of the ordered side of the interphase boundary, causing the experimentally observed interphase boundary behavior in Fig. 2(a). In fact, if one performs an approximate calculation for the average diffusivities and corresponding jump frequencies of the atoms in the ordered and disordered phases, respectively, using the equation $\Gamma = 6D/\lambda^2$, where Γ is the jump frequency (s^{-1}), D is the diffusivity at 305°C of 1.7×10^{-15} m^2/s for the α phase and 7.0×10^{-16} m^2/s for the AuCu-I phase,[25] and λ is the jump distance (3.7×10^{-10} m in AuCu-I), one finds that the jump frequency in the disordered phase is approximately 3×10^5 s^{-1}, as compared to 3×10^4 s^{-1} in the ordered phase, consistent with the considerations mentioned above. However, these frequencies are approximately four orders of magnitude greater than the actual frequencies of interphase boundary movement recorded for the ordered and disordered sides of the interphase boundary, indicating that collective atom volumes, rather than individual atom jumps, are involved in the process. In other words, movement of the interface involves the formation of critical-size fluctuations involving many atoms.[15,17]

An outline of the ordered side of the interface was drawn for a number of individual frames, based on the positions of the bright (001) planes. Figure 3 shows two different interface profiles that were typically found, reflecting the mechanisms by which interface movement (or fluctuation) occurred.[12] Figure 3(a) shows a situation where a perturbation in the ordered phase appeared to nucleate homogeneously on the interface (arrow) away from any other features. This perturbation expanded into and laterally along the order-disorder interface, propagating it forward and resulting in the interface shape outlined in Fig. 3(a). The image in Fig. 3(b) shows a second case, where a fluctuation in the ordered phase nucleated near a boundary between two different ordered domains. This ordered portion then propagated along the order-disorder interface as a series of steps, one (001) plane high (indicated by arrows in Fig. 3(b)), again resulting in net movement forward. In either case, the interface then receded by the opposite movement of such fluctuations (or steps). In most of the frames examined in the video, it was found that the ordered phase advanced into the disordered phase at more than one region along the interface, with nucleation at the domain boundary constituting only a small fraction of the fluctuations. Hence, this intersection did not appear to have a large affect on the overall interface behavior.

Based on Fig. 3, it is clear that the interface does not fluctuate by single-atom jumps, but by a collective mechanism involving the nucleation and spreading of critical-size (001) steps parallel to the interface plane. This is similar to the behavior of highly faceted interfaces in other diffusional transformations such as precipitation and crystallization[26] as shown below, and different from the behavior of fluctuations at solid-liquid interfaces,[27] which are not constrained to a certain height, or energy minimum perpendicular to the interface (e.g. one (001) AuCu-I plane spacing). It also demonstrates that the anisotropy in the interfacial energy of the AuCu-I phase, specifically the (001) interphase boundary energy, is sufficient at

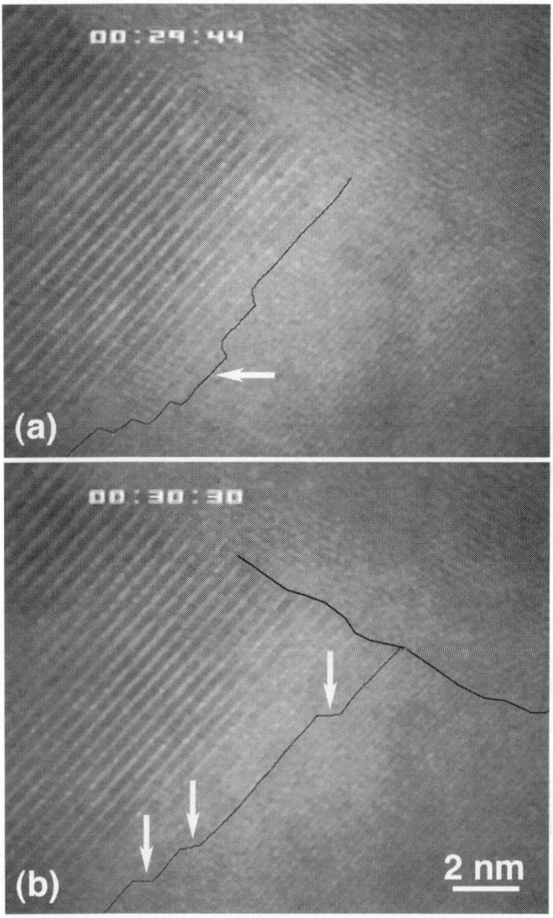

Fig. 3. HRTEM images showing mechanisms of interface movement. (a) A fluctuation nucleates homogeneously on the interface and expands by steps, one (001) AuCu-I plane high, to form a pillbox shape (white arrow). (b) A series of steps (arrows) nucleates at an adjoining boundary and propagates along the interface. The interface profiles are outlined in the figures.

this temperature to influence the morphology and behavior of fluctuations at the diffuse interface. Such behavior is consistent with previous observations of interfaces in other similar systems.[28,29]

Since HRTEM images such as those in Figs. 1 and 3 are projections of the three-dimensional interface structure, the interface fluctuations seen in projection also occur through the specimen thickness along the viewing direction, so that the profile in Fig. 3(a) for example, most likely represents a set of somewhat faceted, concentric pillboxes in three dimensions. It is also quite likely that fluctuations in the interface through the specimen thickness are similar, if not identical, to what is observed in the plane of projection, since the interface plane is (001). The minimum size perturbation (or fluctuation) observed at the interface was one (001) plane spacing high (~0.37 nm) and about 3–4 times this wide (1.2–1.6 nm), so that the critical fluctuation volume is on the order of 0.6 nm^3, or about 40 atoms, assuming a circular pillbox shape, or slightly larger if a square, faceted shape is assumed. Note that this is about the minimum volume of material that one would expect to see embedded in a 20 nm thick specimen.

The interphase boundary thickness and mean position were determined from the data in Fig. 2(a) and these are plotted in Fig. 2(b). The mean interphase boundary position was determined as the midpoint between positions O and D. The average frequency of interphase boundary movement is 1.2 s^{-1}, with a maximum frequency of 7.7 s^{-1} and a minimum frequency of 0.4 s^{-1}. This temporal variation reflects mainly that of the disordered side, which fluctuates at a higher frequency. Also note that the mean interphase boundary position moves up more often than down in Fig. 2(b)), again reflecting the character of the disordered side of the interphase boundary, which tends to move into the disordered phase, expanding the interphase boundary. Thus, the interphase boundary fluctuations are not symmetric about the mean position, but favor the disordered side. The average velocities of the ordered and disordered sides of the interphase boundary were also determined over 60 s using the data in Fig. 2(a) and found to be 0.3 nm/s and 1.3 nm/s, respectively.

The interphase boundary thickness is shown in the bottom graph in Fig. 2(b). The average thickness was approximately 1.7 nm, or the equivalent of 7 {002} planes (or ~4 unit-cells) in the disordered α phase, although the thickness averaged slightly more than 2.0 nm during the latter 20 s of the video sequence, where the disordered side of the interphase boundary moved a few planes further away from the ordered side. The maximum and minimum interphase boundary thicknesses observed were 3.3 and 0.8 nm, respectively, which represent variations in thickness on the order of 100%. The average interphase boundary thickness measured from Fig. 2(b) is comparable to recent studies on strain-induced incomplete wetting above the critical temperature T_c, where the alloy becomes disordered, at AuCu-I (001) surfaces,[24] and is also similar to previous calculations for a Au-Cu alloy[30] and recent atomistic calculations performed on Al-Li and Al-Ag alloys slightly below T_c using embedded atom potentials and Monte Carlo methods.[31,32] In the Al-based alloys, which behave similar to Au-Cu alloy, the calculations indicate that the diffuse interphase boundary should extend over 4–6 unit cells (i.e. $L \sim 1.6$–2.4 nm), and the strain-induced incomplete wetting phenomenon leads to order in AuCu (001) surfaces that extends 5–7 planes (i.e. $L \sim 1.0$–1.4 nm) into the crystal at temperatures much greater than T_c.

3.2. *Partly coherent interfaces in Al-Cu-Mg-Ag alloy*

Recent *in-situ* hot-stage HRTEM studies of precipitate plates in Al-Cu-Mg-Ag and Al-Ag alloys performed both parallel and perpendicular to the plate faces, and comparison of these studies with prior HRTEM and conventional *in-situ* hot-stage TEM investigations, have clearly established the terrace-ledge-kink (TLK) mechanism as the primary atomic mechanism involved in growth and dissolution of faceted precipitates in metal alloys.[2,33,34] This process is illustrated schematically in Fig. 4, which shows a perspective view of a precipitate plate growing by ledges that nucleate on the habit plane and propagate out to the edge, where they stack one above the other. Figure 4 shows two additional views, one perpendicular to the habit plane of the plate, i.e. face-on, and the other parallel to the habit plane, i.e. edge-on. It is clear from the two lower illustrations, that when the plate is viewed edge-on parallel to the facets, it is possible to observe the atomic structure and dynamics of individual ledges and also the motion of the ledges stacked at the plate edges. When the edge of the plate is viewed in the face-on orientation, it is possible to observe kinks that form on and propagate along the ledges. By combining information from these two orientations, it is possible to obtain a three-dimensional description of the atomic mechanisms of interfacial motion, as illustrated by the following data for {111} θ plates in an Al-3.9Cu-0.5Mg-0.5Ag (wt.%) alloy.[15,35]

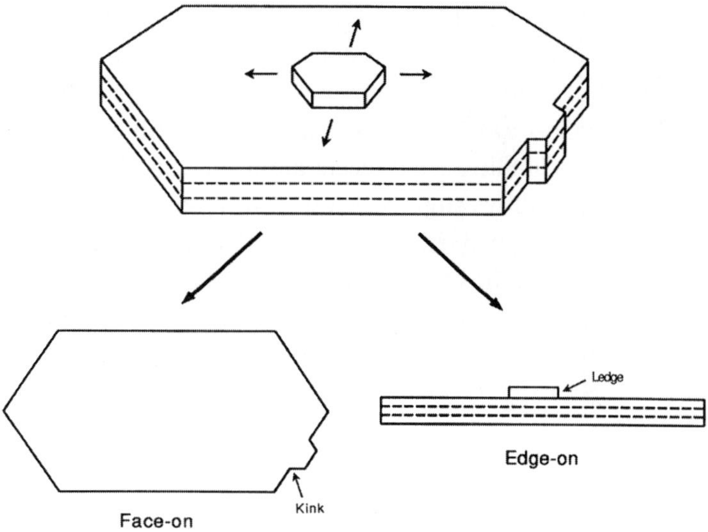

Fig. 4. (Top) Perspective, (bottom-left) face-on and (bottom-right) edge-on views of a precipitate plate growing by a TLK mechanism (from Ref. 26).

3.2.1. *Structural and kinetic analyses in an edge-on orientation*

The orientation relationship of the {111} θ phase with the matrix is $(\bar{1}10)_\theta \| (111)_\alpha$, $[110]_\theta \| [101]_\alpha$ and $[001]_\theta \| [1\,\bar{1}\,1]_\alpha$, which is a low-energy orientation relationship for θ phase designated as a Vaughan II orientation relationship. Figure 5(a) shows a HRTEM image of a ledge on the face of a θ plate viewed edge-on along a $[001]_\theta \| [1\,\bar{2}\,1]_\alpha$ direction, as in the lower-right illustration in Fig. 4. The ledge is approximately two $\{111\}_\alpha$ matrix planes high, or half of a unit cell of the θ structure (0.424 nm). This was the smallest ledge size that was observed on the faces of the θ plates and higher ledges were often observed.[35] The image was taken from the videocassette at about 220°C and the ledge was observed to oscillate several times per second over a distance of about two unit-cells of the θ phase along the precipitate face while moving slowly across the face toward the precipitate edge in the direction indicated by an arrow. *In-situ* experiments performed perpendicular to the plate face indicate that the oscillatory motion is due to the formation and annihilation of kinks along the ledge, as demonstrated in the next section. The videocasette recording also

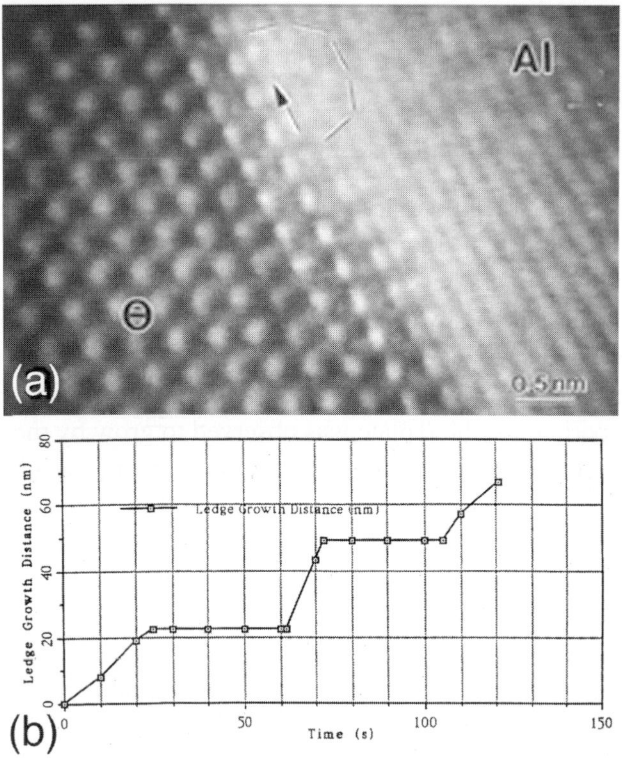

Fig. 5. (a) HRTEM image of a single ledge on a θ plate during growth at about 220°C. (b) Graph of growth distance versus time for the ledge in (a) (from Ref. 35).

revealed direct experimental evidence of enhanced atomic motion in the matrix just ahead of the ledge and this leads to slight blurring in the HRTEM image, which is visible in the enclosed area in Fig. 5(a). It is important to note that the precipitate structure only one unit-cell behind the ledge appears completely transformed, indicating that the structural and compositional changes that are necessary for diffusional growth occur simultaneously within a few atomic distances of the ledge. Although the ledge appeared to move smoothly across the precipitate face over short periods of time, it displayed start-stop behavior when viewed over longer times, as shown in Fig. 5(b). Such periodic lack of mobility during the migration of ledges has been observed previously and attributed to a lack of sites for atomic attachment along the ledges as they align along low-energy matrix directions, in this case $[1\,\bar{2}\,1]_\alpha$.[33,34,36]

3.2.2. *Structural and kinetic analyses in a face-on orientation*

Figure 6 shows a HRTEM image taken at the edge of a θ plate viewed face-on along a $[\bar{1}10]_\theta \| [111]_\alpha$ direction, as in the lower-left illustration in Fig. 4, during an *in-situ* hot-stage experiment at about 275°C. The prominent rectangular pattern of white spots outlined in Fig. 6 with dimensions of 0.244 nm by 0.429 nm relates directly to positions of Cu atoms in the θ structure, as determined by HRTEM image simulation. During the *in-situ* HRTEM experiments, the θ plate was observed to grow by the nucleation of half unit-cell high (0.429 nm) double-kinks along the $(110)_\theta \| (101)_\alpha$ plate edge, which propagated along the edge until they reached the intersecting facet. The arrow in Fig. 6 indicates the end of one such kink. The smallest kinks were one-half of the θ unit cell in height (one rectangular pattern of white spots about 0.429 nm long), but sometimes two or three kinks nucleated and/or dissolved in rapid succession in an oscillatory manner about an average position, similar to the behavior

Fig. 6. An isolated kink (indicated by arrow) traveling along the $[110]_\theta$ facet of a {111} θ precipitate plate during growth at about 275°C (from Ref. 15).

described for the ledge in Fig. 5 and the diffuse interface in Fig. 3. It is important to note that the kink in Fig. 6 is well defined to within two or three half unit-cells of the {111} θ structure. Although the image in Fig. 6 was taken at the edge of the θ plate where several ledges are stacked vertically parallel to the electron beam direction, when this perspective is combined with the one in Fig. 5, it is possible to conclude that the phase transformation is occurring at kinks in ledges on the θ plates and that the transformation is complete within a volume as small as about one unit-cell of the θ phase. This volume contains about four atoms of Cu and eight atoms of Al. Thus, performing *in-situ* hot-stage HRTEM allows observation of the atomic mechanisms of the transformation (the TLK mechanism) as well as the dynamics of transformation. It is also possible to study the dynamics of the kinks at the edges of the plates, as described in detail elsewhere.[15,35]

3.3. *Incoherent interface in Ti-Al alloy*

In contrast to the case of coherent and partly coherent interfaces described above, the atomic mechanisms by which solid-solid incoherent interfaces move are not well understood.[4,37] This is true for both high-angle grain boundaries[1,13,38] and interphase boundaries. During dissolution or at static interfaces when nucleation at the interface is not required for motion, local fluctuations can occur due to thermal effects. *In-situ* HRTEM studies of crystallization and on grain boundaries have revealed such reversible fluctuations.[13,17]

Recent *in-situ* heating HRTEM experiments performed on massive transformation interfaces in TiAl alloy have shown evidence of both continuous and stepwise motion during growth at the atomic level, depending on the orientation relationship and interface plane.[39] These two types of growth behavior were previously observed at much lower levels of resolution during cinematographic studies of the massive transformation in Cu-Ga alloys[40] and during *in-situ* heating TEM experiments of massive interfaces in Ag-Al[41] and Cu-Zn alloys.[42] One problem associated with all of the previous, lower-resolution *in-situ* studies, is that the actual structures of the massive transformation interfaces were not known, but assumed to be incoherent. This assumption has been repeatedly challenged in the literature, particularly since many massive products display definite crystallographic orientation relationships with the parent phase.[4] The following results were obtained from frame-by-frame analysis of *in-situ* HRTEM data obtained on a massive transformation interface in Ti-Al alloy growing by a continuous mechanism, i.e. by atom attachment continuously along the interface rather than at well-defined crystallographic facets or steps in the interface.[43] The results reveal new phenomena associated with continuous growth at solid-solid interfaces that are different from those seen for the previous Au-Cu and Al-Cu interfaces, and are compared with existing experimental and atomistic simulation results on grain and interphase boundaries at the same level of spatial resolution.

Figure 7 shows a HRTEM image of a α_2/γ_m incoherent interface obtained in a Ti-46.5Al (at.%) alloy that was solutionized at 1360°C for 1200 s and

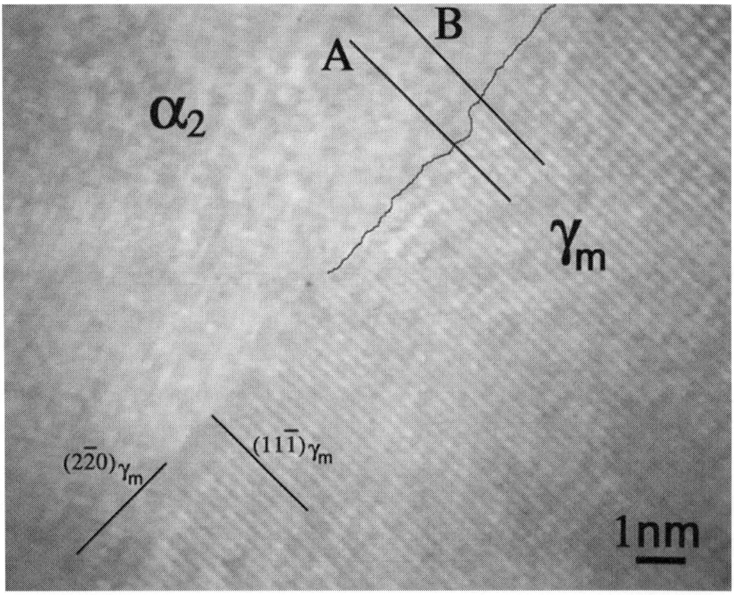

Fig. 7. HRTEM image of a α_2/γ_m incoherent interface obtained after the specimen had been heated for about 5400 s (1.5 hr) at 575°C *in-situ* in the TEM. The $(2\,\bar{2}\,0)\gamma_m$ interface plane and the $(11\bar{1})\gamma_m$ planes in the γ_m phase are indicated. Lines A and B show the positions of two $(11\bar{1})\gamma_m$ planes that were used to obtain Fig. 11 and a typical trace of the α_2/γ_m interface is outlined in the top half of the image (from Ref. 43).

water or ice-brine quenched to nucleate grains of the γ_m (massive) phase in the α (parent) phase, which subsequently orders to form α_2 phase. Figure 7 was taken after heating the specimen for about 5400 s (1.5 hr) at 575°C *in-situ* in the TEM.[39] The interface is edge-on and nearly parallel to a $(2\bar{2}0)_{\gamma_m}$ plane, or roughly perpendicular to the $(11\bar{1})_{\gamma_m}$ planes that are clearly visible and indicated in the γ_m phase in the image. The orientation relationship between the two phases across this interface was approximately $[112]_{\gamma_m} \| [122]_{\alpha_2}$ with about 1.4° tilt between the two zone axes, and $(11\bar{1})_{\gamma_m} \| (\bar{2}23)_{\alpha_2}$, and the viewing direction is approximately $[112]_{\gamma_m} \| [122]_{\alpha_2}$ in Fig. 7. In the video recording, the interface appeared to advance into the α_2 phase by continuous overall motion, without evidence of start-stop growth behavior, which is often observed during motion of coherent and partly coherent interfaces between solid phases with low-index orientations and interface planes by steps or ledges, as shown previously for θ phase in Fig. 5(b). The interface also appeared to undulate slightly as it advanced, rather than remaining planar during growth. These features were examined in detail in the subsequent frame-by-frame analyses over a period of 50 s (or 1500 frames).

A typical trace of the α_2/γ_m interface that was analyzed in this study is indicated by a line in the top-center part of Fig. 7. Figure 8(a) shows the

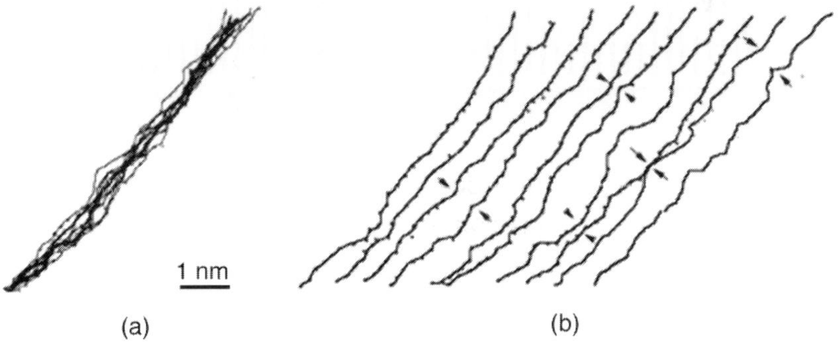

Fig. 8. (a) The positions of 10 traces obtained from each frame of the video recording during the first 0.3 s. (b) The same traces separated by 0.52 nm distance, starting with the original frame on the right, to better reveal their characteristics. The positions of various advancements and recessions in the traces of the interface are indicated by arrows (from Ref. 43).

actual positions of 10 such traces obtained from each frame of the video recording during the first 0.3 s. In Fig. 8(b), the traces were separated by 0.52 nm distance, starting with the original frame on the right, to better reveal their characteristics. The positions of various advancements and recessions in the traces of the interface are indicated by arrows in Fig. 8(b). Initial examination of the traces gives the impression that motion of the interface is highly irregular, although every trace shows at least one location where the interface advances into the α_2 phase. Further examination of the arrows shows that when the interface advances in one location in a frame, it subsequently recedes at the same location in the next frame. This pattern of advancement and recession leads to an oscillatory behavior in the interface position as a function of time, as illustrated using only the $(11\bar{1})$ planes (viewed edge-on) in the γ_m phase in Fig. 9(a). (Note that the regularity of the $(11\bar{1})$ planes shown in Fig. 9(a) is only for purposes of illustration and not meant to imply that the motion of these planes at the interface is so regular.) In addition to the oscillatory behavior caused by the advancements and recessions, movement of the interface appeared to occur by the spreading of certain advancements along the interface.

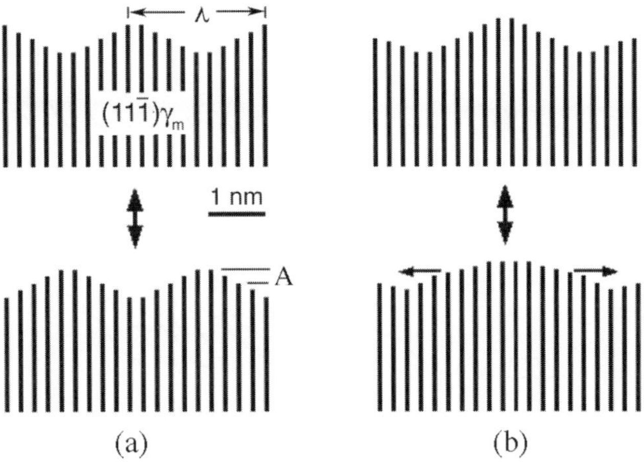

Fig. 9. Illustrations of: (a) the oscillatory behavior of the interface as a function of time, and (b) movement of the interface by the spreading of an advancement in the directions indicated by the horizontal arrows. The amplitude (A) and wavelength (λ) of the fluctuations are indicated in (a) (from Ref. 43).

This behavior is illustrated in Fig. 9(b). Both of these features at the interface appeared to involve the cooperative growth and dissolution among groups of individual $(11\bar{1})$ plane edges in the γ_m phase at the interface.

The distances between advancements in the traces of the interface were measured in each frame and found to display characteristic spacings of approximately 1.15 and 1.61 nm. This feature is illustrated in Fig. 10(a), where the distances between the advancements are indicated by dots. Figure 10(b) shows the same traces with dashed lines draw tangent to the advancements and recessions along the interface traces and a darker, parallel solid line drawn halfway between these. The distances between the dashed lines varied from 0.27 to 0.58 nm, with an average value of 0.41 nm. If the amplitudes and distances of the advancements/recessions at the interface are treated as a wave-like fluctuation, the data in Fig. 10 show that the fluctuation has an amplitude and fundamental wavelength of approximately 0.21 nm and 1.15 nm, respectively. It is interesting to note that the second characteristic wavelength associated with the fluctuation of 1.61 nm is approximately $\sqrt{2}$ times the fundamental wavelength.

The interface dynamics were analyzed over a longer period of time by similarly plotting the interface traces from video frames obtained every 5 s,

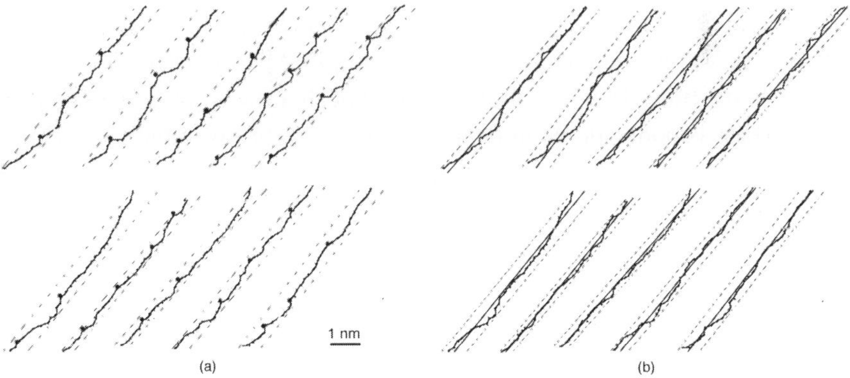

Fig. 10. (a) The same traces as in Fig. 8, with the distances between advancements indicated by dots. These were used to determine the wavelength of the fluctuations. (b) The same traces with dashed lines draw tangent to the advancements and recessions along the interface traces and a darker, parallel solid line drawn halfway between these. These were used to determine the amplitude of the fluctuations (from Ref. 43).

over the duration of the video sequence (50 s). The same interface characteristics were observed over this longer time interval as for the previous short one, indicating that the interface behavior is the same over all time scales within the time resolution of the experiment (i.e. 0.03 s). In the 50 s segment analyzed, the distances between the advancements in the interface were found to be approximately 1.19 and 1.69 nm and the distances between the advancements and recessions of the fluctuation varied from 0.32 to 0.53 nm. These values yield an average amplitude and fundamental wavelength for the interface fluctuation of approximately 0.21 nm and 1.19 nm, respectively, with the higher wavelength of 1.69 nm again being approximately $\sqrt{2}$ times the fundamental wavelength. The same analyses performed at intermediate time intervals yielded similar amplitudes and wavelengths, indicating that they are characteristic of the interface over all time scales of observation from 0.03 to 50 s.

The average velocity of the interface based on the initial and final positions of the interface traces over the 50 s of observation was 0.023 nm/s. In order to further understand the instantaneous velocity/behavior of the interface, the edges of individual $(11\bar{1})_{\gamma_m}$ planes were followed frame-by-frame in two locations at the interface, indicated by lines parallel to the planes labeled A and B in Fig. 7. Figures 11(a) and 11(b) show the resulting plots of these $(11\bar{1})_{\gamma_m}$ plane edges every 30th frame over 1500 frames, or a time of 50 s. These plots clearly illustrate the somewhat irregular but definite oscillatory behavior of the interface, that is, its wave-like fluctuations versus time. In addition, it is apparent that the average velocity of the interface is constant, as indicated by the straight lines that were fitted and superimposed on the plots in Figs. 11(a) and 11(b), although the instantaneous position of the interface oscillates about the average position with time at these locations. The equations for the lines are also given in the plots. This behavior was typical of all such $(11\bar{1})_{\gamma_m}$ planes at the interface and reflects the behaviors illustrated in Figs. 9 and 10. It is important to note that although the average interface velocity was 0.023 nm/s, the instantaneous velocity at any given time can be more than an order of magnitude higher, as evidenced by the slopes of the oscillations in Fig. 11.

The results from the frame-by-frame analyses above indicate that this massive transformation interface, which moves forward continuously as a

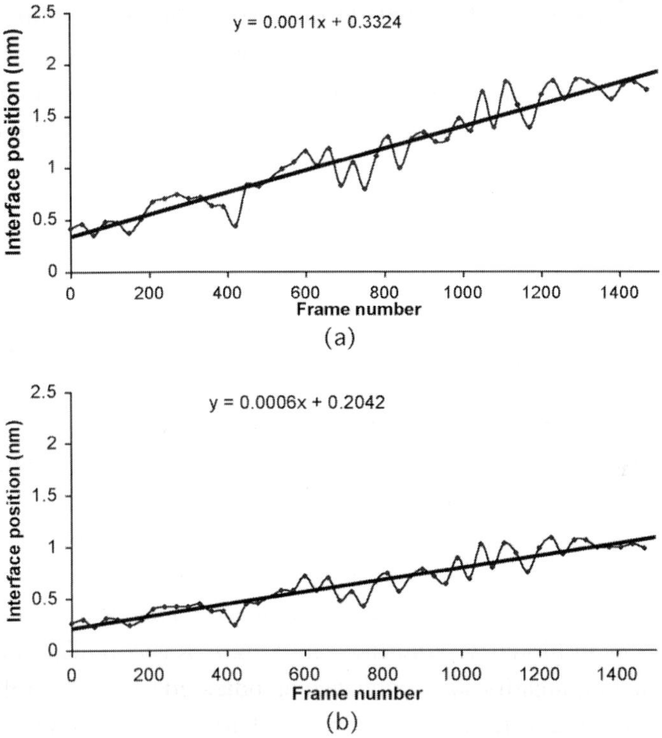

Fig. 11. Frame-by-frame plots of the $(11\bar{1})_{\gamma_m}$ plane edge positions at locations A and B in Fig. 7, with data points indicated every 30th frame over the duration of 1500 frames (50 s). The equations for the straight lines fitted to the data points in the figures are also given (from Ref. 43).

planar interface overall, rather than by an obvious ledge mechanism, displays quasi-periodic fluctuations along its length that can be characterized by a wave-like function. This oscillatory interface behavior, which accompanies the interface as it propagates forward with constant average velocity, is superimposed on the average position of the interface versus time. The oscillations are not regular, but they display a characteristic amplitude and wavelength. There was no obvious connection between either the amplitude or wavelength at the interface with any structural features in the α_2 phase, such as a characteristic ledge spacing or height, although this aspect deserves further examination.

The creation of advancements at the interface and spreading of some of these advancements along the interface indicates that the interface must be overcoming some barrier to migration, rather than by simply moving forward uniformly everywhere. As mentioned at the start of this section, the presence of local barriers to interface migration typically leads to oscillatory behavior at interfaces in diffusional transformations. The present interface is different from the previous Al-Cu-Mg-Ag diffusional interface in that it is a (relatively) planar, incoherent interface between two crystals with a high-index orientation relationship and does not involve long-range diffusion, as opposed to being a partly coherent interface between two crystals with a low-index orientation relationship and cusp-oriented interfaces that grow by a TLK mechanism, involving long-range diffusion. In this regard, it is more like a general, high-angle grain boundary migrating under a driving force.[1,38] Unlike the case of highly faceted cusp-oriented interfaces, it appears that the lack of well-defined structure at this interface allows it to undergo regular wave-like spatial fluctuations in position with time, some of which grow to a critical size and then spread along the interface, moving it forward. The size of the critical fluctuations for this interface appears to be just greater than approximately 0.21 nm high and 1.69 nm wide, since fluctuations larger than these are not commonly observed. In this regard, its motion is similar to that of a planar interface growing by nucleation and spreading of two-dimensional nuclei,[2,35,44] except that the nuclei are not well-defined crystallographically and their location is a stochastic process.

The dynamic characteristics of the present interface are remarkably similar to those observed for the massive transformation in Cu-Ga alloys obtained at a much lower spatial resolution (i.e. at the micrometer scale), but at a higher time resolution (at 64 frames/s) and at much greater velocities (up to 1 mm/s) using cinematographic studies.[40] In fact, if one compares the interface traces in Figs. 8 and 10 with those in Figs. 8 and 9 in Ref. 40, the traces are essentially identical, indicating scaling of certain interface behaviors in these massive transformations over great variations in length, velocity and time. The benefit of the present Ti-Al analysis is that the much higher spatial resolution reveals the previously unknown atomic details of interfacial structure associated with the dynamic behavior. The oscillatory character of the interface and the constant velocity (i.e. the linear displacement versus time in Fig. 11) also compare reasonably well with results from a recent kinetic Monte Carlo study of a massive transformation interface between f.c.c. and b.c.c. crystals (see Figs. 3 and 5 in Ref. 45 for example), although in this study, the low-index close-packed planes in the two crystals were parallel, so that the interface is cusp-oriented and two-dimensional b.c.c. nuclei are clearly present at the interface at low driving forces. It is also interesting to note that some oscillations are present in molecular dynamics simulations of grain-boundary migration in planar, high-angle twist grain boundaries (see Fig. 5 in Ref. 38 for example).

4. Summary and Outlook

The purpose of this article was to demonstrate how *in-situ* hot-stage HRTEM is used to understand the atomic structure and dynamics of interphase boundaries, since these are important in understanding the nature of phase transformations and in utilizing phase transformations to develop the structure and properties of materials. The examples showed that *in-situ* hot-stage HRTEM allows direct observation of the projected atomic structure of interphase boundaries and their mechanisms of motion involving collective atom processes, but not individual atoms or jumps. Since most dynamic processes at interphase boundaries appear to involve collective atom motion, e.g. fluctuations at order-disorder interfaces and growth of precipitate interfaces by a TLK mechanism, not being able to observe individual atom jumps is not a great limitation. The time resolution of dynamic observations is currently limited by a combination of image signal and the image recording devices to ~0.03 s. Even with these limitations, *in-situ* hot-stage HRTEM is able to provide a wealth of information about detailed atomic processes that occur during phase transformations in materials. This situation will continue to improve as rapid progress is being made in increasing the spatial, temporal and chemical resolution of HRTEM's, in the capability of imaging devices, storage and processing, and in precise computer control of the microscope. It is conceivable that it will be possible to image the three-dimensional atomic structures of interfaces in materials using energy-filtering TEM and/or high-angle annular dark-field scanning TEM and image reconstruction methods in the near future.[46] Observing their dynamics will present major additional challenges, but these too may be eventually solved.

Acknowledgments

The author is grateful to the many students and colleagues whose work is referenced herein. This research was supported by the National Science Foundation under Grants DMR-9908855 and DMR-0554792.

References

1. A. D. Sutton and R. W. Balluffi, *Interfaces in Crystalline Materials* (Clarendon Press, Oxford, 1995).
2. J. M. Howe, *Interfaces in Materials: Atomic Structure, Thermodynamics and Kinetics of Solid-Vapor, Solid-Liquid and Solid-Solid Interfaces* (John Wiley & Sons, Inc., New York, 1997).
3. C. M. Wayman, H. I. Aaronson, J. P. Hirth and B. B. Rath (eds.), *Proc. Pacific Rim Conf. on the Role of Shear and Diffusion in the Formation of Plate-Shaped Transformation Products*, *Metall. Mater. Trans.* **25A**, 1781 (1994).
4. H. I. Aaronson, *Metall. Mater. Trans.* **48A**, 2285 (2002).
5. J. W. Martin, R. D. Doherty, and B. Cantor, *Stability of Microstructure in Metallic Systems*, 2nd Ed. (Cambridge University Press, Cambridge, 1997).
6. J. M. Howe, H. I. Aaronson, and J. P. Hirth, *Acta Mater.* **48**, 3977 (2000).
7. J. C. H. Spence, *Experimental High-Resolution Electron Microscopy*, 2nd Ed. (Oxford University Press, Oxford , 1988).
8. B. Fultz and J. M. Howe, *Transmission Electron Microscopy and Diffractometry of Materials*, 2nd Ed. (Springer-Verlag, Berlin, 2002).
9. E. P. Butler and K. F. Hale, *Dynamic Experiments in the Electron Microscope* (North-Holland Publishing Co., New York, 1981).
10. R. Sinclair, T. Yamashita, M. A. Parker, K. B. Kim, K. Holloway, and A. F. Schwartzman, *Acta Cryst.* **A44**, 965 (1988).
11. D. B. Williams and C. B. Carter, *Transmission Electron Microscopy: A Textbook for Materials Science* (Plenum Press, New York, 1996).
12. J. M. Howe, A. R. S. Gautam, K. Chatterjee and F. Phillipp, *Acta Mater.* **55** (2007) 2159.
13. K. L. Merkle, L. J. Thompson, and F. Phillipp, *Phys. Rev. Lett.* **88**, 225501-1 (2002).
14. K. Sasaki and H. Saka, *Phil. Mag. A* **63**, 1207 (1991).
15. J. M. Howe and W. E. Benson, *Interface Sci.* **2**, 347 (1995).
16. R. Sinclair, *Mater. Res. Soc. Bull.* **XIX**, 26 (1994).
17. R. Sinclair, J. Morgiel, A. S. Kirtikar, I.-W. Wu, and A. Chiang, *Ultramicros.* **51**, 41 (1993).

18. J. M. Howe, *Proc. Inter. Conf. Martensitic Transformations '92 (ICOMAT-92)*, C. M. Wayman and J. Perkins (eds.) (Monterey Institute of Advanced Studies, Carmel, 1993), p. 185.
19. Z. Kang and L. Eyring, *Metall. Trans.* **22A** 1323 (1991).
20. H. Gabrisch, L. Kjeldgaard, E. Johnson, and U. Dahmen, *Acta Mater.* **49**, 4259 (2001).
21. J.-G. Lee and H. Mori, *J. Mater. Res.* **20**, 1708 (2005).
22. S. Iijima, *J. Electron Micros.* **34**, 249 (1985).
23. L. D. Marks, *Rep. Prog. Phys.* **57**, 603 (1994).
24. W. Schweika, H. Reichert, W. Babik, O. Klein, and S. Engemann, *Phys. Rev. B* **70**, 041401R (2004).
25. D. B. Butrymowicz, J. R. Manning, and M. E. Read, *J. Phys. Chem. Ref. Data* **3**, 527 (1974).
26. J. M. Howe, W. E. Benson, A. Garg, and Y.-C. Chang, *Advances in Physical Metallurgy — 94*, S. Banerjee and R. V. Ramanujan (eds.) (Gordon and Breach Science Publishers, Amsterdam, 1996), p. 277.
27. J. J. Hoyt, M. Asta, and A. Karma, *Mater. Sci. Eng.* **R41**, 121 (2003).
28. A. Loiseau, C. Ricolleau, L. Potez, and F. Ducastelle, *Proc. Intl. Conf. on Solid-Solid Phase Transformations in Inorganic Mater.*, W. C. Johnson, J. M. Howe, D. E. Laughlin, and W. A. Soffa (eds.), (The Metals, Minerals and Materials Society, Warrendale, 1994) p. 385.
29. K. Chatterjee, J. M. Howe, W. C. Johnson, and M. Murayama, *Acta Mater.* **52**, 2923 (2004).
30. R. Kikuchi and J. W. Cahn, *Acta Metall.* **27**, 1337 (1979).
31. M. Sluiter and Y. Kawazoe, *Phys. Rev. B* **54**, 10381 (1996).
32. M. Asta and J. J. Hoyt, *Acta Mater.* **48**, 1089 (2000).
33. C. Laird and H. I. Aaronson, *Acta Metall.* **17**, 505 (1960).
34. J. M. Howe, U. Dahmen, and R. Gronsky, *Phil. Mag A* **56**, 31 (1987).
35. W. E. Benson and J. M. Howe, *Phil. Mag A* **75**, 1641 (1997).
36. A. Garg, Y.-C. Chang, and J. M. Howe, *Acta Metall. Mater.* **41**, 235 (1993).
37. T. B. Massalski, W. A. Soffa, and D. E. Laughlin, *Metall. Mater. Trans.* **37A**, 825 (2006).
38. B. Schonfelder, D. Wolf, S. R. Philpot, and M. Furtkamp,. *Interface Sci.* **5**, 245 (1997).
39. J. M. Howe, W. T. Reynolds, Jr., and V. K. Vasudevan, *Metall. Mater. Trans* **33A**, 2391 (2002).
40. J. E. Kittl and T. B. Massalski, *Acta Mater.* **15**, 161 (1967).
41. J. H. Perepezko and T. B. Massalski, *Acta Metall.* **23**, 621 (1975).
42. G. Bäro and H. Gleiter, *Acta Metall.* **22**, 141 (1974).
43. N. Raffler and J. M. Howe, *Metall. Mater. Trans.* **37A**, 873 (2006).
44. J. W. Cahn, W. B. Hillig, and G. W. Sears, *Acta Metall.* **12**, 1421 (1964).
45. C. Bos, F. Sommer, and E. J. Mittemeijer, *Acta Mater.* **52**, 3545 (2004).
46. S. J. Pennycook, M. Varela, C. J. D. Hetherington, and A. I. Kirkland, *Mater. Res. Soc. Bull.* **31**, 36 (2006).

CHAPTER 6

IN-SITU TEM OF FILLED NANOTUBES: HEATING, ELECTRON IRRADIATION, ELECTRICAL AND MECHANICAL PROBING

Dmitri Golberg and Yoshio Bando

Nanoscale Materials Center, National Institute for Materials Science
Tsukuba, Ibaraki 305044, Japan

This chapter describes various kinds of *in-situ* TEM experiments performed on filled conventional carbon and non-carbon inorganic nanotubes. The results are summarized for nanotubes filled with liquid and solid metals, ceramics and inorganic compounds. The prospects of functional applications of analyzed nanotubes are particularly highlighted.

1. Introduction

The specific nanotubular shape presumes the natural existence of a tiny, down to 1 nm in diameter or even less, intra-tube channel. This cavity may serve as an important nanotechnology component for making various functional devices or performing multiple designated operations at the nanoscale, all within a nanotube channel. To date, standard C nanotubes, both multi- and single-walled, have been filled with various solids, e.g. metals, alloys, ceramics and salts,[1–4] fluids, e.g. water,[5] and liquid crystals.[6,7] These filled nanotubes might have bright prospects in the realization of revolutionary applications in electronics, optics, nanoscale magnets, energy storage, drug delivery and so forth.

This chapter shows how various *in-situ* TEM experiments may find an interesting use in the case of nanotubes which have been filled with various substances, namely, liquid melts or solid metals, functional ceramics

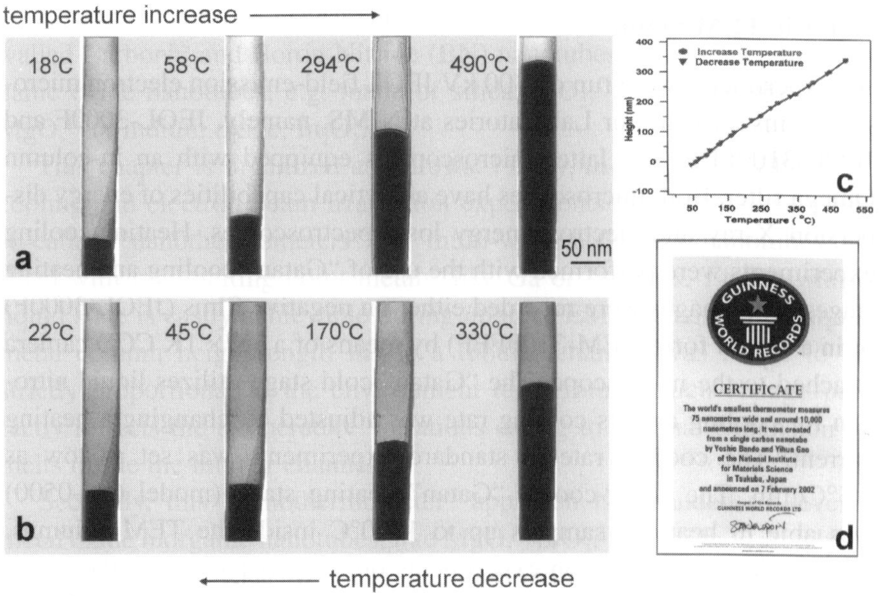

Fig. 1. Consecutive TEM images of a liquid Ga-filled multi-walled C nanotube subjected to thermal heating (a); and cooling (b) inside a transmission electron microscope using a "Gatan" heating stage. The corresponding temperatures are marked on the images; (c) a plot of a measured Ga-column height position within the tube as a function of temperature on heating and cooling; both dependences show a linear character and perfectly match each other thus revealing a perfect "nanothermometer"-like behavior; and (d) an issued *Guinness World Record* certificate for the smallest "thermometer" in the world which measures 75 nm wide and 10.000 nanometers long, and matches the nanotube shown in (a) and (b). Figures 1(a)–1(c) are from Ref. 14.

as documented by the corresponding *Guinness World Record* certificate, Fig. 1(d).

Figure 2 illustrates an analogous multi-walled C nanotube with an encapsulated Ga melt. The tube was cooled below room temperature inside TEM.[16] This particular tube was cooled down to –90°C, then reheated to 90°C. In a liquid state the length of the encapsulated Ga column varied linearly, as expected based on the pioneering experiments shown in Fig. 1. Freezing (solidification) occurred at –80°C. This is evidenced by a sharp Ga volume decrease at this temperature and additionally reconfirmed by the detailed electron diffraction analysis. The sudden sharp volume

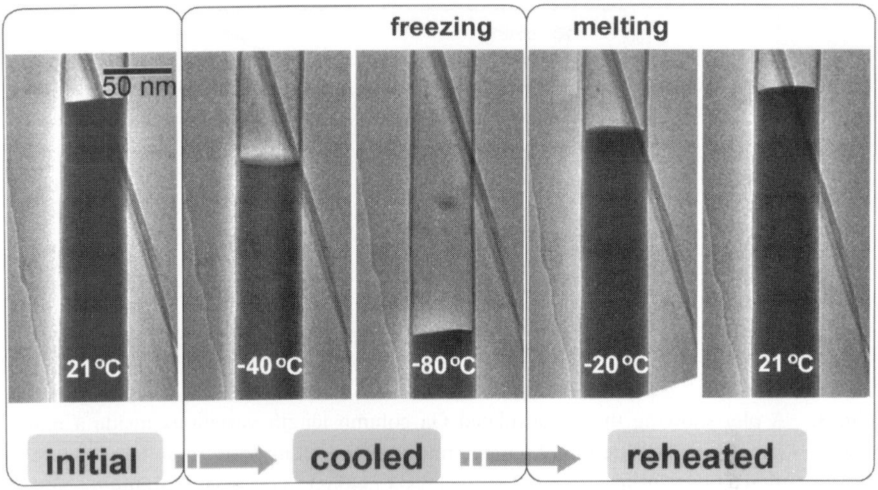

Fig. 2. TEM micrographs revealing the behavior of an encapsulated Ga column in a multi-walled C nanotube during cooling below room temperature (upon freezing and reheating back to room temperature). Ga freezes at −80°C, which is far below the freezing temperatures of any known Ga crystal phases; from Ref. 16.

(length) decrease of the Ga column upon solidification obviously indicates that the present melt is crystallized in another than the most typical α-Ga phase. α-Ga is an ice-type element and it should expand 3.1% while solidifying. The present embedded Ga column remains liquid up to −80°C. This temperature is much lower than the melting points of all known Ga phases.[31–33] For example, the lowest melting point is known for γ-Ga, that is −35.6°C. The plot of Ga-column height as a function of temperature is depicted in Fig. 3. This possesses a marked hysteresis due to supercooling and superheating behaviors of liquid Ga encapsulated in a C nanotube.

Within the same tube different Ga phases may coexist, for example β-Ga and γ-Ga (melting points −17°C and −36°C, respectively), as displayed in Fig. 4. This was established while performing the detailed electron diffraction analyses at various temperatures. The phenomenon leads to various interesting melting/solidification features of various filled parts along the tube as the tube is heated or cooled as a whole.

We suggest that the presently observed Ga melting point deviation from its bulk forms could be due to a combination of several interplaying

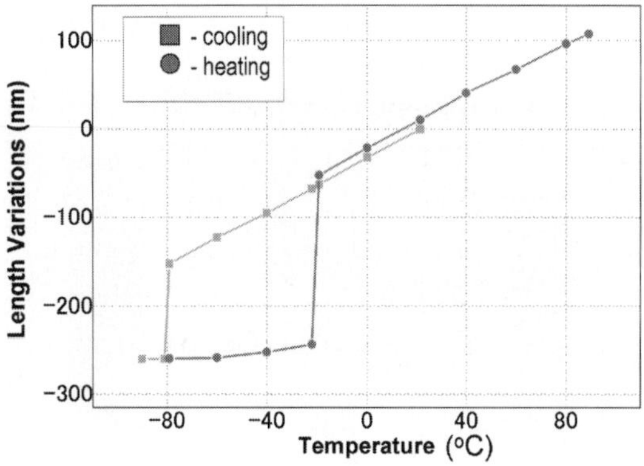

Fig. 3. A plot showing the encapsulated Ga column length variations inside a multi-walled C nanotube as a function of temperature during cooling from room temperature to −80°C, followed by reheating. The plot displays a marked hysteresis implying the supercooled and superheated states of Ga inside the Carbon nanotube; from Ref. 16.

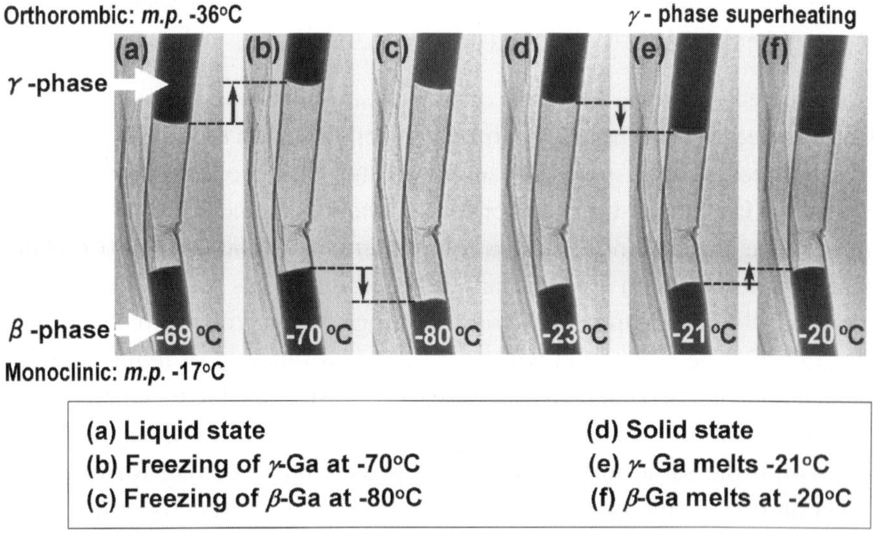

Fig. 4. TEM images illustrating a sequence of phase transformations inside a multi-walled Carbon nanotube within the two Ga fragments possessing various crystal structures. The upper domain corresponds to a γ-phase, while the lower domain to a β-phase. The arrows in figures denote the corresponding changes in the matter states, as seen due to the characteristic drastic changes in filling positions. The legend below the figure specifies the confirmed phase transitions; from Ref. 16.

factors. Most likely, the primary factor is the interactions between Ga and the C nanotube innermost walls. It is worth noting that these interactions should be especially prominent at a low temperature. In fact, it was observed that the curvature of the Ga tip surface is notably changed with temperature. Before solidification, the Ga tip surface is flat rather than has a convex meniscus. The situation is dramatically changed after remelting. Another factor could be the effect of dissolved impurities in the Ga melt, whose level is far beyond the detection limit of all experimental techniques available. The high vacuum in the TEM column may also slightly affect the melting point, but not that much. So-called confinement effect should be definitely ruled out since the present nanotube diameter is rather large (~100 nm) to account for the so-called quantum effects. A series of measurements on C multi-walled nanotubes with different diameters ranging from 100 to 200 nm show no melting point dependence. After ten cycles of Ga freezing/melting the observed nanotube diameter was only of ~96% of its original value. This may reflect some changes in the cross-sectional tube shape due to undefined stress-strain fields under the Ga phase transitions. However, if an encapsulated Ga was cooled down but yet allowed to freeze, its tip surface perfectly returns to the original position when reheated to the starting temperature. Therefore, such unique nanoscale thermometer can be used in a far wider temperature range of –69–500°C than that *a-priori* expected based on the bulk Ga thermal properties (50–500°C). This gives us an opportunity to measure the negative temperatures. It is emphasized here that the upper limit of the present nanotube thermometer utilization is related to fast Ga evaporation in a high-vacuum TEM column when the temperature approaches 550–600°C.

The practical temperature measurements of a given microenvironment using the proposed nanoscale thermometers are not straightforward though and may often require a dedicated TEM facility. One of the possible ways of real temperature recording was originally proposed by Gao *et al.*[15] by noticing that if a C nanotube is open at one of its tip-ends an encapsulated Ga is easily oxidized through this open tip, if the tube is placed in a oxygen-containing environment (e.g. normal air) at a high temperature. Thus a Ga oxide fragment, closer to the open tip, becomes unmovable since it tightly sticks/welds to the innermost C tube walls. This merging than restricts the whole Ga column move. In other words, the

Ga column after the tip part oxidation is frozen and its position remains totally independent of any following-up treatments, e.g. extraction of the thermometer from a hot zone, its cooling to room temperature, further reheating in TEM etc. Having a preliminary calibration of such thermometer performed in TEM, e.g. a Ga column height *versus* temperature (Fig. 1(c)), it becomes possible to estimate the highest temperature experienced by a Ga-filled C nanotube, i.e. the temperature at which the inner Ga melt was actually oxidized.

As a consequence of such approach, very recently Liu *et al.*[17] has documented that if a Ga oxide layer is rather thin it may be spatially separated from the non-oxidized filling part. The Ga column shrinkage remains persistent on cooling from a high-temperature measured. A thin mark made of a Ga oxide layer on the tube walls may be then visible inside TEM. Through using a heating TEM stage, a non-oxidized thermally-expanding Ga column is forced to hit the remaining wall mark. The temperature at which this happens may be considered as the true temperature experienced by a Ga-filled nanotube initially put in a given high-temperature environment.

Another rather smart developed method is based on the electrical calibration of the C nanothermometers,[18] i.e. measuring the overall multi-walled C nanotube electrical resistance. It was noticed that the electrical resistances of a Ga-filled and an empty C nanotube parts markedly differ. The unfilled part shows a relatively high electrical resistance of several tens of kΩ/μm, whereas that of filled one — only several parts of kΩ/μm. The total Ga-filled C nanotube electrical resistance is thus solely proportional to a length/volume ratio of the filled to empty nanotube segments. In such case, a nanotube object mounted on the pre-patterned electrical contacts may be viewed with a scanning electron microscope (SEM) or an atomic force microscope (AFM), and a high-spatial resolution peculiar to TEM and HRTEM techniques is not required anymore.

The performance of a liquid Ga inside a C nanotube may also be used for making reversible on-demand physical contact between an encapsulated Ga melt and an embedded semiconducting nanowire.[34,35] For example, Figure 5 shows an electron beam heating-driven establishment and cutting of such physical contact inside a thin-layered multi-walled Carbon nanotube. It is visible that a Mg_3N_2 semiconducting nanowire

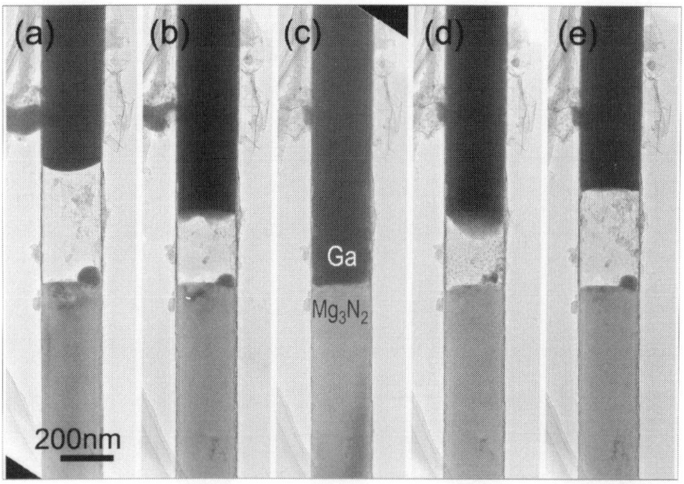

Fig. 5. (a–e) A series of TEM images showing the establishment and a dispatch of a Ga-Mg_3N_2 (metal-semiconductor) junction between the two filled fragments inside a multi-walled C nanotube. The phenomena were recorded during moderate changes in the electron-beam intensity focused on the tube area due to thermal effects of an electron beam; from Ref. 35.

placed inside the tube channel can be delicately contacted by a thermally expanding liquid Ga column on heating (Figs. 5(a)–(c)) and reversely disconnected on cooling (Figs. 5(d), (e)). This leads to a sort of a temperature-driven electrical switch-like action of the nanostructure as a whole.

Galium is not the only low melting point metal which may be used as a thermally-expanding medium. Indium can also be a decent candidate.[20] Usage of In leads to the corresponding changes in a temperature range of the nanothermometer performance due to a higher melting point of In compared to Ga. An In-filled C nanotube is displayed in Fig. 6. Apparently, the change in In column height in a liquid state is monotonic and linear. The sudden jumps in height take place at the temperatures corresponding to In melting/freezing at approximately 156°C. Since Ga and In create a continuous row of alloy solid solutions, a design of intra-tube liquid media with tunable melting points and ranges of practical temperature recording is practically achievable.

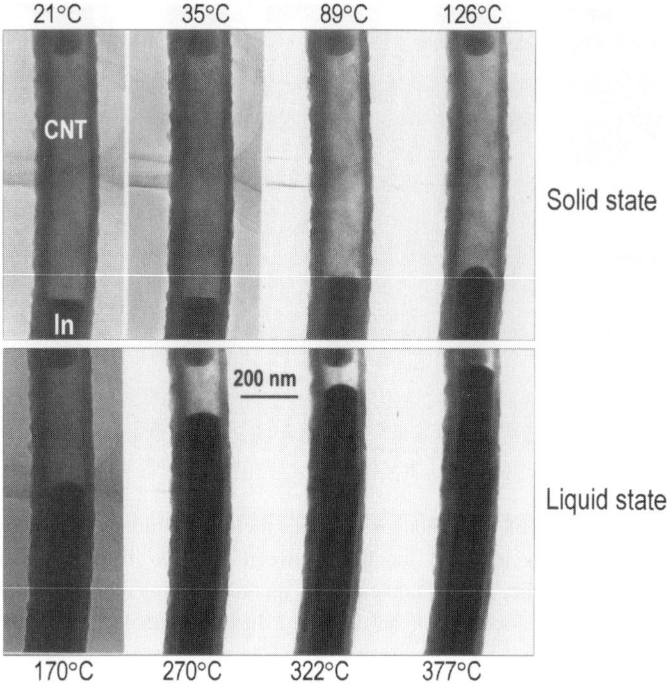

Fig. 6. Consecutive TEM images of an In-filled multi-walled Carbon nanotube upon thermal heating inside a TEM column using a heating stage. The corresponding temperatures are marked near the images. The upper row of the micrographs corresponds to a solid In state, whereas the lower row shows In behavior after its melting inside the tube at approximately 156°C. The filling length in the liquid state changes linearly against temperature similar to the above-described case of Ga-filled Carbon nanotubes; from Ref. 20.

3.2. *Alternative filled inorganic nanotubes performing as nanothermometers: in-situ TEM heating*

Carbon nanotubes are rather easily burned in air at temperatures not exceeding 450–500°C. Thus their applications for practical needs have serious limitations. However, the same "thermometer" approach does work for a liquid Ga if it is placed in the cores of inorganic oxide nanotubes. Such nanotubes are typically less prone to oxidation than conventional multi-walled C nanotubes and may withstand much higher temperatures. Figure 7 depicts consecutive TEM images of a Ga-filled

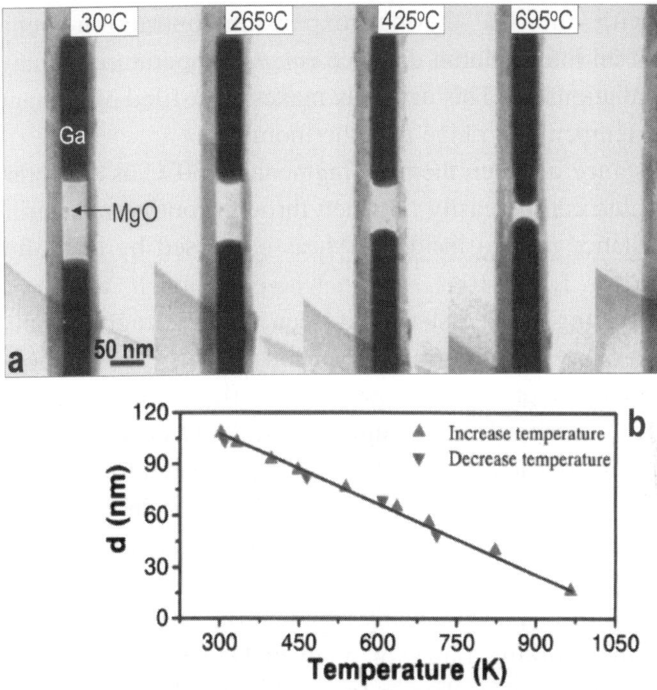

Fig. 7. (a) Consecutive TEM images of a MgO nanotube which possesses two encapsulated liquid Ga fragments. The fragments approach each other upon nanostructure heating inside TEM, so that a measured gap between them fits the line *versus* the experimental temperature (b) inside the TEM holder. The columns move away from each other on cooling, completely restoring the former positions inside the tube channel at each given temperature. This reveals the possibility of a multiple high-temperature usage of the present inorganic nanotube thermometer; from Ref. 13.

MgO nanotube which was heated inside TEM.[13] It is worth noting that Magnesium oxide is one of the most stable refractory compounds, as proved by its numerous high-temperature applications. TEM images of a Ga-filled MgO nanotube were recorded in the temperature range of 30–695°C. The two Ga fragments, exhibiting a 1.94 μm entire length were sealed inside the MgO tube and became separated by a distance of 109 nm at 30°C, leaving a central tube part blank. The distance between the tips of the two fragments decreases with increase in temperature during heating, and increases with decrease in temperature during cooling, in accordance with the Ga thermal expansion/contraction behavior. Dependence of an inter-column distance *versus* temperature displays the perfect linear relationship. This naturally makes a Ga-filed MgO nanotube a proper wide-temperature range nanothermometer.

Using a distance between the two fragments at 30°C as a reference, a temperature value can be easily obtained through routine measuring the changeable distance d (nm) inside TEM, as expressed by the following formula, $T = T_o + 1.39 \times 10(d_o - d)/L_o$, where L_o and d_o are the entire length of a Ga filling and a distance between the two fragments at T_o. If we consider possible linear thermal expansion of a MgO nanotube cavity during heating and adopt the corresponding coefficient of the bulk matter (0.14×10^{-4} K^{-1}) as the reference value for the MgO nanotubes, a linear volume expansion coefficient of 1.14×10^{-4} is calculated for the encapsulated Ga column. This is slightly higher that the value in a macroscopic state of liquid Ga (1.01×10^{-4} K^{-1}). Such difference may originate from a smaller thermal expansion of a MgO nanotube itself (compared to the bulk state) due to its characteristically higher surface to volume ratio.

Generally, the working range of a given Ga-filled MgO nanotube solely depends on L_o and d_o. For instance, the one shown in Fig. 7 can effectively work up to ~800°C, Fig. 7(b). We note here that the MgO nanotubes normally exhibit not circular, but square-like cross-sections, due to a cubic lattice of the constituting MgO compound. However, this fact does not negatively affect the reliability and reversibility of a Ga-column move inside a MgO tube in a strict proportion to a temperature inside TEM (Fig. 7(b)).

The sole drawback of the above-described Ga-filled MgO nanothermometers is the fact that once the two columns had been merged inside the tube, they could not be separated again due to the high-surface tension of liquid Ga. Therefore, such merging would eventually kill the thermometer, if overheated. In such a case, a MgO thermometer may be used only for a single (not cyclic) temperature recording. However, unless the environment temperature remains lower than that at which the two neighboring Ga columns merge, the thermometer is effective for a multiple use with no failure.

Another working example of an inorganic nanotube that may decently serve for the temperature measurements is a SiO_2 nanotube.[21] Silica

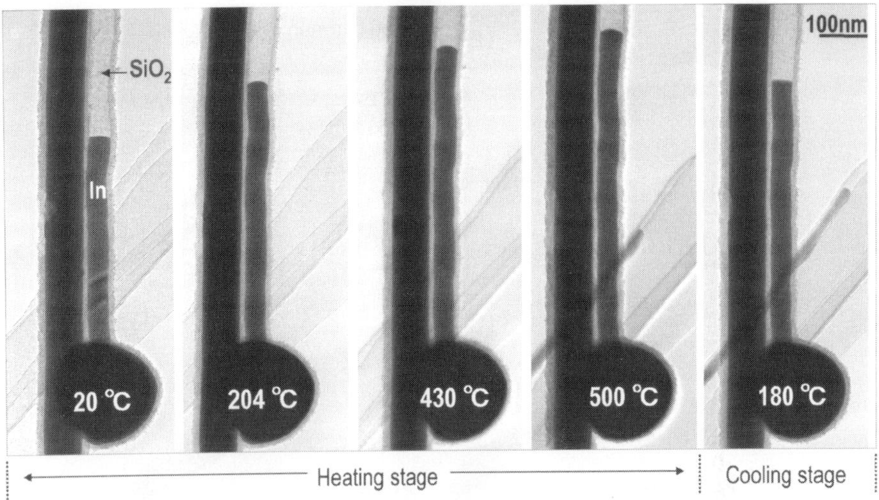

Fig. 8. A series of TEM images of an In-filled silica nanotube during thermal heating and cooling inside a microscope. The changes in a In-filling height are proportional to the temperature inside the TEM heating/cooling holder. A large In ball providing the feedstock material for In thermal expansion is located at the bottom of the SiO_2 nanotube. The In movement is fully reversible upon cooling, as shown on the right-hand-side image; from Ref. 21.

nanotubes have been found to be highly valuable as nanoscale host materials in bioanalysis, biocatalysis etc.[11,12] These nanotubes have also been probed by us for hosting metallic In. Representative TEM images of an In-filled SiO_2 nanotube in a temperature range of 20–500°C are shown in Fig. 8. The results demonstrate that the melting point of an In nanocolumn confined with a silica tube is ~152°C. That is slightly lower than the standard value of ~156°C, characteristic of a macroscopic In state.

During heating, melting of In cores at 152°C is always accompanied by a drastic jump in the filling height, similar to the case of C nanotube in Fig. 6. This is obviously caused by the corresponding density difference between the solid and liquid In phases. The length of an In melted column monotonically increases on heating and decreases on cooling in accord with the standard thermal expansion behavior of In. A plot of length change *vs* temperature fits a straight line, thus again displaying a thermometer-like behavior.

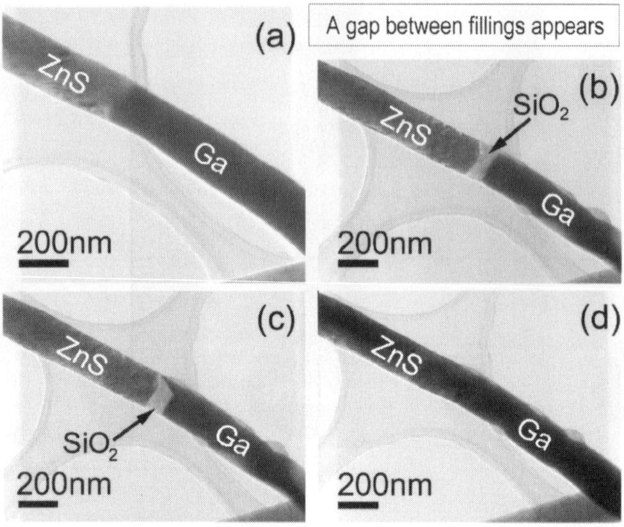

Fig. 9. TEM images illustrating the possibilities of manipulation with a physical contact between an encapsulated liquid Ga column (a metal) and an embedded solid ZnS nanowire (semiconductor) inside a silica nanotube by using the thermal effect of an electron beam. Through varying the beam intensity in (b–d) a Ga melt may be moved backward and forward with respect to the semiconducting wire, whose position remains unchanged in the tube; from Ref. 36.

The silica nanothermometers were judged to be liquid-proof and tough enough for the use as fluidic nanochannels due to their characteristic pin-hole-free amorphous-like walls. Interestingly, a silica nanotube may be used to merge encapsulated liquid metal, i.e. Ga, and a semiconducting nanowire, ZnS,[36] in a way analogous to that described previously for a Carbon nanotube. Figure 9 displays the sequence of transformations at the metal-semiconductor boundary; the two domains may be delicately joined and separated by using electron-beam-induced heating inside TEM.

4. *In-Situ* Electron Irradiation Experiments on Filled Nanotubes

4.1. *Filled carbon nanotubes*

We found that the changes within Ga-fillings in multi-walled C nanotubes may be related to not only thermal effects but also to the electron-beam

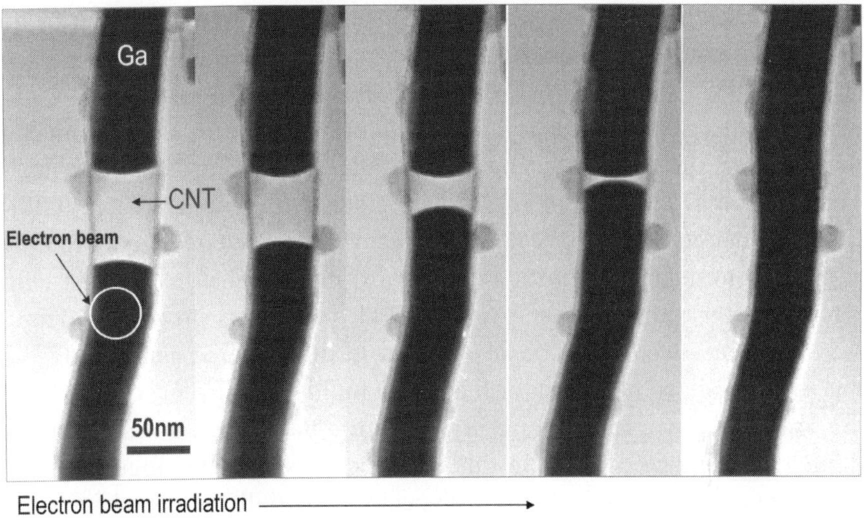

Fig. 10. TEM images showing the selective irradiation-induced movements of a lower Ga-segment inside a multi-walled Carbon nanotube, while the upper Ga-segment position remains unaffected. The actual beam size and its position on the tube are shown on the left-hand-side image. The two domains may be entirely merged within the tube channel through the irradiation, as illustrated in the right-hand-side image; from Ref. 19.

irradiation phenomena inside TEM. Under room temperature TEM irradiations of numerous Ga-filled C nanotubes we estimated that the temperatures could easily rise to around 100–150°C, which is far above the melting point of most commonly seen α-Ga (29.8°C). For example, Fig. 10 displays the consecutive stages of Ga expansion under the electron-beam irradiation owing to the considerable expansion coefficient (1.015×10^{-4} T) and good thermal conductivity of liquid Ga.[19] We note that a wetting angle between a Ga melt and a C tubular sheath slightly changes under irradiation. Most likely, this change may be caused by the thermal effect of irradiation. Interestingly, the top Ga segment remains unchanged in all images in Fig. 10. This implies that the thermal conductivity of a C tubular shield is negligible compared to the metallic Ga.

Therefore, it should be concluded that the thermal effect of an electron beam only comes into effect within the sole irradiated region marked with a circle in Fig. 10. The expansion of the lower Ga segment is an indicator of the sufficient electron-beam-induced heating. Moreover, if the beam intensity reaches the threshold it may cause the damage of an irradiated target.

Generally, the beam damage of a graphitic C and metals is dominated by the direct atomic displacements, so-called knock-on displacements.[37] This effect may create numerous point defects within an irradiated multi-walled C nanotube. Along with the defect creation, we realized a unique possibility to manipulate with a Ga-filled C nanotube: it may be cut with a beam of the sufficient energy. Figure 11 shows this process in detail. A convergent electron probe of ~10 nm in diameter, as marked on the image, was moved back and forth across a multi-walled C nanotube. After the irradiation over 2 min (Fig. 11(c)) the irradiated region first becomes thinner. The image contrast in this area becomes lighter, suggesting that metallic Ga (darker optical contrast specie) is partially removed from the region due to the evaporation by high-energy electrons. Further irradiation

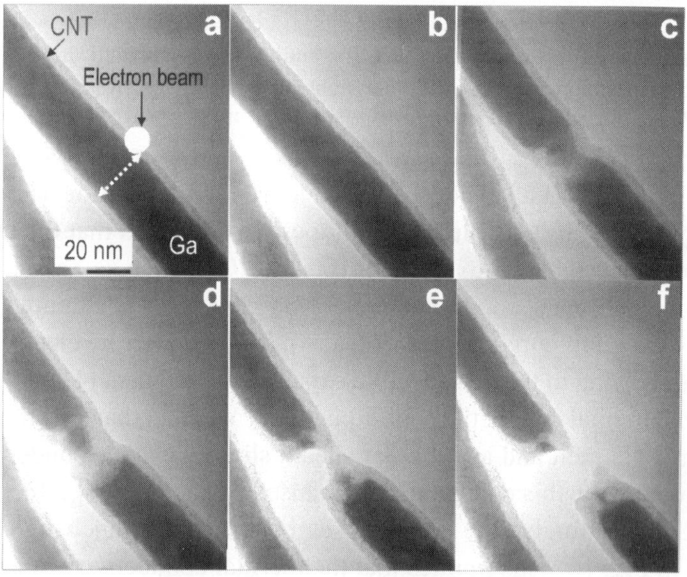

Fig. 11. (a–f) Consecutive TEM images illustrating the possibility of cutting apart a Ga-filled C nanotube using a focused electron probe which is slowly scanned across the tube (along the dashed line marked in (a)); from Ref. 19. The actual size of the probe is marked with a white circle.

over next 2 min, Fig. 11(d), completely removes Ga from the tube making a prominent internal gap within the tubular sheath. Then, the remaining carbon material is gradually shrunk. The entire process resembles a necking phenomenon during a metal rod/wire failure under a tensile test. As a result of the whole electron beam irradiation process, the two isolated independent Ga-filled and entirely sealed multi-walled C nanotubes are formed, Fig. 11(f).

We finally note here that an encapsulated Ga melt does not hinder the reconstruction of a tubular shape in the newly formed composite structures. Once the C sheath shrinks, the liquid Ga is ready to recede and separate into two domains due to its liquid status. It is noteworthy that the applied electron irradiation simultaneously anneals the pre-existing defects within the carbon tubular sheath and causes the knock-on damage of the nanostructure. During both processes the self-organization phenomena take place, that yield many spherical C onion-like nanoparticles at the tips of separated tubular domains.[19,37]

4.2. *Filled boron nitride nanotubes*

A Boron Nitride nanotube is the well-known analog of a standard C nanotube in which the alternating B and N atoms entirely substitute for C atoms in a honeycomb graphene-like network.[10] However, the BN nanotubes have been found to be much more stable to oxidation in air compared to their C counterparts thus revealing excellent thermal and chemical stabilities.[27] BN tubes withstand heating in air up to 800–900°C without visible deterioration. These properties of BN nanotubes make them particularly useful as far as high-temperature or high-energy experiments are concerned. In addition, the electrical response of BN nanotubes is independent of their atomic structures. This is totally opposed to C nanotubes. The BN tubes are always electrically insulating independent of helicity, number of tubular layers and diameters.[38]

Among numerous *in-situ* TEM observations performed on multi-walled BN nanotubes in recent years within our Groups (NIMS) the one deserves a special attention. This is related to Mg-compound filled BN nanotubes and their behavior under a TEM electron beam.[25,26] In contrast to the above-mentioned liquid phase filling of a nanotube, the present section deals with the first example of a solid phase nanotube filling.

It is known that Mg forms several oxides, e.g. MgO, MgO_2 and MgO_4.[39] The last oxide is highly unstable at atmospheric pressure and temperatures just above ~30°C. MgO is one of the most stable inorganic compounds with a melting point of ~2800°C. By contrast, MgO_2 is an unstable material exhibiting melting at just ~88°C, followed by gradual decomposition, which in turn generates molecular oxygen and stable MgO. We note that such temperature is easily achievable inside TEM.

In this section we show how it is possible to observe the regarded phase transformations in the Mg-O system compounds, which are put in a nanotubular BN container and *in-situ* TEM electron-beam irradiated.

When a BN nanotube had been filled with an oxygen-release compound MgO_2 some prominent changes within the fillings were recorded during *in-situ* TEM. First of all, the fillings easily melted and fragmented into numerous tiny pieces under irradiation even at room temperature, Fig. 12. The horizontal line in Fig. 12 marks the initial level of filling height before the irradiation started. During the irradiation the filling is gradually consumed and numerous voids appear in it, implying a corresponding volume and density change within the filled matter. Multiple tiny fragments, appearing as a result of the oxygen-rich filling deterioration (the upper parts of the images), possess a well-defined faceted morphology peculiar to cubic crystal lattices. Diffraction analysis of these fragments and the remaining filling parts after melting, deterioration and consumption indeed revealed a highly-crystalline cubic Mg-based phase with the lattice parameters similar to the standard MgO.

The phenomenon was further verified during a test irradiation run at a slightly decreased electron dose (in order to slow down irradiation-induced damage) which was approximately 20% higher of that found in normal imaging conditions within TEM (1–2 A/cm^2), Fig. 13. Even under such moderate irradiation the marked changes occur within an oxygen-rich core of a BN nanotube. Dark contrast (stronger scattering) precipitates become visible after the initial flash irradiation. These precipitates are likely depleted in oxygen, as compared to the core matrix, as the spatially-resolved oxygen map tells, Fig. 13(h). In fact, the oxygen distribution within the irradiated region becomes highly

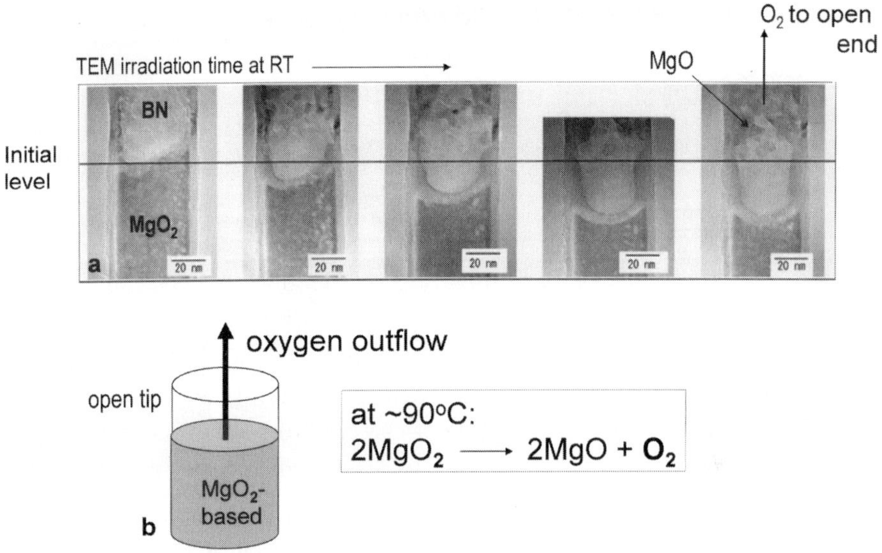

Fig. 12. (a) TEM images showing the gradual deterioration and consumption of a starting MgO_2 filling inside a multi-walled BN nanotube under the electron beam irradiation at room temperature. The particles possessing the square-like morphology are peculiar to cubic MgO crystals. They become well visible after the prolonged irradiation, as marked with an arrow on the right-hand side TEM image. (b) a proposed sketch of the on-going process within the tube channel and an expecting chemical reaction involved. The reaction yields an outflow of a pure molecular oxygen from the open tube end; from Ref. 26. An electron beam current density on the TEM screen during the experiment was ~50 pA/cm^2.

inhomogeneous. By contrast, the B, N and Mg maps look rather uniform along the tube.

Keeping in mind that the melting point of MgO_2 is surely achieved during the present irradiation runs, we may conclude that along with the filling decomposition and transition to a stable MgO oxide a pure molecular oxygen outflow should be taken place from the open BN nanotubular channel.

In other words, the present MgO_2 encapsulated multi-walled BN nanotube illustrates an original nanoscale oxygen burner and/or generator. It may be widely used, if properly manipulated, for the needs of localized metal oxidation at the high spatial precision to produce well-defined quantum dots, or for the analysis of living and deterioration conditions of different viruses or bacteria in nanobiology, and medical fields.

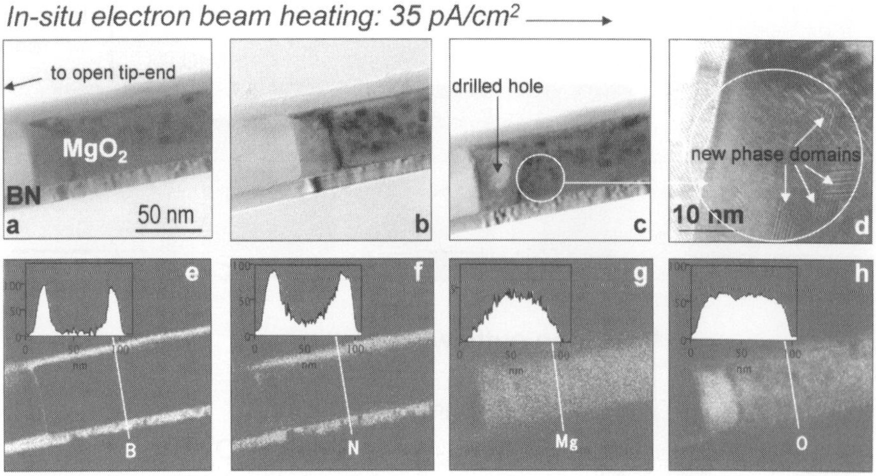

Fig. 13. (a–d) Morphological changes of a MgO_2 filling inside a multi-walled BN nanotube recorded under the low dose electron beam irradiation at room temperature. Appearance of the dark-contrast phase domains corresponding to stable MgO nanocrystals is well visible in (b),(c) and (d); (e–h) elemental maps of the constituting B, N, Mg and O species acquired at the intermediate irradiation stage, i.e. corresponding to (b). A highly inhomogeneous oxygen map within the filling is highlighted in (h) and additionally confirmed by showing the cross-sectional contrast intensity in the inset; from Ref. 25. An electron beam current density is shown, as measured on a TEM screen.

4.3. *Filled silica nanotubes*

Harada and Adachi[11] first developed a surfactant-mediated technique for the synthesis of SiO_2 nanotubes. Filling of these tubes with liquid metals, for example In, in addition to the thermometer-like behavior, described in Sec. 3.2, may also lead to some interesting electrical-driven effects due to purely insulating nature of a silica tube and associated charging effects inside a TEM column.

Figure 14 displays an entirely sealed silica nanotube. The tube initially has an In nanowire core. Under the electron beam irradiation, we surprisingly observed the fast cyclic nucleation, growth and abrupt disappearance (in an explosive-like fashion) of liquid In balls on the SiO_2 tube periphery. The ball-like shapes of the In particles indicate that they are indeed molten. In fact, detailed electron diffraction and high-resolution TEM

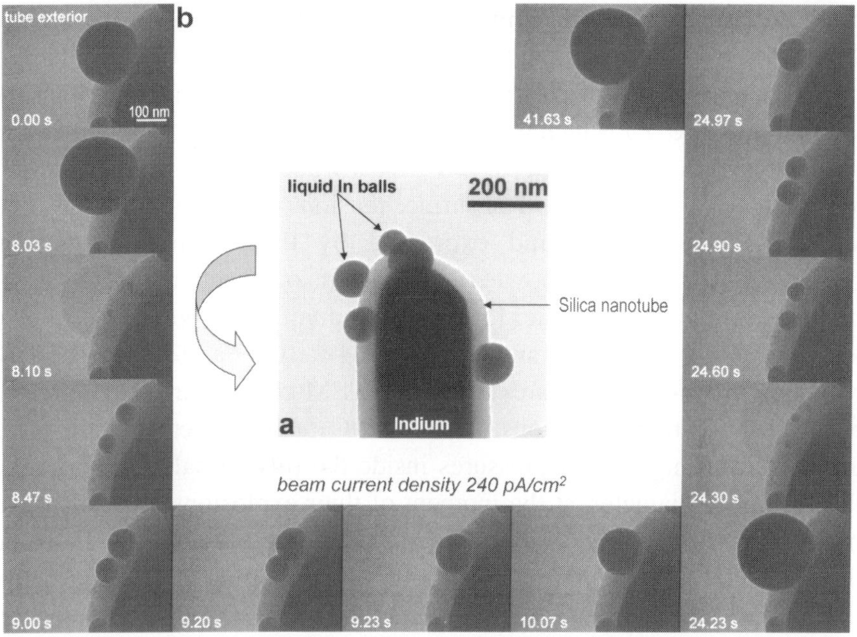

Fig. 14. (a) A general view of an In-filled silica nanotube subjected to electron beam irradiation; and (b) consecutive video frames verifying the cyclic appearance and disappearance in an-explosion-like fashion of numerous liquid Indium balls at the tube periphery under electron beam irradiation. The running irradiation time is marked on the images; from Ref. 23. An electron beam current density is shown, as measured on a TEM screen.

studies performed on these particles have shown neither clear diffraction spots nor crystalline lattice fringes. The In nuclei in Fig. 14 always appear at nearly similar sites on the amorphous SiO_2 tube shield, followed by abrupt explosion of a larger liquid In ball.

Clearly, when heated, an In volume should increase if allowed the room for expansion.[40,41] However, SiO_2 nanotubes are very rigid and tough material to break. Instead, the amorphous silica sheath appears to serve as a sieve, or mesh, through which the In atoms squeeze towards the tubular periphery. As the Indium flux toward the tube surface persists, liquid balls gradually grow. The ball cross-section for the capture of high-energy 300 kV electrons also increases. Accumulated energy leads to liquid In boiling inside the balls.[40] It may be assumed that when the vapor pressure inside the balls becomes high enough to prevail over the surface tension of an In liquid film a ball explodes.

In the framework of classical physics it can be demonstrated that the internal pressure *(P)*, which is required to break a given In vapor-filled liquid In sphere may be written as: $P = 2\ \sigma/R$, where σ is the In melt surface tension and R is a ball radii. The surface tension of liquid Indium has been studied experimentally and expressed by the empirical equation: $\sigma = 566.7 - 0.0666(T - T_m)$, where the units of σ are mN/m, and T_m is the melting point of a bulk In, i.e. 156°C. The Indium vapor pressure is known to be relatively low, for example, only 8.34×10^{-3} Pa at 725°C. For the realistic range of temperatures inside the TEM column under the present electron irradiation conditions (~150–200°C) a simple calculation gives unrealistically high vapor pressures inside the tiny In balls, only several dozens nm in diameter, at the moment of their explosions: 10^6–10^7 Pa or more (this corresponds to the temperature of more than 2000°C inside them). Therefore, apparently, in addition to the thermal effects of irradiation, some other effects, namely, knock-on atomic collisions, electron charge distribution and so forth should be taken into account and assigned for the In ball boiling. Most likely, the driving force behind the phenomenon is related to the well-established electromigration phenomena.[42]

In order to support such claim we have additionally confirmed that an In mass transport may even take place within a single SiO_2 nanotube channel. Figure 15 shows how some In liquid balls inside a channel grow at the expense of the others. The balls grow and shrink in a "flip-flop"-like fashion, while exchanging with the positions within the channel. It worth noting that some other balls do not change at all (Fig. 15), implying that the observed fast mass transport is very sensitive to the local electro-thermal conditions within a given inter-channel site.

All observed phenomena have a sort of lifetime. They typically cease after 3–4 min since an electron beam was focused on a given fragment of a nanostructure. Thus there is a distinct underlying physical process leading to the structure change, stabilization and/or freezing. Unfortunately, this process can be hardly modeled quantitatively, since the electrical/thermal status of the silica nanotubes inside TEM is not known and depends on many unset parameters: electrical and thermal conductivities at each particular point of a multi-crossed/linked nanotube array

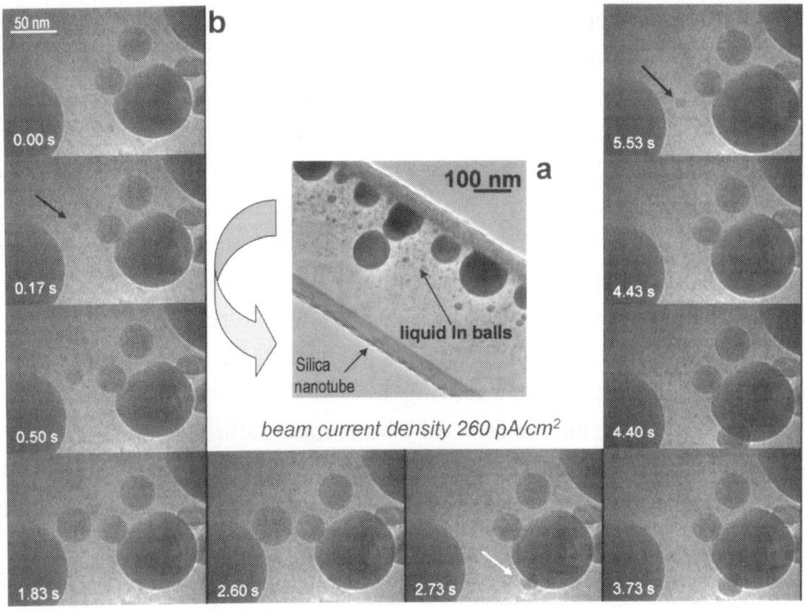

Fig. 15. (a) A general view of a SiO_2 nanotube filled with numerous In particles subjected to electron beam irradiation; and (b) consecutive video frames displaying the In mass transfer between the two distinct regions inside the SiO_2 nanotube channel (shown with the black and white arrows) under the irradiation. The running irradiation time is marked on the images; from Ref. 23. An electron beam current density measured on a TEM screen is shown.

residing on an amorphous C grid, the intimacy of the nanotube/grid contacts at each particular place chosen for the beam focusing and so forth. However, at least qualitatively, it may be though that a SiO_2 nanotube should be definitely charged up under electron irradiation with 300 kV electrons. The charge distributes rather irregularly within the nanotube set-up, and even within an individual nanotube. The same applies to the distribution of thermal fields caused by the electron irradiation.

It is known that atoms of a metal under a gradient of electrical potential and/or temperature are submitted to a force, which has a double origin.[42] This force is called "direct" in the electric case, or "intrinsic" in the thermal one. The "direct" force is due to the unscreened action of the electric field on the ionic charge, and the "intrinsic" contribution corresponds to the enthalpy transfer due to an atomic jump. Notable migration under external forces, mainly an electric field, takes place also at surfaces. The innermost and/or outermost shell of an amorphous SiO_2 nanotube represents such a model potential surface. It has been established that the diffusivity on the surfaces is much higher than in the bulk and may proceed in a temperature range where normal bulk diffusion is negligible. An additional difficulty for the interpretation of the present results stems from the fact that diffusion is expected to be highly anisotropic on non-perfect surfaces, i.e. amorphous silica. Diffusion is thought to take place through individual atom jumps only at very low temperatures ($T < 0.15\ T_m$). At higher temperatures, matching the case of melted In, several new mechanisms have been proposed.[42] During intense electron irradiation, vacancies and interstitials, and/or ions may be created within SiO_2 shells and In cores and/or particles, either as isolated Frenkel pairs or as complex cascades. The number of such defects is roughly given by $N_d = kT_d/2E_d$, where T_d is the elastic energy given to a material, E_d — the displacement threshold energy, and K — an efficiency factor. Such defects may dramatically enhance the In-diffusion. For example, for In on a Si(001) surface very fast surface electromigration (~8000 μm/min) towards the cathode was found.[43] Some observations also show that the In surface electromigration is related to an ordered 4 × 3 In phase together with a two-dimensional In gas phase over the 4 × 3 phase and an In-disordered phase at the front end of the In flux. The electromigration suddenly set in at 450°C in the latter case. The driving force for In surface electromigration has yet been ascertained. Neither the structure of the 4 × 3 phase nor the properties of the 2D In gas have been established. We also note that the velocity of In on another Si-plane, (111), was measured to be ~10–100 μm/min [one of the fastest velocities of diffusion on Si(111)].[43]

It has been stated that a voltage gradient, rather than a thermal gradient primarily determines the direction of In transport on the surface of conducting multi-walled C nanotubes.[44] Regan *et al.*[44] have observed that an In mass always transports to the cathode part in the D.C.-assisted experiments. We note that the latter fact is inconsistent with our observations of the present In drive towards the external tube surface (Fig. 14) which is thought to be positively charged under the utilized electron irradiation and thus the resultant picture seems to be more complex than in the case of sole *d.c.* driven experimental runs.

5. Electrical Probing Experiments on Filled Nanotubes

5.1. *Ferromagnet-filled carbon nanotubes*

Since recently there has been a growing interest in nanotube electrical probing and manipulation.[45–61] The usage of piezo-driven stages within the standard TEM setups is thought to be one of the efficient routes to accomplish such goals.[45,48,50–52,56–61] The prime advantage of this technique, as opposed to the commonly-used transport measurements using nanotube containing pre-formed electrical circuits, is the possibility to thoroughly characterize a nanotube prior, during and after electrical measurements. All standard TEM operations, e.g. high-resolution imaging at the atomic resolution, spatially-resolved nanobeam electron diffraction and chemical composition analysis using electron energy loss or X-ray dispersion spectrometers could be selectively performed on exactly the same nanotube that is electrically probed.

Most of other popular pre-existing experimental setups, like AFM- or SEM-based techniques, suffer from limited magnifications and spatial resolutions. Even if the time- and effort-consuming electrical tests on an individual nanotube had been successfully accomplished, there was always a huge degree of uncertainty as to what particular structure, morphology and/or nanotube chemical composition they were related to and how the given nanostructure was affected during the data collection.

The series of the *in situ* TEM highly informative experiments using piezo-stages have already been performed on standard carbon nanotubes.[45,48,50–52] However, we noticed here that there have been far fewer such works performed on metal-filled nanotubes.

We performed *in-situ* TEM experiments on such nanotubes by means of a "Nanofactory Instruments" piezo-holder inserted into a JEOL-3000F 300 kV field emission high-resolution transmission electron microscope. The electronics and software from the "Nanofactory Instruments" were used. The inertial sliding mechanism consists of a sapphire ball rigidly attached to a piezo tube and a movable part with six springs that embraces the sapphire ball.[57] The experimental setup within the holder is sketched in Fig. 16. Prior to the electrical measurements the nanotubes were thoroughly analyzed using selected area electron diffraction, high-resolution lattice imaging and spectroscopic techniques.

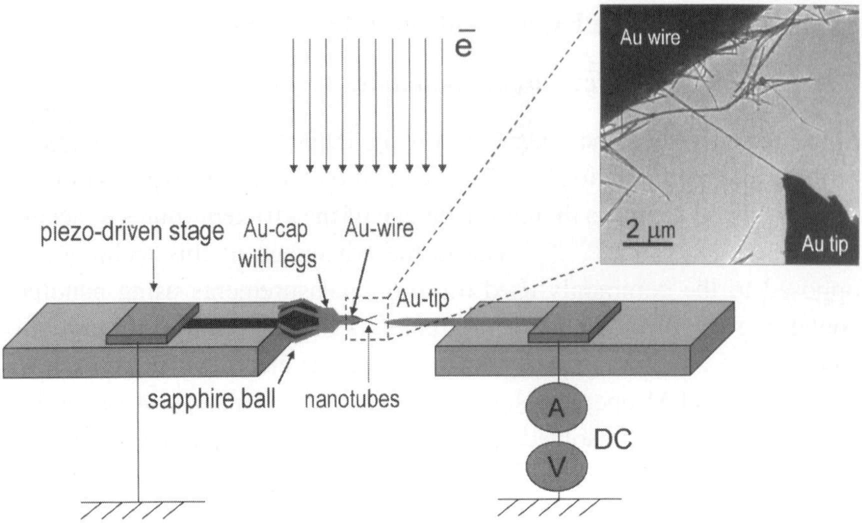

Fig. 16. A sketch summarizing the experimental setup under the utilization of a "Nanofactory Instruments" piezo-driven STM-TEM holder for the *in-situ* TEM analysis of the nanotube electrical and mechanical properties. The inset on the top-right-side displays the same setup, as seen inside TEM; from Ref. 30.

A sample containing free-standing Fe-filled C nanotubes was mechanically attached to a grounded gold wire (0.25 μm in diameter) using a graphite paste (sapphire ball side), while the opposite gold tip (an etched 0.25 μm gold wire) was biased during the electrical measurements and manipulation. The two-terminal current-voltage (I–V) curves were measured using a "Nanofactory Instruments" D.C. power supply (maximum voltage ± 140 V). The position of a sample wire could be easily adjusted in the X, Y and Z directions within a wide range (several hundreds of μm).

Figure 17 illustrates the overall structural and chemical characteristics of a Fe-filled CNT that has been tested and its representative low-resistant I–V curve. The sharp gold tip was sweeping along the tube axis while measuring the numerous I–V curves at various locations. At all occasions we observed that the nanostructure is truly metallic, independent of the contact point. The recorded I–V curve is nearly linear thus revealing a good Ohmic contact between the gold tip and the outermost Fe-filled

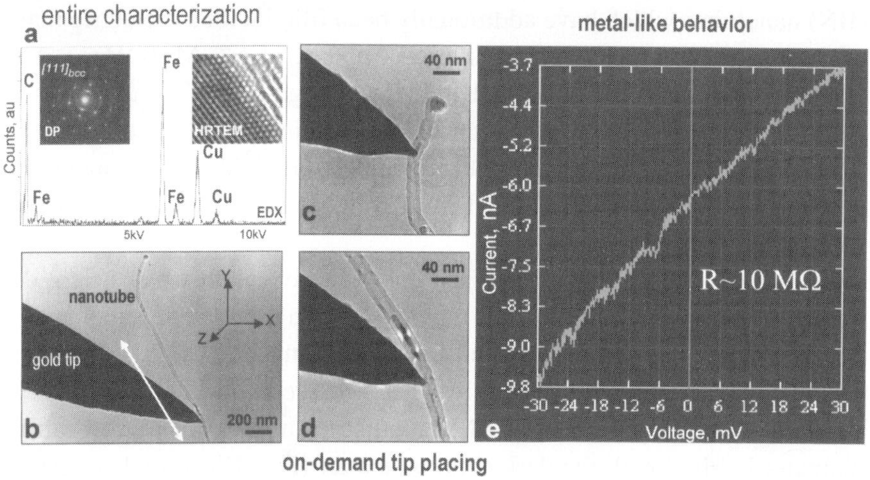

Fig. 17. A layout illustrating the entire chemical composition (a); and morphological, and structural characterization (a–d) of a Fe-filled multi-walled Carbon nanotube inside TEM prior to its electrical probing using the "Nanofactory Instruments" holder. The insets in (a) display the electron diffraction pattern (left) and high-resolution images (right) of the regarded nanotube. The filled tube reveals the characteristics of a metal independent of the contact site and possesses a resistivity of ~10 MΩ,, as evidenced by a representative *I–V* curve in (e).

C nanotube shells. The measured electrical resistance on Fe-filled nanotubes always falls in the range of 10–300 MΩ.

5.2. *Ceramic-filled BN nanotubes*

The electrical measurements on alternative non-carbon inorganic nanotubes, e.g. BN, have been lingering far behind of CNTs. Only very recently, Cumings and Zettl[58] have first evaluated field emission and *I-V* properties of multi-walled BN nanotubes using an *in-situ* TEM manipulation stage basically similar to the one we used. Surprisingly, BN nanotubes were found to possess stable field emission currents, albeit they were detected to be insulating at a low bias. The BN tubes showed stable, reversible breakdown current at a high bias of several tens volts.

In this section we present the first two-contact *I–V* measurements and *in-situ* electrically-driven TEM manipulation with insulating boron nitride

(BN) nanotubes which have additionally been filled with a dielectric magnesium-oxygen-hydrogen containing phase, i.e. $Mg(OH)_2$. The "Nanofactory Instruments" TEM holder was again used to accomplish this work. In the course of the electrical measurements we serendipitously realized a rare opportunity to delicately adjust the dielectric BN nanotube position and/or deflection within TEM through smooth tuning of a bias on a gold counter-tip.

Magnesium hydroxide-filled nanotubes were synthesized using the experimental procedure which has previously been used in our Laboratories for the synthesis of carbon-free multi-walled BN nanotubes.[62,63]

Figure 18 presents the synopsis of the obtained TEM and *I–V* data. It is visible that a representative BN nanotube of approximately 40 nm in diameter is filled with the dark contrast matter (Fig. 18(a)) and is capped at the tip-end. Detailed electron diffraction (Fig. 18(b)) and chemical composition analysis (Fig. 18(c)) verify that the filling is made of insulating hexagonal magnesium hydroxide $Mg(OH)_2$. While looking at the

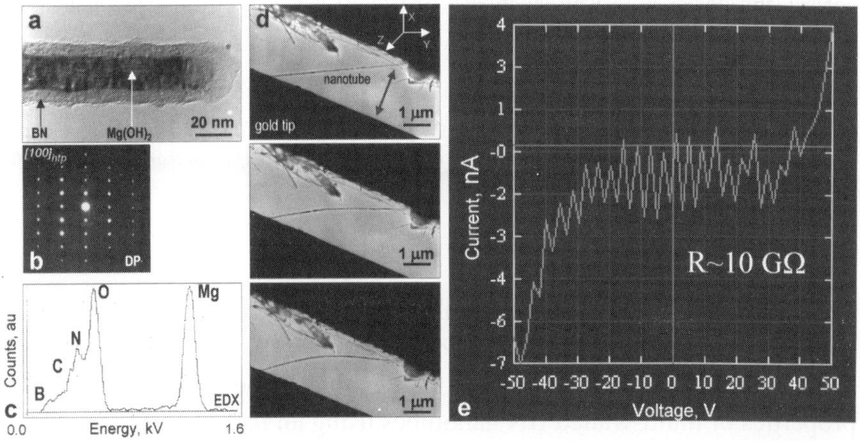

Fig. 18. A layout showing the entire structural (a,b) and chemical composition (c) characterization of a $Mg(OH)_2$-ceramic filled multi-walled BN nanotube slightly squeezed between the two gold contacts (d) and electrically probed inside TEM using a two-terminal scheme within the "Nanofactory Instruments" holder. The nanotube possesses the characteristics of an electrical insulator with a resistivity exceeding ~10 GΩ,, as evidenced by a representative *I–V* curve shown in (e); from Ref. 28.

TEM images, it is clear that the nanotube surface is free from a distinct amorphous-carbon based residue, which may negatively affect the electrical measurements and lead to artifacts.

The two probe *I–V* curves were taken on the filled tube which was preliminary gently squeezed between the two gold electrodes (Figs. 18(d)) in order to improve the physical contact between the tube and the gold leads. The resultant *I–V* curve is depicted in Fig. 18(e). At a low bias there is no conduction at all (except a noise current). The regarded *I–V* curve is highly serrated showing the current discrepancy in the range of ~1 nA. The voltage can be increased to up to ± 30 V with not detectable current pass or tube burning-out/deterioration. In contrast, it has been known that when just a ~2–4 V bias is applied to a standard multi-walled C nanotube it is typically burned-out or collapsed.[50,51,58] Thus the present magnesium hydroxide filled BN nanotube is a perfect insulator up to the voltages of approximately ± 25–30 V. The major contribution of a nanotube-gold contact resistance to the measured extremely high-resistant *I–V* curves could be surely ruled out. We have performed many test experiments on pure carbon nanotubes (with and without fillings) using the exactly similar setup: the latter nanotubes start to conduct just at a several mV bias already revealing passing currents of 10 nA or even more.

At voltages of more than ± 30 V the filled BN nanotubes exhibited passing currents exceeding several dozens of nA. The electron transport is reversible and does not lead to any morphological destruction of the nanotubes, similarly to the case of unfilled pure BN nanotubes.[58] The breakdown occurs at nearly ± 25–30 V, which is slightly higher the values reported for the empty BN nanotubes (12–25 V). The marked serrations on the *I–V* curves (Fig. 18(e)) are thought to illustrate the specific varying electrostatic interactions between the nanotube and the gold counter electrode during bias sweeping.

Finally, the Figs. 19(a) and 19(b) demonstrate two representative experiments in which the on-demand dielectric BN nanotube manipulation is thoroughly demonstrated. Initially the nanotubes were kept out of contact with the gold tip before applying the bias voltage to it. Applying of a negative bias to the tip effectively pulls-up the BN nanotubes from a nanotube debris on the sample side towards the gold tip-side. Applying of a positive bias pushes the nanotubes out of the tip. Both modes are

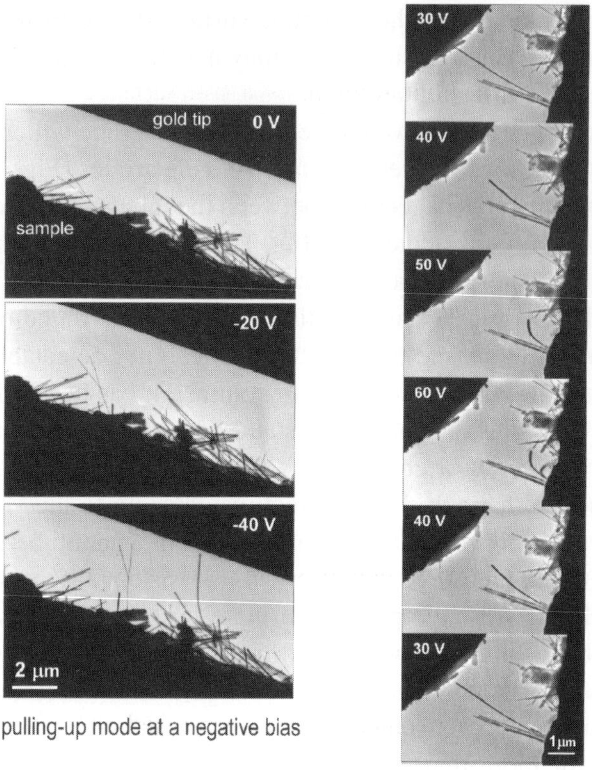

a) pulling-up mode at a negative bias

b) pushing-down mode at a positive bias

Fig. 19. TEM images illustrating the possibilities of a BN multi-walled nanotube debris pulling-up (a) and pushing-down (b) through smooth tuning of a bias voltage on the gold tip of the "Nanofactory Instruments" TEM holder. The nanotubes are attracted to the electrode at a negative bias and repulsed from it at a positive bias due to the prominent electrostatic interactions between the positively charged BN nanotubes (under 300 kV electron irradiation) and the gold electrode; from Ref. 28.

fully reversible and a given BN nanotube may be repeatedly manipulated in a circle-like fashion, as shown in Fig. 19(b). It is worth noting that the present dielectric nanotubes can be gently pulled-up, pushed-down and even bent reversibly (Fig. 19(b)) over many circles through smooth tuning of the applied bias voltage on the counter gold electrode. The present filled BN nanotubes appear to be very flexible, tough and elastic to a large degree of deformation, seemingly higher than standard unfilled BN nanotubes, as will thoroughly discussed in the following-up Section.

It should be admitted that sometimes a bias-assisted deflection was noticed for carbon nanotubes during *in-situ* experiments, as well.[64] However, in the latter case the deflections are of sporadic character, since they solely relate to the bad nanotube-grounded holder electrical contact (which is unpredictably changeable and is difficult to control) rather than to natural stable dielectric nanotube performance itself.

The mechanism lying behind the regarded effective manipulation is suggested to be very simple and is related to the electrostatic repulsive and attractive forces. Under the electron beam of the microscope the dielectric nanotubes on the grounded stage of the holder are positively charged. Applying a positive bias to a gold counter-electrode would repulse the nanotubes, whereas a negative bias would attract them.

The phenomenon described in this Section may allow one to reliably manipulate with any dielectric nanotubes inside TEM. The present effect has been verified but not limited to the magnesium oxide/hydroxide filled BN nanotubes. It makes possible precise positioning and/or deflection of a given dielectric nanotube, e.g. BN, SiO_2, MgO *etc.* inside TEM and designing multifunctional mechanical actuators based on them.

6. Mechanical Deformation of Filled BN Nanotubes

It is known that a carbon nanotube possesses the extremely high Young's modulus and yield strength. The same is true for a BN nanotube.[10]

The experimental studies[65–69] and theoretical estimates[70–73] of C nanotube deformation mechanics and physics have been performed. Several experiments have attempted to measure the elastic modulus of C nanotubes through vibrating/blurring of individual CNTs under an alternating current (A.C.) and/or thermal field, or by direct measuring of a reaction force for an imposed displacement,[65–67,74] and by nanoindentation. Generally, a nanotube has been treated as a homogeneously-structured solid cylinder and its specific structural details, like defects, possible fillings *etc.*, have not been taken into account at all. On the other hand, the importance of specific CNT structural features for the deformation behavior[68,69] and electronic structure changes caused by deformation[52] have been well illustrated in several TEM studies. Such works related to BN nanotubular counterparts have been not available in the literature as yet.

A BN nanotube possess a ~1.2 TPa Young modulus, as was measured in a pioneering work by Chopra and Zettl[74] using a thermal vibration amplitude technique. To the best of our knowledge this work has been the only experimental research devoted to the BN nanotube mechanical response.

In this Section we show how by using the "Nanofactory Instruments" STM-TEM holder it is possible to physically deform a filled multi-walled BN nanotube under carrying out all conventional TEM and HRTEM operations, like high-resolution imaging, spatially-resolved electron diffraction, elemental mapping using electron-energy loss (EELS) and Energy dispersion X-ray (EDX) spectroscopies.

We again emphasize here that the appealing advantage of the present experiments over the pre-existing mechanical measurements on nanotubes (performed in AFM or SEM set-ups is a significantly improved spatial resolution. This allows us to carry out a precise analysis of deformation structural features and to properly understand the mechanics/physics behind the observed deformation-driven phenomena.

In addition, the regarded Section particularly highlights the prime importance of detailed microscopic study of a nanotube for the understanding of the observed mechanical behavior. Prior to bending inside TEM a BN nanotube was firstly analyzed in details with respect to its morphology, e.g. diameter, number of layers, structural inter-shell and intra-tube defects, chemical composition and electrical properties. Only then, the nanotube was mechanically tested under continuous tracing of all above-mentioned parameters. Such careful control gave us a true insight into microscopic peculiarities of a filled BN nanotube deformation. This is very important as far as the incorporation of BN nanotubes into future nanotechnological materials and devices, like polymeric nanotube composites, sensors, nanocables and transistors are concerned.

Multi-walled BN nanotubes used for mechanical deformation were produced through the procedure described by us in Refs 62 and 63. A selected tube was contacted by a sharp gold electrode through adjusting a relative height of the two holder sides inside the TEM pole piece using the X-wobbler function. Then, the piezo-driven part was delicately moved in the *X*, *Y* and *Z* directions so that the nanotube was firstly physically contacted by the sharp tip and, then, gradually squeezed between the two gold leads/contacts. Preliminary, the nanotube low- and high-magnification images, electron diffractions, and EDX and EELS spectra were taken in various domains on the tube. In addition, its starting insulating performance, highlighted in the previous Section, was documented by the tip bias sweeping from −140 V to + 140 V in a contact mode. No meaningful current (except a noise current of less than ~1 nA) was recorded during this operation. The nanotube was then reversibly bent at least twenty times between the two contacts with a special attention paid to an alternating change in bending direction. At all stages of deformation the morphological, diffraction and chemical control of the nanotube was carried out.

Firstly, the detailed electron diffraction study was performed on all BN nanotubes (around twenty), protruding from the gold wire edge in order to select a representative crystal structure. Albeit various types of the diffractions patterns were observed, the majority of BN nanotubes (~90%) had the diffraction features peculiar to zig-zag or nearly zig-zag orientation of the tubular shells with respect to the tube axes, i.e. the [10-10] direction of the graphitic shell was eventually parallel to the tube axes in accord with our previous observations over the years.[75–77] The longest nanotube of such type was selected for the bending experiments inside TEM. This particular nanotube is shown in Fig. 20. The nanotube originally has two distinct contrast regions marked in Fig. 20, namely 1 and 2. In domain 1 the nanotube appears to be hollow, Fig. 20(b), whereas in domain 2 it has some dark-contrast core filling, Fig. 20(c). Electron diffraction analyses of these domains, Figs. 20(d) and 20(e), display a DP of domain 2 exhibiting some additional features absent in domain 1, namely, polycrystalline rings with slightly larger *d*-spacings compared to those peculiar to a standard hexagonal BN (d_{100} = 2.1 Å, as marked with an ***a*** arrow). Analysis of the diffraction data gave the most intense (first) ring *d*-spacing (marked with ***b*** on Fig. 20(e)) for this phase of 2.43 Å. This parameter does not fit the known lattice constants either for a stoichiometric layered BN (hexagonal or rhombohedral phases) or for a high-pressure cubic BN phase. However, the EEL spectra recorded from these regions revealed the sole B and N element presence with no other species detected. Precise crystallography analysis of the core phase is rather difficult due to the preliminary probed electrically and its chemical composition was verified. It exhibited normal insulating behavior, Fig. 21(b), with a resistivity exceeding ~10 GΩ, as we *a priori* expected for a stoichiometric layered Boron Nitride compound, Fig. 21(c).

It is noted that the nanotube bending-relief can be performed at least twenty times without tube failing. An in-plane measured bending angle reached ~70° or even more. This strongly contradicts the pre-existing belief about BN nanotube brittleness due to the partial ionic character of chemical bonding between the B and N atoms. Another appealing issue is that the nanotube bending is always occurred at the analogous kink (or neck) point which is located in the vicinity of the originally filled nanotube domain. This feature was persistent and independent of a cycle number and mode of bending deformation (direct or reverse).

High-resolution images of a deformation kink are shown in Fig. 22. The insets to Fig. 22 illustrate the corresponding low-magnification images which give an idea of the corresponding high degrees of nanotube bending. The images clearly imply that in spite of the heavily deformed

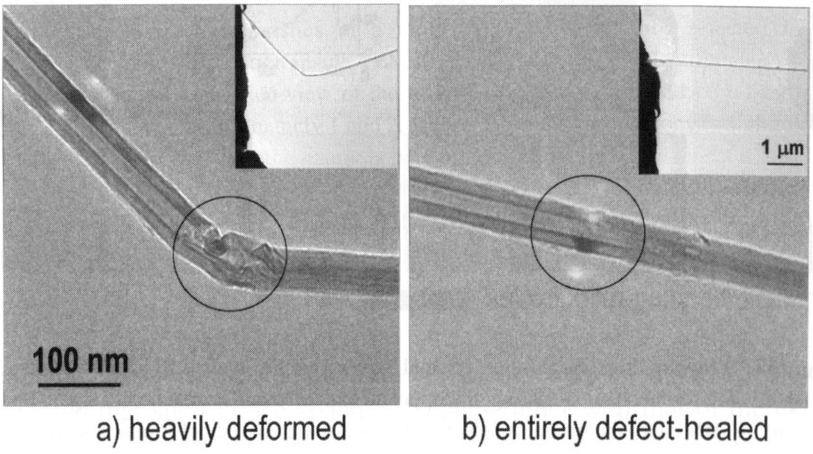

Fig. 22. TEM images of a BN nanotube heavily deformed with a maximum applied degree of bending (a); and that fully structurally recovered after reloading (b). Note that the deformation always occurs in the vicinity of a BN nanotube segment, which is originally filled with a BN matter; from Ref. 30.

and corrugated tubular shells in the vicinity of the kink, the nanotube fully restores its original structure after reloading. Therefore, a BN multi-walled nanotube is remarkably flexible — a striking result which was not expected by us *a priori* (if one keeps in mind a profound ionic-like bonding component in a layered BN). The pre-existing nanotube helicities (close to the zig-zag orientation) and the overall B/N stoichiometry were also not affected by the deformation. The flexibility of a BN nanotube rivals that of a standard C nanotube.[80]

The only change in the nanotube morphology was that after numerous deformation cycles the former nanotube polycrystalline BN filling was entirely graphitized (hexagonal-like BN), resembling a flattened BN tube and displaying the DPs peculiar to a normal layered BN. The basal planes of this layered BN became oriented in parallel with the tube axis and perpendicular to a bending plane under the elastic deformation.

Thus the set of data presented here implies that it is imperative to analyze the particular nanotube morphology and atomic structure, while studying its macroscopic deformation. The detailed comparative studies of the mechanical behavior of BN nanotubes exhibiting various structural features (filled *versus* empty, zig-zag *versus* arm-chair *etc*) are the subject of the on-going research in our Laboratoty.

To sum up this section, we describe the morphological peculiarities on individual multi-walled BN nanotube deformation. The nanotube exhibited the superb flexibility which was not expected *a-priori* for a hexagonal BN compound. The reversible bending deformation was found to be truly elastic. No traces of residual plastic deformation up to the bending angles exceeding 70° (in alternating directions) were observed. The highly corrugated BN tubular layers in the vicinity of a reproducible kink entirely restored their original structure after reloading. The kink position on the nanotube was persistent, and corresponded to an area in the vicinity of a non-stoichiometric B_xN phase-filled short nanotube domain.

7. Concluding Remarks

The inorganic nanotubes made of graphitic carbon, boron nitride and metal oxides filled with various matters represent an important domain of the nanotube family. This chapter clearly showed how various *in-situ* TEM experiments, including thermal heating/cooling, electron irradiation and electrical and mechanical probing, may shed an additional light on the chemistry, physics and mechanics of such novel and practically important nanotubes. This work is particularly important and necessary as far as prospective engineering, electrical and functional applications of filled nanotubes, and their integration into modern nanotechnology are concerned.

Acknowledgments

The authors thank many researchers and post-doctoral fellows at the Nanoscale Center of the National Institute for Materials Science (NIMS), Tsukuba, for their contribution to some of the *in-situ* TEM experiments summarized in the this chapter. Special thanks to P.M.F.J. Costa, M. Mitome and K. Kurashima (NIMS) and O. Lourie (Gatan, USA), for their participation in the "Nanofactory Instruments" holder-oriented experiments. The extensive synthetic and TEM work by C.C. Tang, Y.C. Zhi, J.H. Zhan, Y.H. Gao, Y.B. Li, J.Q. Hu, L.W. Yin (all NIMS), X.D. Bai (Chinese Academy of Sciences) and Z.W. Liu (University of Sydney) is also highly appreciated. The long-standing experimental support of Y. Uemura is particularly acknowledged.

References

1. P. M. Ajayan and S. Iijima, *Nature* **361**, 333 (1993).
2. E. Duajardin, T. W. Ebbesen, H. Hiura, and K. Tanigaki, *Science* **265**, 1850 (1994).
3. S. C. Tsang, Y. K. Chen, P. J. F. Harris, and M. L. H. Green, *Nature* **372**, 159 (1994).
4. E. Philip, J. Sloan, A. I. Kirkland, R. R. Meyer, S. Friedrichs, and J. L. Hutchinson, *Nature Mater*. **2**, 788 (2003).
5. M. P. Rossi, H. Ye, Y. Gogotsi, S. Babu, P. Ndungu, and J. C. Bradley, *Nano Lett*. **4**, 989 (2004).
6. T. Jin, G. P. Craford, R. J. Crawford, S. Zumer, and D. Finotello, *Phys. Rev. Lett*. **90**, 015504 (2003).
7. H. J. Shah, A. K. Fontecchio, M. P. Rossi, D. Mattia, and Y. Gogotsi, *Appl. Phys. Lett*. **89**, 043123 (2006).
8. A. Oberlin, M. Endo, and T. Koyama, *J. Cryst. Growth* **32**, 335 (1976).
9. S. Iijima, *Nature* **354**, 56 (1991).

10. N. G. Chopra, R. J. Luyken, K. Cherrey, V. H. Crespi, M. L. Cohen, S. G. Louie, and A. Zettl, *Science* **269**, 2045 (1995).
11. M. Harada and M Adachi, *Adv. Mater.* **12**, 839 (2000).
12. R. Fan, Y. Y. Wu, D. Y. Li, M. Yue, A. Majumdar, and P. D. Yang, *J. Am. Chem. Soc.* **125**, 5254 (2003).
13. Y. B. Li, Y. Bando, D. Golberg, and Z. W. Liu, *Appl. Phys. Lett.* **83**, 999 (2003).
14. Y. H. Gao and Y. Bando, *Nature* **415**, 599 (2002).
15. Y. H. Gao, Y. Bando, Z. W. Liu, and D. Golberg, *Appl. Phys. Lett.* **83**, 2913 (2003).
16. Z. W. Liu, Y. Bando, M. Mitome, and J. H. Zhan, *Phys. Rev. Lett.* **93**, 095504 (2004).
17. Z. W. Liu, Y. Bando, J. Q. Hu, K. Ratinac, and S. P. Ringer, *Nanotechnology* **17**, 3681 (2006).
18. P. Dorozhkin, S. Tovstonog, D. Golberg, Y. H. Zhan, Y. Ishikawa, M. Shiozawa, H. Nakanishi, K. Nakata, and Y. Bando, *Small* **1**, 1088 (2005).
19. J. H. Zhan, Y. Bando, J. Q. Hu, and D. Golberg, *J. Phys. Chem. B*, **109**, 11580 (2005).
20. Y. H. Gao, Y. Bando, and D. Golberg, Appl. Phys. *Lett.* **81**, 4133 (2002).
21. Y. B. Li, Y. Bando, and D. Golberg, *Adv. Mater.* **16**, 37 (2004).
22. Y. B. Li, Y. Bando, and D. Golberg, *Adv. Mater.* **15**, 581 (2003).
23. D. Golberg, Y. B. Li, M. Mitome, and Y. Bando, *Chem. Phys. Lett.* **409**, 75 (2005).
24. D. Golberg, F. F. Xu, and Y. Bando, *Appl. Phys. A.* **76**, 479 (2003).
25. D. Golberg, Y. Bando, K. Fushimi, M. Mitome, L. Bourgeois, and C. C. Tang, *J. Phys. Chem.B,* **107**, 8726 (2003).
26. D. Golberg, Y. Bando, M. Mitome, K. Fushimi, and C. C. Tang, *Acta Mater.* **52**, 3295 (2004).
27. D. Golberg, Y. Bando, K. Kurashima, and T. Sato, *Scripta Mater.* **44**, 1561 (2001).
28. D. Golberg, M. Mitome, K. Kurashima, C. Y. Zhi, C. C. Tang, Y. Bando, and O. Lourie, *Appl. Phys. Lett.* **88**, 123101 (2006).
29. X. D. Bai, D. Golberg, Y. Bando, C. Y. Zhi, C. C. Tang, M. Mitome, and K. Kurashima *Nano Lett.* **7**, 632 (2007).
30. D. Golberg, X. D. Bai, M. Mitome, C. C. Tang, C. Y. Zhi, and Y. Bando, *Acta Mater.* **55**, 1293 (2007).
31. L. Bosio, *J. Chem. Phys.* **68**, 1221 (1978).
32. L. Bosio, A. Defrain, H. Curien, and A. Rimsky, *Acta Crystallogr. B*, **25**, 995 (1972).
33. L. Bosio, A. Rimsky, M. Dupont, and H. Curien, *Acta Crystallogr. B*, **28**, 1974 (1972).
34. J. Q. Hu, Y. Bando, J. H. Zhan, C. Y. Zhi, and D. Golberg, *Nano Lett.* **6**, 1136 (2006).
35. J. Q. Hu, Y. Bando, J. H. Zhan, C. Y. Zhi, and D. Golberg, *Adv. Mater.* **19**, 1342 (2007).
36. J. Q. Hu, Y. Bando, J. H. Zhan, and D. Golberg, *Adv. Mater.* **17**, 1964 (2005).
37. F. Banhart, *Rep. Prog. Phys.* **62**, 1181 (1999).
38. X. Blasé, A. Rubio, S. G. Louie, and M. L. Cohen, *Europhys. Lett.* **28**, 335 (1994).
39. V. G. Samsonov, *The Oxide Handbook*, Plenum Press, New York, p. 89 (1973).
40. H. W. Koren and R. J. Lang, *Rev. Sci. Instr.* **41**, 468 (1970).

41. B. B. Alchagirov, A. G. Mozgovoi, and D. I. Kurshev, *Russ. J. Phys. Chem.* **78**, 2298 (2004).
42. J. L. Boucquet, G. Brebec, and Y. Limoge, *Physical Metallurgy*, R. W. Cahn and P. Haasen (eds.) (North-Holland Phys. Publ., New York, 1983).
43. S. Kono, T. Goto, Y. Ogura, and T. Abukawa, *Surf. Sci.* **420**, 200 (1999).
44. B. C. Regan, S. Aloni, R. O. Ritchie, U. Dahmen, and A. Zettl, *Nature* **428**, 924 (2004).
45. P. Poncharal, Z. L. Wang, and W. A. De Heer, *Science* **283**, 1513 (1999).
46. P. Kim and C. M. Lieber, *Science* **286**, 2148 (1999).
47. S. Paulson, M. R. Falvo, N. Snider, A. Helser, T. Hudson, A. Seeger, R. Taylor, R. Superfine, and S. Washburn, *Appl. Phys. Lett.* **75**, 2936 (1999).
48. Z. L. Wang, P. Poncharal, and W. A. De Heer, *J. Phys. Chem. Sol.* **61**, 1025 (2000).
49. M. F. Yu, O. Lourie, M. J. Dyer, K. Moloni, T. F. Kelly, and R. S. Ruoff, *Science* **287**, 637 (2000).
50. J. Cumings and A. Zettl, *Science*, **289**, 602 (2000).
51. P. C. Collins, M. S. Arnold, and P. Avouris, *Science* **292**, 706 (2001).
52. K. Suenaga, C. Colliex, and S. Iijima, *Appl. Phys. Lett.* **78**, 70 (2001).
53. B. Q. Wei, R. Spolenak, P. Koehler-Redlich, M. Ruhle, and E. Arzt, *Appl. Phys. Lett.* **74**, 3149 (1999).
54. D. Golberg, P. Dorozhkin, Y. Bando, M. Hasegawa, and Z. C. Dong, *Chem. Phys. Lett.* **359**, 220 (2002).
55. P. Dorozhkin, D. Golberg, Y. Bando, and Z. C. Dong, *Appl. Phys. Lett.* **81**, 1083 (2002).
56. D. Golberg, M. Mitome, K. Kurashima, and Y. Bando, *J. Elec. Microsc.*, **52**, 111 (2003).
57. K. Svensson, Y. Jompol, H. Olin, and E. Olsson, *Rev. Sci. Instrum.* **74**, 4945 (2003).
58. J. Cumings and A. Zettl, *Sol. St. Comm.* **129**, 661 (2004).
59. K. Svensson, H. Olin, and E. Olsson, *Phys. Rev. Lett.* **93**, 145901 (2004).
60. B. C. Regan, S. Aloni, K. Jensen, R. O. Ritchie, and A. Zettl, *Nano Lett.* **5**, 1730 (2005).
61. M. Sveningsson, K. Hansen, and E. Olsson, *Phys. Rev. B.* **72**, 085429 (2005).
62. C. C. Tang, Y. Bando, T. Sato, and K. Kurashima, *Chem. Comm.* 1290 (2002).
63. C. Y. Zhi, Y. Bando, C. C. Tang, and D. Golberg, *Sol. St. Comm.* **135**, 67 (2005).
64. Z. L. Wang, in *Electron Microscopy of Nanotubes*, Z. L. Wang and C. Hui (eds.) (Kluwer Academic Publishers, Boston/Dordrecht/London, 2003), pp. 174–177.
65. M. M. Treacy, T. W. Ebbesen, and J. M. Gibson, *Nature* **381**, 678 (1996).
66. M. F. Yu, O. Lourie, M. J. Dyer, K. Moloni, T. F. Kelly, and R. S. Ruoff, *Science* **287**, 637 (2000).
67. E. W. Wong, P. E. Sheehan, and C. M. Lieber, *Science* **277**, 1971 (1997).
68. C. Bower, R. Rosen. L. Jin, J. Han, and O. Zhou, *Appl. Phys. Lett.* **74**, 3317 (1999).
69. O. Lourie, D.M. Cox, and H.D. Wagner, *Phys. Rev. Lett.* **81**, 1638 (1998).
70. B. I. Yakobson, C. Brabec, A. Maiti, and J. Bernholc, *Phys. Rev. Lett.* **76**, 2511 (1996).
71. M. Buongiorno Nardelli, B. I. Yakobson, and J. Bernholc, *Phys. Rev. B.* **57**, R4277 (1998).
72. A. Maiti, *Chem. Phys. Lett.* **331**, 21 (2000).

73. A. Pantano, M. C. Boyce, and D.M. Parks, *Phys. Rev. Lett.* **91**, 145504 (2003).
74. N. G. Chopra and A. Zettl, *Sol. St. Comm.* **105**, 297 (1998).
75. D. Golberg, Y. Bando, M. Eremets, K. Takemura, K. Kurashima, and H. Yusa, *Appl. Phys. Lett.* **69**, 2045 (1996).
76. D. Golberg, W. Han, Y. Bando, L. Bourgeois, K. Kurashima, and T. Sato, *J. Appl. Phys.* **86**, 2364 (1999).
77. D. Golberg, Y. Bando, K. Kurashima, and T. Sato, *Sol. St. Commun.* **116**, 1 (2000).
78. K. Ploog, P. Rauh, W. Stoeger, and H. J. Schmidt, *Cryst. Growth* **13**, 350 (1972).
79. A. Celik-Atkas, J. M. Zuo, J. F. Stubbins, C. C. Tang, and Y. Bando, *Acta Cryst. A.* **61**, 533 (2005).
80. M. S. Wang, L. M. Peng, J. Y. Wang, and Q. Chen, *Adv. Funct. Mater.* **16**, 1462 (2006).

CHAPTER 7

IN-SITU ION AND ELECTRON BEAM EFFECTS ON THE FABRICATION AND ANALYSIS OF NANOMATERIALS

Kazuo Furuya*,‡, Minghui Song* and Masayuki Shimojo*,†

**High Voltage Electron Microscopy Station*
National Institute for Materials Science
3-13 Sakura, Tsukuba, 305-0003, Japan

†Advanced Science Research Laboratory, Saitama Institute of Technology
1690 Fusaiji, Fukaya, 369-0293, Japan
‡Furuya.Kazuo1@nims.go.jp

Ion implantation causes the formation of nano-phases as well as radiation damage. "*In-situ*" observation in a TEM is a unique technique to clarify such phenomena. One example is Xe nanocrystals embedded in a metal matrix. HRTEM observations revealed the atomic structures and the motion of atoms in a Xe nanocrystal. Electron beam-induced deposition is another technique to fabricate nano-structures. Nanostructures having desired shape and size can be obtained. Metal atoms are deposited using focused electron beam irradiation under the presence of a small amount of precursor gas molecules on the substrate. The details of these ion and electron beam effects are reviewed.

1. Introduction

Beam irradiation to materials causes structural evolution associated with atomic processes such as radiation damage, chemical decomposition, sputtering, phase-separation and the nucleation of precipitates and so on. The resultant structures are mainly composed of a variety of "nano-phases", and are needed to be analyzed with transmission electron microscopy (TEM) in atomic scale for utilizing and controlling the property of materials.

The nanostuctures by beam effect are formed either "embedded into materials" or "on the surface of materials", or both. One of the typical example is the energetic ion implantation to materials. When the energy of ions is low enough to stay near surface, the reaction between host materials and irradiated atoms occurs on the surface. As the energy of ions increased, the reaction region reaches deep inside the materials, depending upon the range distribution of ions, and embedded nanostructures are formed. Since irradiated ions are generally foreign elements to host materials, atomic interaction causes "second phase formation" in nano-scale. This energy driven process is sometime non-thermodynamic and causes "embedded non equilibrium nano-phases".

The electron beam effect is rather simple unless foreign materials are intentionally added in the area of irradiation. The typical example is electron-beam induced deposition (EBID), in which a precursor gas is introduced in a chamber when the electron beam is turned on. In general, the gas molecules decompose and solid elements are deposited on the surface of substrates. If the electron beam is focused to nanometers in diameter, the size and position of the deposits can be easily controlled.

To investigate the above mentioned processes of ion and electron beam effects on the fabrication of nanomaterials, we use "*in-situ*" transmission electron microscopy with ion implantation and gas introduction interfaces, which is an unique and powerful tool, to directly analyze the structural evolution of nanostructures both "embedded into materials" and "on the surface of materials".

2. *In-Situ* TEM Studies of Ion Implanted Structures

2.1. *Introduction of In-situ ion implantation experiments*

Ion implantation results in the formation of nano-phases as well as radiation damage. Understanding the process of the nano-phase formation is clearly related to those of atomic process of defect clusters. Therefore, *in-situ* TEM studies initially focused on issues with radiation damages of materials. However, the field has extended to characterize the materials with nanometer-sized structures and phases, which show different mechanical, electronic, magnetic, optical, or thermodynamic properties from those of a bulk material. One of the examples for the nano-phases introduced by ion implantation is the inert gas nano-precipitates. They were investigated as the by-products of nuclear fission reactions in radiation damage studies, but they are now attracting scientists as typical meta-stable nanocrystals, by which atomic behavior in nano-scale range can be understood.

In-situ TEM, especially a high voltage transmission electron microscope (HVEM) with ion beam interface allows the use of the same electron beam to produce atomic displacements, and to image the process directly with TV cameras at atomic scale in high-resolution TEM (HRTEM) mode. Furthermore, the use of analytical tools, such as electron energy loss spectroscopy (EELS) and energy dispersive X-ray spectroscopy (EDS) provides accurate chemical information of meta-stable nano-phased materials.

2.2. *Instruments with ion beam interfaces*

The motivation of connecting TEMs, including HVEMs, to ion implanters was to introduce ion irradiation cascade damage. To obtain this approach, researchers in the world have being made a great deal of effort to develop ion implanter-interfaced TEMs. Several papers have reviewed the instrument and the relevant research.[1–3] Two classes of *in-situ* ion beam irradiation systems interfacing to TEM have been developed: small guns mounted directly on the microscope[4]; and large freestanding accelerators that also serve additional target chambers.[5–8] The interfacing of the ion-irradiation system involves a relatively simple modification of the TEM in the vicinity of the objective lens without significant degradation of microscope resolution. The critical issues for the ion-implanter interfaced TEM are the vibration isolation of the TEM and the accurate measurement of electron and ion dosimetry. The anti-vibration can be achieved by connecting the ion-implanter and the TEM with soft and elastic bellows. Faraday cups located close to the TEM specimen are usually used to monitor ion beam.

The *in-situ* ion-irradiation facility at Argonne National Laboratory was initially installed in 1978 and upgraded in 1995. In the upgraded facility, an intermediate-voltage TEM (IVEM), a Hitachi H-9000NAR, was interface to a 650-kV accelerator and a 2-MV tandem accelerator.[7–9] The TEM voltage ranges from 100 to 300 kV. The gap of the objective lens pole-piece of the IVEM was 11 mm. To permit ion-beam incidence on the specimen at 30° with respect to the electron beam, the bore of the upper pole-piece was increased to 11 mm in diameter. Then, the Cs for the TEM was 2.8 mm. The guaranteed point-to-point resolution at 300 kV was 0.25 nm with a line resolution of 0.14 nm. These resolutions were also achieved with the ion beam line attached.

A schematic drawing of the ion acceleration and transport system attached to an HVEM at National Institute for Materials Science is shown in Fig. 1.[10] This ION/HVEM system consists of a 1000 kV HVEM (JEM-ARM1000, JEOL Ltd), and dual ion implanters of 200 and 30 kV with a hollow cathode and a RF discharge ion source, respectively. Ions can be accelerated and introduced into the HVEM column through an electrostatic beam control system, which is operated with network coupled PC-based computers from the HVEM operation console. This is necessary

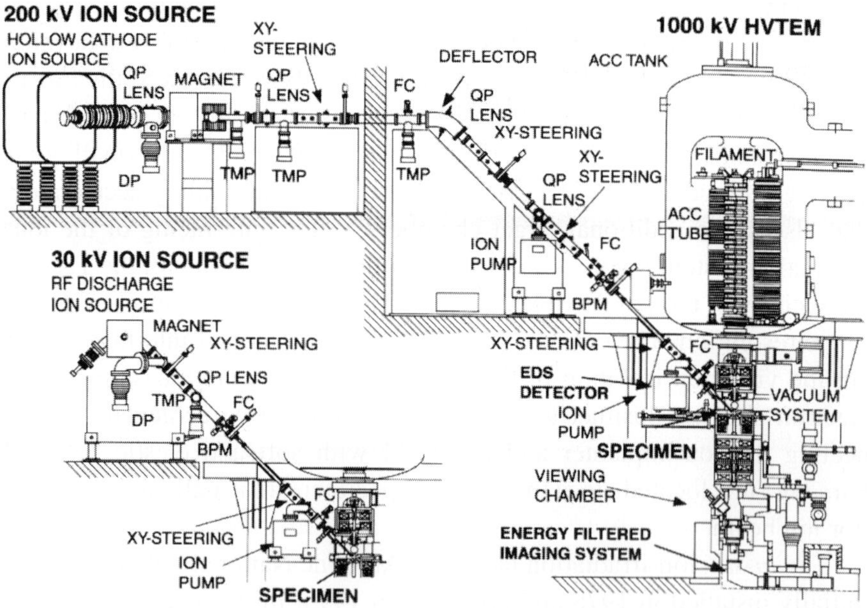

Fig. 1. A schematic drawing of the ION/HVEM system, consisting of a high-resolution high-voltage transmission electron microscope (HVEM, JEM-ARM1000), and 200 and 30 kV dual ion implanters.[10]

when *in-situ* implantation is carried out with HRTEM. Electron and ions are irradiated with the angles of 90 and 45 degrees to the specimen surface, respectively. A fixed aperture of 2 mm in diameter at the entrance of the HVEM controls an ion beam position and shields uncollimated ion beams. The beam profile and total intensity of ion beams can be measured by the micro-Faraday cup holder with 0.25 mm-diameter detection-hole and by the analytical holder before and during the implantation, respectively. The mechanical modification was made by drilling several holes with 2 mm in diameter through the pole-piece of the objective lens for the introduction of ion beams. The resolution of the HVEM was measured under the irradiation of 15 keV He^+ to Al TEM specimens at room temperature, and turned out to be better than 0.23 nm in the dynamic observation using videotage.[11] The typical ion beam current density of ions is $0.01 \sim 5 \times 10^{-1}$ A m^{-2}.

2.3. *Irradiation induced phase transformations*

The atomic displacement by both electron and ion irradiation causes phase transformation in materials. The typical type of those are crystalline to amorphous or the reverse, ordered phase to disordered one, segregation, solid phase transformation, etc.[12–15] One of the examples of ion-irradiation induced phase transformation is that in TiAl alloys. Due to their high strength-to-weight ratio and good resistance to temperature variations, intermetallic TiAl alloys are supposed to be used in airplane and spaceship manufacturing. Additional potential application includes first-wall material for fusion reactors. Thus, knowledge of the stability of these alloys under irradiation is very important.

During the low-temperature (i.e. $T \leq 40$ K) Kr irradiation of Al_3Ti, the structure became fully amorphous after a dose of 1 dpa. On the other hand, light ion irradiation, 40 keV He, leads to only partial chemical disordering and strong lattice distortion, but not to complete amorphization after doses as high as 2 dpa.[16]

Room-temperature heavy-ion irradiation (50 keV Xe or 25 keV Ar) of γ-Ti49-Al51 alloys[17,18] to a dose of 2.3 dpa induced a new phase that had sizes up to about several tens of nanometers (Fig. 2). This phase has a hexagonal structure with $a = 0.286$ nm and $c = 0.462$ nm. The crystallographic

Fig. 2. The irradiation induced phase in γ-TiAl irradiated with 50 keV Xe to a dose of 2.2×10^{19} ions m^{-2} at room temperature. (a) An HRTEM image of the Xe induced phase; (b) an enlargement of region E in (a); (c) a simulated image of the induced phase in [100]; (d) a simulated image of the γ-TiAl in [011]; (e) an SAD pattern corresponding to the area including (a); (f) a simulated diffraction pattern of the induced phase.[17]

orientation relationship between the induced phase and the γ-TiAl matrix is: $(001)_P//(1\bar{1}1)_\gamma$ (subscript P denotes the induced phase) and $[100]_P//[011]_\gamma$. On the other hand, no phase transformation was induced in γ-TiAl by light ion irradiation (15-keV He) up to 4.6 dpa.[19] These results suggest that the phase transformation depends on displacement cascade formation rather than the total irradiation dose.

Another example of the phase transformation by ion irradiation is the surface modification of austenite steels. The first evidence for the occurrence of a martensitic phase transformation from face-centered cubic (fcc) to body-centered cubic (bcc) in austenitic stainless steels after ion implantation was obtained with electron microscopy from 18/8 and 316 steels implanted with phosphorus.[20,21] Further research revealed that the phase transformation was also induced by implantations with other energetic ions, such as Sb,[22] the constituent element ions (Fe, Ni, and Cr) of the austenitic stainless steels,[23] and inert gas ions (He, Ar, Kr, Xe).[23–25] The orientation relationship of the transformed bcc phase to the fcc matrix were observed to agree mostly with the Nishiyama-Wassermenn rule, $(110)_\gamma//(111)_\gamma$ and $[1\bar{1}0]_\gamma//[\bar{2}11]_\gamma$.[22,23] A relationship agreeing with the K-S rule, $(01\bar{1})_\gamma//(1\bar{1}1)_\gamma$ and $[111]_\gamma//[011]_\gamma$, was also reported for 100 keV Xe ion irradiation of 304 stainless steels.[26] Studies on the depth distribution of transformed martensite in Xe-implanted stainless steels have revealed that the distribution of the martensite is consistent with the range of implanted ions.[27] This suggests that the strain in the matrix induced by implanted ions or agglomerated clusters is an important driving force for the phase transformation.

2.4. *Nano-inclusions in materials*

Nanoscale solid inclusions in matrix, for example, nano-solids of Pb, Sn and Tl in Al[28–30] and alloy nano-particles of Pb-Cd in Al and in Si,[31,32] have increasing fundamental and applied interest in materials research. *In-situ* TEM is particularly important in studying the behavior of such buried nanophase under ion irradiation or during heating.

Johnson and Dahmen carried out *in-situ* TEM observations of alloying of nanoscale Pb inclusions in Al by implantation with Cd ions.[33] Both Pb and Cd elements are insoluble in Aluminum but together they form a simple eutectic system. Ion implantation of Pb was first performed to form nanoscale Pb inclusions in Al, and then ion implantation of Cd was performed *in-situ* in order to observe the alloying process of Cd elements into Pb nanoparticles. Implantations below the lead/cadmium eutectic temperature of 521 K showed nucleation and growth of solid cadmium slabs on $(111)_{Pb}$/Al interfaces, while implantations above the eutectic temperature induced gradual melting of the inclusions after the solubility of cadmium in lead was exceeded. Melting was seen initially to nucleate on $(100)_{Pb}$/Al interface facets, followed either by continuous melting or by melting in abrupt steps within a fairly limited concentration range.

Other *in-situ* TEM observations during annealing were carried out to observe the melting and solidification of Pb-Cd alloy nanoscale inclusions in real time.[34] The same dose of 5×10^{19} m^{-2} of Pb^+ and Cd^+ were implanted sequentially to form Pb-Cd inclusions containing approximately equal amounts of Pb and Cd. Figure 3 shows a series of micrographs from an *in-situ* heating/cooling sequence recorded both on negatives and video tape. The solid inclusion, which is imaged close to the $\langle 110 \rangle$ matrix direction, contains a Cd slab on one of the edge-on $\{111\}_{Pb}$ facets. The ratio between the projected areas of Cd and Pb is about 2:3 corresponding approximately to a composition of 50% Cd and 50% Pb, i.e. a composition that is Cd-rich relative to the eutectic composition.

2.5. *Xe and inert gas in metals*

Inert gas is essentially insoluble in metals, therefore it tends to precipitate, and usually forms nanometer sized particles in gaseous state (bubble) or solid state (crystal or amorphous). The morphology of the inert gas precipitates depends on the matrix, gas species, temperature, and size of the particle itself.[35–42] Small Xe, Kr, and Ar precipitates in Al can be in solid state at room temperature. The basic reason for their solidification in metals is due to a high pressure in GPa order produced by the suppression of the matrix.[35] Typically, Xe nanoprecipitates in metals has been known to be in an fcc structure in fcc metals and to have the same orientation as the host fcc metals from electron diffraction analysis and TEM observations.

2.5.1. *Nucleation and fluctuation of Xe nanocrystals*

The direct observation of nucleation and growth of Xe precipitates was performed *in-situ* with HREM.[43] The Xe ions were implanted at a low flux

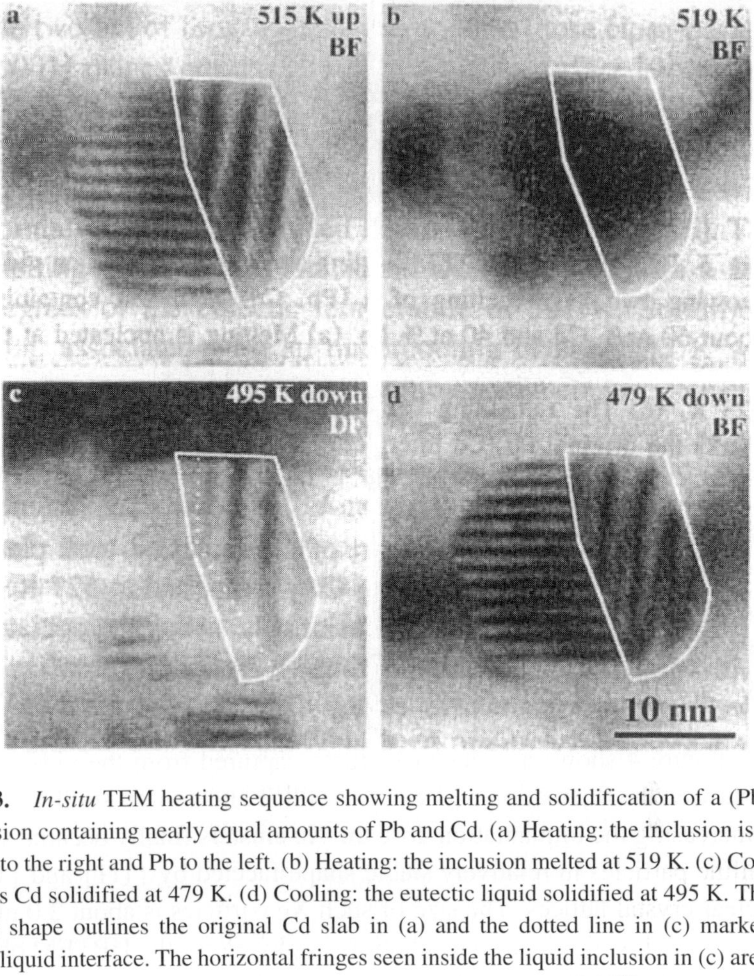

Fig. 3. *In-situ* TEM heating sequence showing melting and solidification of a (Pb, Cd) inclusion containing nearly equal amounts of Pb and Cd. (a) Heating: the inclusion is solid, Cd is to the right and Pb to the left. (b) Heating: the inclusion melted at 519 K. (c) Cooling: excess Cd solidified at 479 K. (d) Cooling: the eutectic liquid solidified at 495 K. The full white shape outlines the original Cd slab in (a) and the dotted line in (c) marked the solid/liquid interface. The horizontal fringes seen inside the liquid inclusion in (c) are from an overlapping solid inclusion. (BF) bright-field, and (DF) dark-field.[34]

below 2.8×10^{15} ions m^{-2} s^{-1}, at which the drift of the specimen during ion implantation is small enough for atomic recording with HREM. The observations were carried out with an off-axis imaging technique, in which a few degrees of tilt away from the Al $\langle 110 \rangle$ zone axis and non-Scherzer defocus are employed to reduce the image contrast from low-index Al reflections.[44] Below the dose of about 2.2×10^{19} ions m^{-2}, the regions with black contrast in size of about 1~2 nm which were considered to be Xe

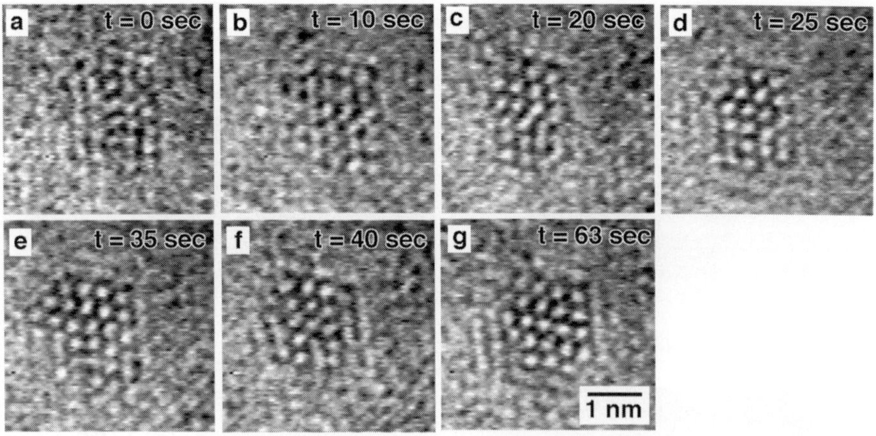

Fig. 4. A series of captured HRTEM micrographs from the video tape recorded during Xe ion implantation at room temperature showing the nucleation and grwth of a solid Xe precipitate. The relative time is shown in the figures.[43]

clusters in non-crystalline state were seen, but they were not stable in shape and contrast. With the further implantation, the Xe clusters began to show regular arrangement of atom rows, which were not stable at the earlier period. Figure 4 shows a series of pictures captured from the video tape. They showed the fluctuation between crystalline and non-crystalline states. With increasing the implantation dose the Xe clusters finally became small crystalline particles in relatively stable shape faceted by {111} and {100} with clear crystal lattices. The size of such Xe particles is about 2.0 nm.

Under the 1 MeV electron irradiation employed for the HRTEM observation at room temperature, the nanoprecipitates exhibited a number of readily observed phenomena including migration within the matrix, changes in shape, melting, recrystallization, associated with changes in interface structure.[45,46] The possible mechanisms for these structure changes were suggested to be beam heating and changes in precipitate volume associated with fluctuations in arrival of Al vacancies and interstitials, or with transport of Xe to and from the Al matrix due to electron irradiation damage in the Al and/or the Xe precipitate.

The changes in structure became drastic with small Xe particles. It was observed that a Xe particle about 1 nm in size changed structure from an

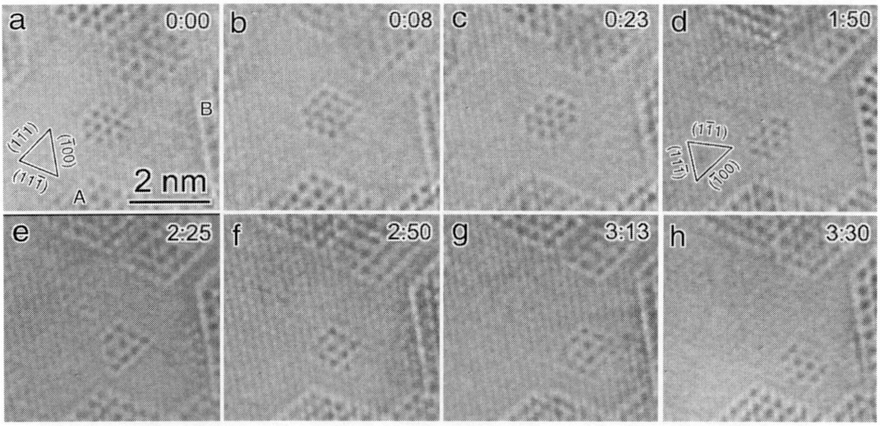

Fig. 5. The HRTEM images of Xe particles embedded in an Al matrix, showing changes in structure with time. The relative time of m:s are shown in figures.[47]

fcc to another one, and changed the orientation refer to the matrix but still in an fcc structure (Fig. 5).[47] However, the 1 nm sized Xe particle was hardly observed to be in amorphous state in the observation by a video system with time resolution of 1/30 second. The lattice parameters of Xe particles from about 1 to 6 nm in sizes were measured based on the observation.[48] Those of Xe crystals in size from about 2 to 6 nm were almost the same as obtained from diffraction results. But the lattice parameter of a 1 nm fcc Xe particle can be about 20% smaller than the average value obtained from electron diffraction, implying that the particle was compressed by about 80%.

2.5.2. *Motion of atoms in a Xe nanocrystal*

The structure change happened in a Xe nanocrystal can be observed in atomic resolution level *in-situ* using the off-axis observation technique.[44] One of the examples is the observation of defect introduction into a Xe nanocrystal and the recovery process under the irradiation of 1 MeV electron beam.[49,50] The observation was performed with an HREM conjunction with a video system. It was observed in real time that an intrinsic stacking fault was introduced into a Xe nanoprecipitate on {111} with the electron beam irradiation. The defect extended from the precipitate vertex

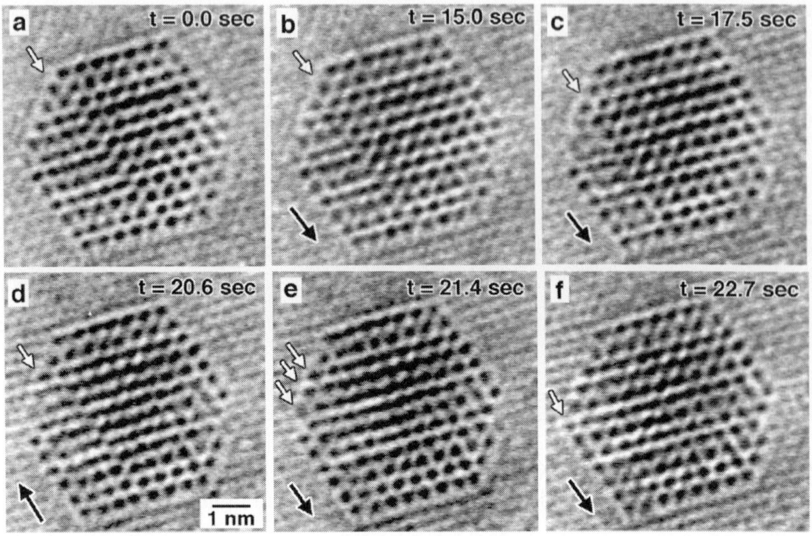

Fig. 6. The successive images of the defect recovery. The white arrow indicates the layer in which the recovery occurred, and the black arrow indicates the direction of movement. The number indicates the time in seconds measured from the first observation of the defect.[49]

and terminated at a Shockley partial. The defect moved out of the crystal by successive shuffling of {111} layers, as shown in Fig. 6. Detailed analysis revealed that the displacement of Xe atom columns approximately corresponded to the projected displacement produced by a Shockley partial dislocation, bounding an intrinsic stacking fault, but the displacements were reduced at the precipitate/matrix interface, implying that there would be some compromise between the strain field of the partial dislocation and the interface compatibility requirement.

2.5.3. *Coalescence of Xe nanocrystals*

Under the irradiation of 1 MeV electron beam, Xe nanoparticles in Al change structures and grow larger. One of the processes is the coalescence of the Xe nanoparicles. *In-situ* HREM observation using the off-axis technique makes it possible to reveal the process in real time.[44] The coalescence of Xe nanoparticles was observed at room temperature with an

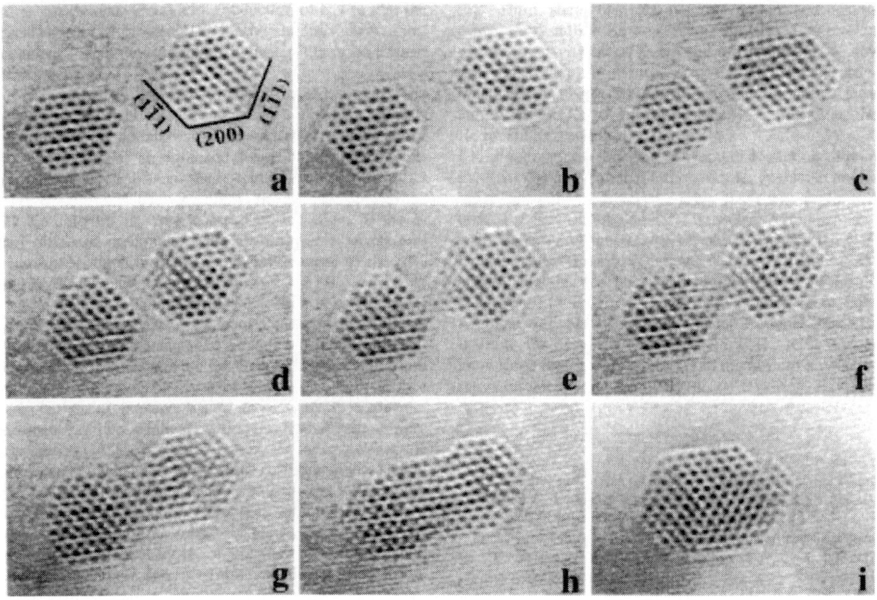

Fig. 7. Motion and coalescence of two isolated crystalline Xe precipitates during continuous 1 MeV electron irradiation. Measured from the first image, the elapsed times at which video frames were recorded are (a) 0, (b) 101, (c) 418, (d) 549, (e) 550, (f) 551, (g) 561, (h) 584, and (i) 727 seconds. Traces of crystallographic planes are indicated in frame (a).[51]

HVEM, JEM-ARM1000 operated at 1 MeV as shown in Fig. 7.[51] It was revealed that atomic-level fluctuations of cavity facets resulted in shape changes and precipitate motion leading to coalescence, and that there was no apparent elastic interaction between precipitates separated by as little as 0.5 nm. After coalescence, crystalline Xe conformed by plastic deformation without melting to change in cavity shape. Cavity volume, not surface area, was conserved during coalescence, implying that cavity pressure was not determined solely by the interface tension.

2.5.4. *Ordering in a fluid Xe inclusion contained in Al metal*

Theoretical investigations of the structure of liquids at solid interfaces have indicated that density modulation perpendicular to the interface is likely to occur.[52–56] Such layering is expected to occur even in simple hard-sphere systems and results from the geometrical constraining effect of the solid surface on the atoms or molecules of the liquid.[57] However, the observational evidence of such interfacial layering has been ambiguous. The inert gas in metals has a tendency to cluster together to form small solid or liquid bubbles with radii of a few nanometers. The small inert gas-filled cavity in a thin metal film is a useful high-pressure cell, permitting the studies of the interface between inert-gas and solid surface. Donnelly *et al.* took advantage of such tendency of inert gases, clearly observed the layering structure of liquid Xe gas on surface of Al metal.[58] The HRTEM image in Fig. 8 shows two large faceted cavities containing fluid Xe. The cavities are seen in $\langle 110 \rangle$ projection with Al $\{111\}$ and $\{100\}$ planes approximately normal to the plane of the image with the off-axis technique.[44] Fringes are visible in the cavities on the $\{111\}$ facets. It was confirmed by MD simulations that the fringes resulted from the layering of Xe atoms in the fluid at the interface. The results provide conclusive evidence for layering at a fluid/solid interface in a simple liquid.

3. *In-Situ* Studies of Electron Beam-Induced Deposition (EBID)

3.1. *Introduction of EBID*

Electron-beam induced deposition (EBID) is a versatile nanofabrication technique in which a precursor gas is introduced in the chamber. The precursor gas adsorbs on the substrate and is decomposed by focused electron beams. As electron beams can be focused to a nanometer-sized region using modern electron microscopes, a nano-sized deposit is formed in the irradiated regions on the surface of the substrate. A schematic illustration is shown in Fig. 9(a). For example, when the vapor of tungsten hexacarbonyl, $W(CO)_6$ is used as a precursor gas, the electron irradiation decomposes the gas into W, CO, C, O and various molecular species. W atoms are deposited on the substrate, while the volatile materials are pumped out. The deposit grows upwards from the surface when the beam is stationed at one point, and a line is produced by moving the beam position. Accordingly, more complicated shape can be formed if the position of the beam is controlled by deflecting the beam using a computer. A schematic illustration of an arbitrary nano-fabrication using

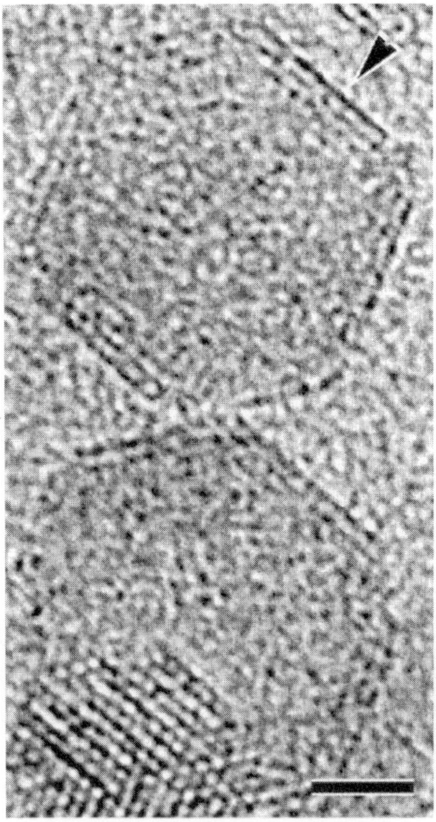

Fig. 8. High-resolution TEM image of faceted, Xe-containing cavities in Al at room tempaerature. The Xe in the two large cavities is fluid; in the small cavity at the bottom of the figure, whose image overlaps one of the large cavities, Xe is solid. The image was recorded at an objective lens defocus of +72 nm. The contrast transfer function of the microscope under these conditions, combined with a small degree of tilt, results in very little intensity from the Al lattice in the micrograph, enabling layering in the fluid Xe to be clearly seen. Bar, 2.5 nm.[58]

an EBID technique is shown in Fig. 9(b). The electron beam was positioned on the edge of the substrate and then moved to empty space, which resulted in the formation of a freestanding rod. The growth of the nanorod follows the position of electron beam so that the shape of the freestanding rod is controlled arbitrarily.

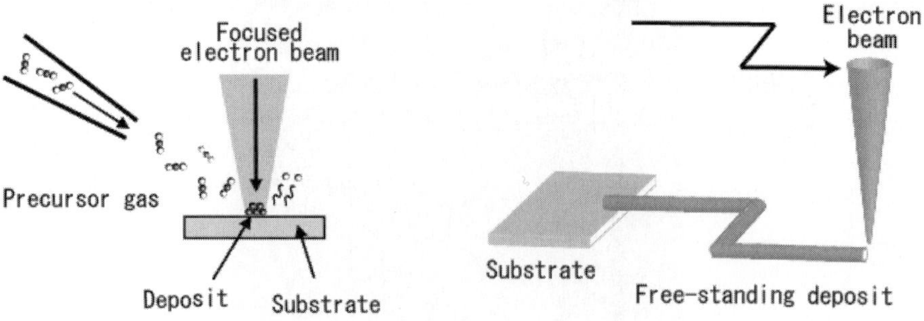

Fig. 9. Schematic illustrations of (a) deposition mechanism and (b) an arbitrary-shaped free-standing structure formation.

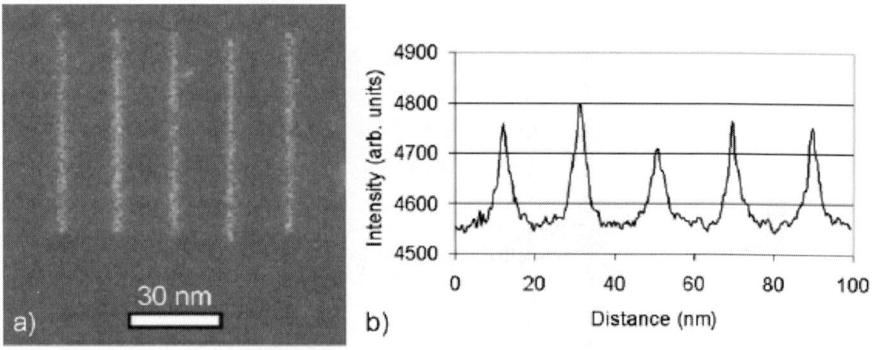

Fig. 10. Line drawing on a substrate by EBID. (a) HAADF-STEM image and (b) their line profiles, showing a line width of about 3.2 nm at full width at half maximum.[59]

Here, some examples from previous studies are shown to demonstrate the flexibility of EBID in a nanoscale. Figure 10 shows an example of line drawing on a substrate and line profiles indicating a line width of 3.2 nm at FWHM made by Van Dorp *et al.*[59] Figure 11 demonstrates the characters writing on a substrate, which consists of hundreds of dots formed by EBID, and a freestanding nano-ring using a $W(CO)_6$ precursor gas. Figure 12 shows a nanoscale map of the world drawn using EBID with a $W(CO)_6$ precursor on a silicon nitride substrate fabricated by Crozier and Van Dorp.[60] EBID is now recognized as a very promising nanofabrication technique, and the amount of works devoted to this field is rapidly increasing. In addition to the flexibility in shaping, a wide variety of materials

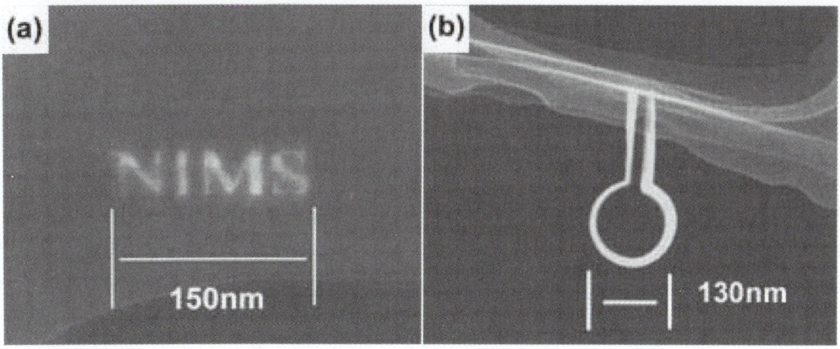

Fig. 11. Examples of (a) character writing on a Si substrate by placing dots with 3.5 nm in diameter using a $W(CO)_6$ precursor and (b) a free-standing structure grown from an edge of a carbon film using a $W(CO)_6$ precursor.

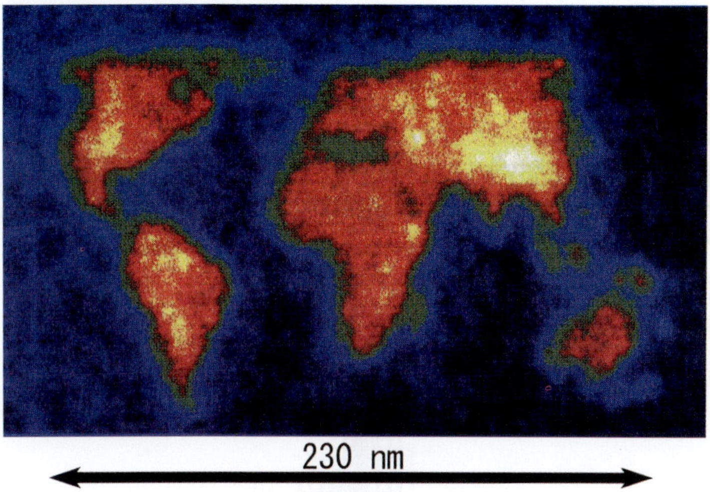

Fig. 12. Nanoscale map of the world drawn by EBID with a $W(CO)_6$ precursor on a silicon nitride substrate.[60]

can be deposited. In this section, deposition mechanisms, deposited materials, and applications of EBID techniques are reviewed.

3.2. *Mechanisms of EBID*

The precursors are decomposed or dissociated by electron beams in the EBID process. The dissociation cross-section, which indicates the probability of dissociation or decomposition as a function of electron energy, is an important factor for EBID. The dissociation cross-section curves are known for simple gases, such as, H_2, N_2, CH_4 etc.,[61] but the data for the gases used in EBID is still limited.[62–64] A dissociation cross-section curve generally has a peak at a low energy region around 100 eV. Thus, low-energy electrons, such as secondary electrons, play an important role in the EBID process, and the knowledge of the secondary electron distribution is essential. Secondary electron distribution has been studied using a Monte Carlo method[65] described below.

Primary electrons injected into a substrate are scattered elastically and/or inelastically. The electron scattering processes in a substance are complicated so that the simulation of these processes is usually performed by a Monte Carlo method using the fast secondary model.[65] A dynamic Monte Carlo profile simulation was proposed by Silvis-Cividjian *et al.*[66,67] to simulate deposition by including the electron scattering inside the already-grown deposit. Figure 13 shows a schematic illustration of the program display window during calculation. By these simulations, it was recognized that the distribution of primary electrons determined the shape of the deposit.[68]

It is argued by many authors[69] that the deposition is mainly caused by secondary electrons generated by primary electrons. This is reasonable in

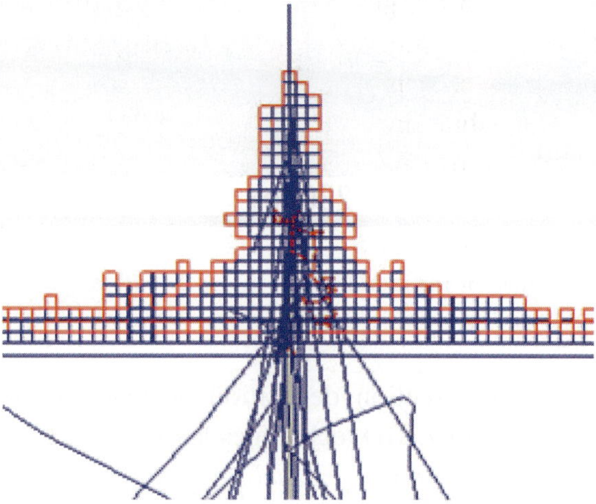

Fig. 13. Schematic illustration of Monte Carlo simulation, indicating the profile of the deposit and electron scattering inside the already-grown deposit.

view of the energy dependence of dissociation cross section. However, such low energy electrons do not travel a long distance due to their short mean free paths, although they have a high dissociation cross-section. Therefore, the overall distribution, which is generated by low energy secondary electrons, is almost overlapped by the distribution of primary electron trajectories. This means that the shape of the deposit can be roughly estimated by the trajectories of the high energy primary electrons.

3.3. *Resolution limit*

The main tool used for EBID has been scanning electron microscopes (SEMs) with an accelerating voltage up to 30 kV. For a pillar structure which is fabricated by stationing a beam at a point, the lateral size (the radius of the pillar) is dictated by the escape distance of the secondary electrons, which is several tens of nanometers. A nanometer-sized electron probe can easily be obtained by field emission (FE) SEM. Therefore, such an instrument is a reasonable choice for EBID.

Using transmission electron microscopes (TEMs), an electron beam can be focused to less than 0.2 nm at the sample position. Some groups used TEM or scanning transmission electron microscopes (STEMs) to explore the resolution limit of EBID. Tanaka *et al.*[70] used an ultra high-vacuum TEM having a FE gun with a careful control of precursor gas pressure, and obtained a minimum size of a tungsten dot of 1.5 nm in diameter. The dots are shown in Fig. 14. The dots are located in the intersecting points of the white lines in a TEM bright field image in Fig. 14(a). They are so small that it is hard to distinguish them from the amorphous substrate. Figure 14(b) shows a high angle annular dark field (HAADF)-STEM image of the dots. In this way, dots are managed to be observed. Recently, Van Drop[59] obtained a minimum deposit size of 0.7 nm at full-width half maximum (FWHM) using a STEM machine.

3.4. *Materials and precursors for EBID*

3.4.1. *Overview of precursors for metal deposition*

A wide variety of organic precursors have been used for EBID. Many organic compounds are commercially available, which are mostly used in

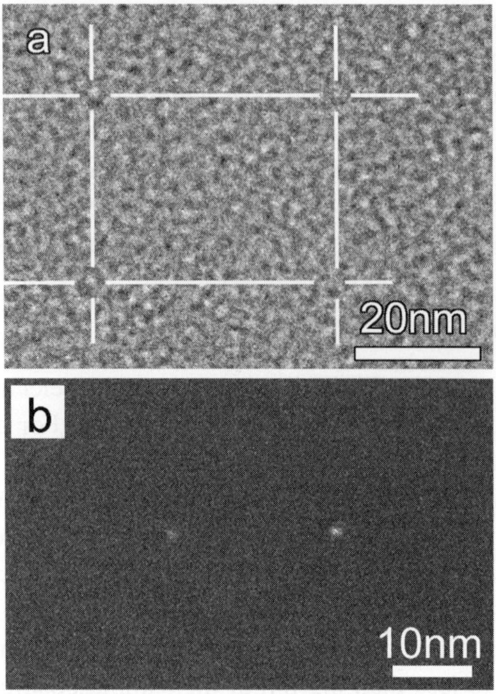

Fig. 14. (a) Bright field TEM and (b) HAADF-STEM images of dots having a diameter of 1.5 nm. The dots are so small that they cannot be seen by conventional bright field TEM imaging, but they are managed to be observed using HAADF-STEM imaging.

chemical vapor deposition techniques. Some inorganic substances are also used as precursors for EBID. In an early stage of the EBID history, carbon contamination which often occurs in electron microscopes is used to make dots and structures.[71] Such carbon contamination occurs due to the dissociation and deposition of hydrocarbons that exist on uncleaned specimens or atmosphere from the diffusion pump oil, etc. Then, researchers have introduced the vapor of many substances intentionally into the chamber so that many materials have been deposited so far.

Tungsten hexacarbonyl, $W(CO)_6$ has been used for EBID in many studies.[72–74] Tungsten hexacarbonyl is a white crystalline solid and sublimates slightly even at room temperature so that the vapor can be transferred into the chamber. Han *et al.*[75] characterized nanowires using a nano electron diffraction technique and revealed that the deposits from this precursor contained tungsten, tungsten carbide, tungsten oxides, and amorphous carbon.

Cyclopentadienyl platinum trimethyl, $CpPtMe_3$, is often used for platinum deposition.[76] It was reported that the deposits contained Pt nanocrystals in a carbon-containing amorphous matrix. The metal content changed with changing accelerating voltage and beam current. Similar compounds, such as trimethyl methylcyclopentadeinyl platinum, were also used to deposit Pt.[77]

Several organic compounds have been used as precursors for gold deposition, which are dimethylacetylacetonato gold, $Me_2Au(acac)$, dimethyltrifluoroacetylacetonate gold, $Me_2Au(tfac)$, and dimethylhexafluoroacetylacetonato gold, $Me_2Au(hfa)$. However, the content of gold in the deposits was not large and some researchers have tried to increase the gold content. Weber *et al.*[78] used $Me_2Au(tfac)$ and obtained the deposits containing 3–10 at% of Au, depending on the beam current, etc. Folch *et al.*[79] increased the Au content to 50% by using $Me_2Au(hfac)$ with oxygen or water vapor. Molhave *et al.*[80] introduced $Me_2Au(acac)$ with water vapor in an environmental SEM and obtained a deposit with a dense gold core surrounded by a crust of gold nanocrystals in amorphous carbon. Recently, Botman *et al.*[81] increased to 60% by heating in a reactive atmosphere of oxygen after deposition.

One of the most serious drawbacks of EBID is carbon contamination in the deposits. The carbon atoms come from the dissociated ligands of organic molecules. Thus, it is considered that the carbon contamination can be minimized if inorganic compounds are used as precursors. Hiroshima *et al.*[82,83] used tungsten hexafluoride, WF_6, and fabricated a highly conductive deposit. A more complex precursor, $[RhCl(PF_3)_2]_2$ was used to deposit rhodium by Cicoira *et al.*[84] PF_3AuCl was used to deposit gold.[85] Crozier[86] used perdeuterated gallium azide, D_2GaN_3, to produce GaN dots. The purity of the deposits from these precursors may be high but it should be noted that some of these precursors are corrosive or explosive.

3.4.2. *Iron and iron-compound deposition*

The production of nanometer-sized magnetic materials or structures is of significant interest due to the potential application in novel magnetoelectronic devices as well as fundamental understanding of nanomagnetics.

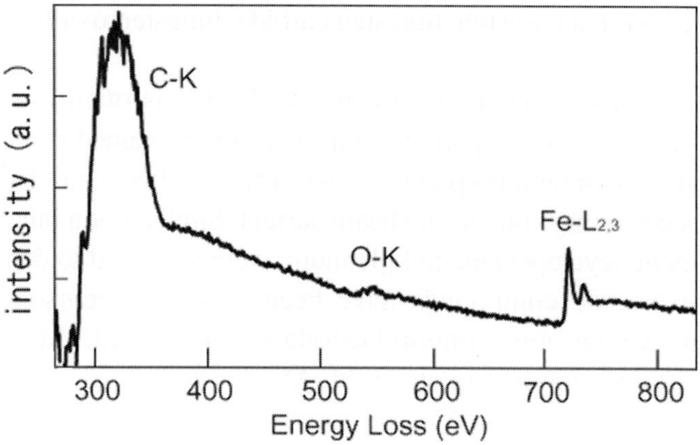

Fig. 15. EELS spectrum of a deposit from $Fe(CO)_5$. The deposit contains iron, carbon and a small amount of oxygen.

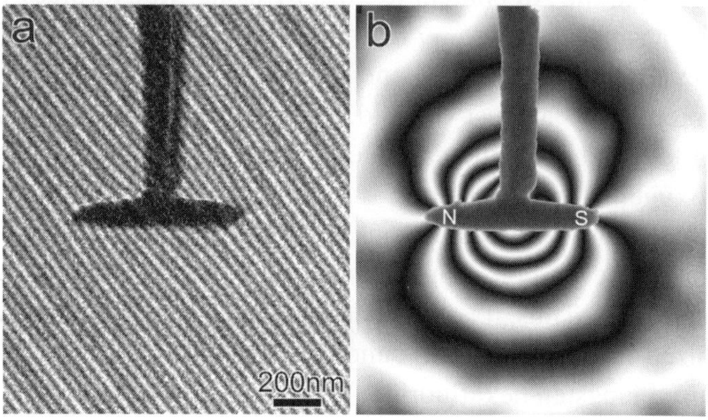

Fig. 16. Electron hologram and reconstructed phase image of a nano-magnet made at the apex of a sharp tungsten tip, indicating the magnetic flux around the feromagnetic nano-deposit.

Biscyclopentadienyl iron (Ferrocene), is an orange-colored solid and generates Fe containing deposits,[87] though the content of Fe compared to carbon is low. Deposits from iron pentacarbonyl, $Fe(CO)_5$, which is a liquid at room temperature and has a vapor pressure of about 5 kPa, contain more Fe, which is about 60 at%.[88] Figure 15 shows an electron energy loss spectroscopy (EELS) spectrum from the deposit using $Fe(CO)_5$. As the deposit contains a large amount of Fe atoms, it shows ferromagnetism.

Figure 16 shows an electron hologram and a reconstructed phase image of a nano-magnet made at the apex of a tungsten tip, indicating magnetic flux around the nano-magnet. It was also reported that the deposit from $Fe(CO)_5$ was transformed to bcc-Fe by post-deposition heat-treatment in vacuum.[88]

Che *et al.*[89] formed FePt intermetallic compound nanostructures using a mixture of $Fe(CO)_5$ and cyclopentadienyl platinum trimethyl as a precursor and heated in vacuum after deposition. FePt has a higher residual magnetization than pure Fe. Thus, this technique is important for the formation of magnetic nanostructures.

Deposition at high temperatures induces the reaction of Fe with the substrate. Tanaka *et al.*[90,91] reported the formation of iron silicide using an $Fe(CO)_5$ precursor on a cleaned Si substrate.

Shimojo *et al.*[92] added water vapor to $Fe(CO)_5$ vapor and used it as a precursor for EBID. The deposit was identified to be crystalline Fe_3O_4 magnetite. No carbon was detected in the deposit by EELS. Figure 9 shows a high resolution TEM image of an Fe_3O_4 nanowire. This is the first carbon-free crystalline oxide formation in EBID performed at room temperature.

3.5. *Applications of EBID*

3.5.1. *Mask repair and device fabrication*

EBID has many advantages, such as, (i) the minimum deposit size is a few nanometers, (ii) nano-structures having desired shape at desired position can be formed, and (iii) three-dimensional growth is possible. Using these advantages, many applications have been made so far. In this section, some examples of applications are described.

Mask repair is a process of editing pattern structures to make a correction of a defective mask in lithography techniques. As the size of integrated circuits is decreasing, the mask repair tools with a size and placement accuracy on the order of 10 nm or less are required. EBID techniques are suitable to repair defective masks because EBID has a "point-and-shoot" capability with an accuracy of 1 nm. Koops *et al.*[93] have extensively studied and developed a photomask repair system including electron beam-induced deposition and etching, which meets the 65 nm combined with "*in-situ*" TEM is one of leading technologies in top-down manner to manufacture materials with complete control and to understand atomic structures precisely.

Acknowledgments

This work was supported by "R&D program of new device materials" of the National Institute for Materials Science (NIMS), Japan and by "Nanotechnology Support Project" and "R&D program for atomic level analysis of nuclear materials" of the Ministry of Education, Culture, Sports, Science and Technology (MEXT), Japan. We would like to thank those who provided preprints or unpublished material that was used in this article, and M. Takeguchi, M. Tanaka, K. Mitsuishi at NIMS, R. C. Birtcher, C. W. Allen at Argonne National Laboratory, USA and S. E. Donnelly at the University of Salford, UK for their valuable discussion on the entire works.

References

1. S. Ishino, *J. Nucl. Mater.* **206**, 139 (1993).
2. C. W. Allen, S. Ohnuki, and H. Takahashi, *Trans. Mat. Res. Soc. Jpn.* **17**, 93 (1994).
3. R. C. Birtcher, M. A. Kirk, K. Furuya, G. R. Lumpkin, and M-O. Ruault, *J. Mat. Res.* **20**, 1654 (2005).
4. K. Hojou, S. Furuno, H. Ohtsu, K. Izui, and T. Tsukamoto, *J. Nucl. Mater.* **155–157**, 298 (1988).
5. M.-O. Ruault, M. Lerme, B. Jouffrey, and J. Chaumont, *J. Phys. E: Sci. Instrum.* **11**, 1125 (1979).
6. M-O. Ruault, J. Chaumont, and H. Bernas, *Nucl. Instrum. Methods Phys. Res. B* **209–210**, 351 (1983).
7. A. Taylor, C. W. Allen, and E. A. Ryan, *Nucl. Instrum. Methods Phys. Res. B* **24–25**, 598 (1987).
8. C. W. Allen, L. L. Funk, E. A. Ryan, and S. T. Ockers, *Nucl. Instrum. Methods Phys. Res. B* **40–41**, 553 (1989).
9. C. W. Allen, *Ultramicroscopy* **56**, 200 (1994).
10. K. Furuya, K. Mitsuishi, M. Song, and T. Saito, *J. Electro. Micros.* **48**, 511 (1999).
11. N. Ishikawa and K. Furuya, *Ultramicroscopy* **56**, 211 (1994).
12. A. T. Motta, *J. Nucl. Mater.* **244**, 227 (1997).
13. K. Tsuchiya and K. Marukawa, *J. Electron Microsc.* **48**, 375 (1999).
14. T. Nagase and Y. Umakoshi, *Mater. Trans.* **47**, 1469 (2006).
15. H. Wollenberger, *J. Nucl. Mater.* **216**, 63 (1994).

16. C. Jaouen, M. F. Denanot, and M. F. Riviere, *Nucl. Instrum. Methods Phys. Res. B* **80**, 386 (1993).
17. M. Song, K. Mitsuishi, M. Takeguchi, K. Furuya, T. Tanabe, and T. Noda, *Philos. Mag. Lett.* **80**, 661 (2000).
18. M. Song, K. Mitsuishi, M. Takeguchi, K. Furuya, T. Tanabe, and T. Noda, *J. Nucl. Mater.* **307–311**, 971 (2002).
19. M. Song, K. Furuya, T. Tanabe, and T. Noda, *J. Nucl. Mater.* **271–272**, 200 (1999).
20. E. Johnson, T. Wohlenberg, and W. A. Grant, *Phase Transitions* **1**, 23 (1979).
21. E. Johnson, T. Wohlenberg, W. A. Grant, P. Hasen, and L. T. Chadderton, *J. Microsc.* **116**, 77 (1979).
22. E. Johnson, U. Littmark, A. Johansen, and C. Christodoulites, *Philos. Mag. A* **45**, 803 (1982).
23. E. Johnson, A. Johansen, L. Sarholt-Kristensen, L. Gråbæc, N. Hayashi, and I. Sakamoto, *Nucl. Instrum. Methods Phys. Res. B* **19–20**, 171 (1987).
24. N. Hayashi and T. Takahashi, *Appl. Phys. Lett.* **41**, 1100 (1982).
25. I. Sakamoto, N. Hayashi, B. Furubayashi, and H. Tanaue, *J. Appl. Phys.* **68**, 4508 (1990).
26. G. Xie, M. Song, K. Mitsuishi, and K. Furuya, *J. Nucl. Mater.* **281**, 80 (2000).
27. E. Johnson, E. Gerritsen, N. G. Chechenin, A. Johansen, L. Sargolt-Kristensen, H. A. A. Keetels, L. Gråbæc, and J. Bohr, *Nucl. Instrum. Methods Phys. Res. B* **39**, 573 (1989).
28. E. Johnson, A. Johansen, U. Dahmen, S. Chen, and T. *Fujii, Mater. Sci. Eng. A* **304–306**, 187 (2001).
29. E. Johnson, C. R. H. Bahl, V. S. Touboltsev, and A. Johansen, *Nucleation and Growth Processes in Materials*, A. Gonis, P. E. A. Turchi, and A. J. Ardell (eds.), *MRS Symp. Proc.* **580**, 177 (2000).
30. A. H. Sørensen, E. Johnson, K. K. Bourdelle, A. Johansen, H. H. Andersen and L. Sarholt-Kristensen, *Philos. Mag. A* **75**, 1533 (1997).
31. L. Sarholt, E. Johnson, A. Johansen, T. B. Stibius Jensen, A. B. Stibius Jensen, and U. Dahmen, *Mater. Sci. Forum* **312–314**, 299 (1999).
32. L. Sarholt, A. S. B. Jensen, V. S. Touboltsev, A. Johansen, and E. Johnson, *Mater. Sci. Forum* **360–362**, 283 (2001).
33. E. Johnson and U. Dahmen, *Microsc. Microanalysis* **3**, 409 (1997).
34. E. Johnson, V. S. Touboltsev, A. Johansen, U. Dahmen, and S. Hagège, *Nucl. Instrum. Methods Phys. Res. B* **127–128**, 727 (1997).
35. A. Vom Felde, J. Fink, T. H. Muller-Heinzerling, J. Pfluger, B. Scheerer, and H. Linker, *Phys. Rev. Lett.* **53**, 922 (1984).
36. J. H. Evans and D. J. Mazey, *J. Phys. F: Met. Phys.* 15, L1-6 (1985); *Scripta Metallurgica* **19**, **621–623** (1985).
37. C. Templier, R. J. Gaboriaud, and H. Garem, *Mat. Sci. Eng.* **69**, 63 (1985).
38. C. Templier, H. Garem, and J. P. Riviere, *Philos. Mag. A* **53**, 667 (1986).
39. C. Templier, H. Garem, J. P. Riviere, and J. Delafond, *Nucl. Inst. and Meth. B* **18**, 24 (1986).
40. R. C. Birtcher and W. Jäger, *Ultramicroscopy* **22**, 267 (1987).
41. R. J. Cox, P. J. Goodhew, and J. H. Evans, *Acta Metall.* **35**, 165 (1987).

42. M. Song, K. Mitsuishi, and K. Furuya, *Micron* **30**, 129 (1999).
43. M. Song, K. Mitsuishi, and K. Furuya, *Mater. Sci. Eng. A* **304–306**, 135 (2001).
44. K. Furuya, N. Ishikawa, and C. W. Allen, *J. of Microsc.* **194**, 152 (1999).
45. C. W. Allen, R. C. Birtcher, S. E. Donnelly, K. Furuya, N. Ishikawa, and M. Song, *Appl. Phys. Lett.* **74**, 2611 (1999).
46. C. W. Allen, R. C. Birtcher, S. E. Donnelly, M. Song, K. Furuya, N. Ishikawa, and K. Mitsuishi, *J. Electro. Microsc.* **51**, S175 (2002).
47. M. Song, K. Mitsuishi, K. Furuya, C. W. Allen, R. C. Birtcher, and S. E. Donnelly, *Appl. Surf. Sci.* **241**, 96 (2005).
48. M. Song, K. Mitsuishi, K. Furuya, C. W. Allen, R. C. Birtcher, and S. E. Donnelly, *J. Microsc.* **215**, 224 (2004).
49. K. Mitsuishi M. Song, K. Furuya, R. C. Birtcher, C. W. Allen, and S. E. Donnelly, *MRS Symp Proc* **504**, 417 (1998).
50. K. Mitsuishi, M. Song, K. Furuya, R. C. Birtcher, C. W. Allen, and S. E. Donnelly, *Nucl. Inst. and Meth. in Phys. Res. B* **148**, 184 (1999).
51. R. C. Birtcher, S. E. Donnelly, M. Song, K. Furuya, K. Mitsuishi, and C. W. Allen, *Phys. Rev. Lett.* **83**, 1617 (1999).
52. A. Bonissent and B. Nutaftschiev, *Philos. Mag.* **35**, 67 (1977).
53. J. W. Cahn and R. Kikuchi, *Phys. Rev. B* **31**, 4300 (1985).
54. B. B. Laird and D. J. Haymet, *Chem. Rev.* **92**, 1819 (1992).
55. F. Spaepen, *Sloki State Phys.* **47**, 1 (1994).
56. M. Plischke and D. Henderson, *J. Chem. Phys.* **84**, 2846 (1986).
57. J. N. Israelachvili, *Internolecular Forces and Surface Forces*, p. 261 (Academic Press, San Diego, CA, 1992).
58. S. E. Donnelly, R. C. Birtcher, C. W. Allen, I. Morrison, K. Furuya, M. Song, K. Mitsuishi, and U. Dahmen, *Science* **296**, 507 (2002).
59. W. F. Van Drop, B. Van Someren, C. W. Hagen, P. Kruit, and P. A. Crozier, *Nano Lett.* **5**, 1303–1307 (2005).
60. P. A. Crozier and W. F. Van Drop, (Private communication in 2006).
61. H. F. Winters and M. Inokuti, *Phys. Rev.* **A25**, 1420–1430 (1982).
62. P. Seuret, F. Cicoira, T. Ohta, P. Doppelt, P. Hoffmann, J. Weber and T. A. Wesolowski, *Phys. Chem. Chem. Phys.* **5**, 268–274 (2003).
63. T. Ohta, F. Cicoira, P. Doppelt, L. Beitone, and P. Hoffmann, *Chemical Vapor Deposition* **7**, 33–37 (2001).
64. J. D. Fowlkes, S. J. Randolph, and P. D. Rack, *J. Vac. Sci. Technol.* **B23**, 2825–2832 (2005).
65. D. C. Joy, *Oxford Series in Optical and Imaging Sciences*, Oxford University Press (1995).
66. N. Silvis-Cividjian, C. W. Hargen, P. Kruit, M. A. J. v.d. Stam, and H. B. Groen, *Appl. Phys. Lett.* **82**, 3514–3516 (2003).
67. N. Silvis-Cividjian, C. W. Hargen, and P. Kruit, *J. Appl. Phys.* **98**, 84905-1-12 (2005).
68. K. Mitsuishi, Z. Q. Liu, M. Shimojo, M. Han, and K. Furuya, *Ultramicrosocopy* **103**, 17–22 (2005).
69. T. E. Allen, R. R. Kunz, and T. M. Mayer, *J. Vac. Sci. Technol. B* **6**, 2057–2060 (1988).

70. M. Tanaka, M. Shimojo, M. Han, K. Mitsuishi, and K. Furuya, *Surf. Interface Anal.* **37**, 261–264 (2005).
71. A. N. Broers, W. W. Molzen, J. J. Cuomo, and N. D. Wittels, *Appl. Phys. Lett.* **29**, 596 (1976).
72. H. W. P. Koops, R. Weiel, D. P. Kern, and T. H. Baum, *J. Vac. Sci. Technol. B* **6**, 477 (1988).
73. K. T. Kohlmann-von Platen, L. -M. Buchmann, H. -C. Petzold, and W. H. Brunger, *J. Vac. Sci. Technol. B*, **10**, 2690 (1992).
74. P. C. Hoyle, J. R. A. Cleaver, and H. Ahmed, *J. Vac. Sci. Technol. B*, **14**, 662 (1996).
75. M. Han, K. Mitsuishi, M. Shimojo, and K. Furuya, *Phil. Mag.* **84**, 1281 (2004)
76. H. W. P Koops, A. Kaya, and M. Weber, *J. Vac. Sci. Technol. B* **13**, 2400 (1995).
77. D. Beaulieu, Y. Ding, Z. L. Wang, and W. J. Lackey, *J. Vac. Sci. Technol. B* **23**, 2151 (2005).
78. M. Weber, M. Rudolph, J. Kretz, and H. W. P. Koops, *J. Vac. Sci. Technol. B* **13**, 461 (1995).
79. J. Tejada, C. H. Peters, and M. S. Wrighton, *Appl. Phys. Lett.* **66**, 2080 (1995).
80. K. Molhave, D. N. Madsen, A. M. Rasmussen, A. Carlsson, C. C. Appel, M. Brorson, C. J. H. Jacobsen, and P. Boggild, *Nano Lett.* **3**, 1499 (2003).
81. Botman, J. J. L. Mulders, R. Weemaes, and S. Mentink, *Nanotechnology* **17**, 3779 (2006).
82. H. Hiroshima and M. Komuro, *Nanotechnology* **9**, 108 (1998).
83. H. Hiroshima, N. Suzuki, N. Ogawa, and M. Komuro, *Jpn. J. Appl. Phys.* **38**, 7135 (1999).
84. F. Cicoira, K. Leifer, P. Hoffmann, I. Utke, B. Dwir, D. Laub, P. A. Buffat, E. Kapon, and P. Doppelt, *J. Cryst. Growth* **265**, 619 (2004).
85. Utke, P. Hoffmann, B. Dwir, K. Leifer, E. Kapon, and P. Doppelt , *J. Vac. Sci, Technol. B* **18**, 3168 (2000).
86. P. A. Crozier, J. Tolle, J. Kouvetakis, and C. Ritter, *Appl. Phys. Lett.* **84**, 3441 (2004).
87. R. C. Che, M. Takeguchi, M. Shimojo, W. Zhang, and K. Furuya, *Jpn. J. Appl. Phys.* **45**, 5582 (2006).
88. M. Takeguchi, M. Shimojo, and K. Furuya, *Nanotechnology* **16**, 1321 (2005).
89. R. C. Che, M. Takeguchi, M. Shimojo, W. Zhang, and K. Furuya, *Appl. Phys. Lett.* **87**, 223109 (2005).
90. M. Tanaka, F. Chu, M. Shimojo, M. Takeguchi, K. Mitsuishi, and K. Furuya, *Appl. Phys. Lett.* **86**, 183104 (2005).
91. M. Tanaka, F. Chu, M. Shimojo, M. Takeguchi, K. Mitsuishi, and K. Furuya, *J. Mater. Sci.* **41**, 2667 (2006).
92. M. Shimojo, M. Takeguchi, and K. Furuya, *Nanotechnology* **17**, 3637 (2006).
93. K. Edinger, H. Becht, J. Bihr, V. Boegli, M. Budach, T. Hoffmann, H. W. P. Koops, P, Kuschnerus, J. Oster, P. Spies, and B. Weyrauch, *J. Vac. Sci. Technol. B* **22**, 2902 (2004).
94. T. Liang, E. Frendberg, B. Lieberman, and A. Stivers, *J. Vac. Sci. Technol. B* **23**, 3101 (2005).
95. G. Boero, I. Utke, T. Bret, N. Quack, M. Todorova, S. Mouaziz, P, Kejik, J. Brugger, R. S. Popovic, and P. Hoffmann, *Appl. Phys. Lett.* **86**, 042503 (2005).

96. M. Komuro and H. Hiroshima, *J. Vac. Sci. Technol. B* **15**, 2809 (1997).
97. H. W. P. Koops, C. Schossler, A. Kaya, and M. Weber, *J. Vac. Sci. Technol. B* **14**, 4105 (1996).
98. K. Murakami and M. Takai, *J. Vac. Sci. Technol. B* **22**, 1266 (2004).
99. X. Yang, M. L. Simpson, S. J. Randolph, P. D. Rack, L. R. Baylor, H. Cui, and W. L. Gardner, *Appl. Phys. Lett.* **86**, 183106 (2005).
100. J. Fujita, M. Ishida, T. Ichihashi, Y. Ochiai, T. Kaito, and S. Matsui, *Jpn. J. Appl. Phys.* **43**, 3799 (2004).
101. C. H. Jin, J. Y. Wang, Q. Chen, and L.-M. Peng, *J. Phys. Chem. B* **110**, 5423 (2006).
102. I.-C. Chen, L.-H. Chen, C. Orme, A. Quist, R. Lal, and S. Jin, *Nanotechnology* **17**, 4322 (2006).

CHAPTER 8

ELECTRON IRRADIATION OF NANOMATERIALS IN THE ELECTRON MICROSCOPE

Florian Banhart

Institut de Physique et Chimie des Matériaux, Université de Strasbourg
23, rue du Loess, 67034 Strasbourg, France
banhart@ipcms.u-strasbg.fr

Electron irradiation of specimens is generally inevitable in electron microscopy. The energetic beam in the electron microscope can displace atoms in the specimens, and this may lead to visible structure transformations. Although radiation damage is often an unwelcome artefact, some classes of nanomaterials show the formation of novel phases or morphologies when subjected to electron irradiation. The advantage of *in-situ* electron microscopy is that structural transformations can be induced and imaged with the same electron beam and studied in real time with atomic resolution. This chapter summarizes the physical principles of irradiation phenomena with emphasis on effects in nanoparticles. In an experimental section the techniques of *in-situ* electron microscopy and electron irradiation at high specimen temperatures are treated. Examples of electron irradiation phenomena in graphitic nanoparticles such as carbon nanotubes or fullerene-like structures are given. Graphitic materials have a unique ability to reconstruct after atom displacements and thus show a great variety of structural transformations under the electron beam.

1. Introduction

The formation of an image in an electron microscope requires the interaction of the object with an energetic electron beam, therefore electron irradiation of the specimens is inevitable in electron microscopy. In typical transmission electron microscopes (TEM), acceleration voltages in the range of 100–300 kV (in a few instruments up to 1–3 MV) are used nowadays, and the energies of the electrons are sufficiently high to alter the structure of the specimens. Electronic excitations, the breaking of bonds between the atoms, or ballistic atom displacements cause damage in the material under the beam. Normally this is not the aim of TEM studies because artefacts may appear in the images, reflecting features that have not been in the pristine materials. An explanation of these phenomena is generally desirable not only because radiation effects have to be avoided. Electron irradiation in the electron microscope can also be used to simulate the behavior of materials under particle bombardment, for example, in nuclear fission or fusion reactors. This has been a subject of intense research since the 1960s. Furthermore, the applicability of sensitive materials in space where a considerable flux of energetic cosmic particles prevails, can be simulated by exposing the materials to an electron beam. The particular advantage of these experiments is that the evolution of radiation defects can be studied *in-situ* and recorded in real time at high spatial resolution. Of course, the electrons causing radiation-induced alterations and the electrons for imaging in the microscope are not the same. Structural changes in the material go along with an energy loss of the electrons, and these electrons cannot be used for imaging. But it is the same electron beam and the same instrument that can be used for both at the same time.

Nowadays, modern TEMs offer the possibility to study materials with lattice resolution down to the scale of 0.1 nanometers or even below and enable us to study radiation effects on the atomic scale. Furthermore, the electron beam in TEMs with a field emission gun can be focused onto spots of less than 1 nm in diameter (below 0.1 nm with aberration-corrected condensers) so that deliberate beam-induced structural modifications of the specimen can be carried out on the scale of few or even single atoms. Hence, it is now possible to create and image defect structures on the smallest scale. The most important types of radiation defects are interstitials and vacancies which are commonly denoted as point defects. With modern techniques of electron microscopy the imaging of individual point defects in the TEM appears feasible (see Chapter 9). Aberration-corrected instruments promise further advances in this direction which would be of great benefit for the study of radiation effects.

Electron irradiation of bulk specimens has been carried out since decades, and the effects are, at least in some common materials (metals, semiconductors), well understood.[1–5] With the increased interest in nano-sized objects in the last years, nanoparticles have been subject of intense research by electron microscopy, and novel effects, induced by electron irradiation, have been discovered. Due to the large fraction of surface atoms and the general proximity of surfaces, radiation effects in nanoparticles are different from those in macroscopic materials. In crystals of a few nanometers in size, defect annealing at surfaces is generally facilitated, therefore nanocrystals often show increased stability against irradiation. Of particular interest are nanoparticles based on graphite, e.g. carbon nanotubes or fullerene-like structures, not only due to their technical importance but also because graphite has a unique ability to reconstruct and anneal after atom displacements. Many unexpected phenomena were observed in graphitic nanoparticles under electron irradiation and will be treated in detail in the following sections.

Irradiation drives a system away from thermal equilibrium. In some cases, the degree of dissipation in the systems can be very high and, with appropriate "boundary conditions", self-organized structure formation can occur. A crystalline structure with radiation defects is generally in a metastable state after irradiation, and radiation defects, once created, tend to vanish by annealing when equilibrium is restored at high temperature. The temperature of the specimen is generally an important parameter that has to be adjustable in electron irradiation studies by using dedicated heating stages. The knowledge of defect dynamics as a function of temperature during and after irradiation is of great importance but can often not be understood without the help of atomistic simulations, e.g. by molecular dynamics.

This chapter will concentrate on electron irradiation effects that are caused by ballistic atom displacements. Electronic damage (bond breaking, excitations) do not show phenomena of comparable interest and will not be treated in detail. The main focus will be on the author's own work on graphitic nanoparticles because of the great variety of irradiation effects in graphitic structures. A concise review of all types of nanoparticles under irradiation is beyond the frame of this chapter. Another aspect that cannot be treated here is contamination as a result of electron irradiation of specimens in the electron microscope. For a recent review about electron and ion irradiation effects in macroscopic systems, the reader is referred to the article by Birtcher *et al.*[5] Reviews about irradiation effects in carbon nanoparticles can be found in the literature.[6–9]

Section 2 gives a short treatment of the physics behind electron irradiation. The experimental techniques of *in-situ* electron irradiation studies in the TEM are treated in Sec. 3. Section 4 describes examples of irradiation-induced structural modifications in nanoparticles.

2. Irradiation of Solids with Electrons

2.1. *Mechanisms of electron-solid interaction*

In a transmission electron microscope the specimen is exposed to irradiation with electrons whose energy is given by the acceleration voltage of the instrument (typically 100–300 kV). The scattering of the beam in the specimen may be elastic or inelastic, i.e. without or with energy loss of the electrons.[10] Elastic scattering does not alter the specimen but can be used for the formation of the image. Inelastically scattered electrons have transferred energy (resp. momentum) to the nuclei or electrons in the specimen and cannot be used for high-resolution imaging. Due to momentum conservation, the energy transfer from the energetic beam electrons to the electrons in the specimen is high whereas the heavy nuclei can only take up a small fraction of the impact energy. The importance of a scattering event is described quantitatively by the scattering cross-section σ which is given in units of barns (1 barn = 10^{-28} m^2). The probability of a scattering event $p = \sigma j$ is given by the product of the scattering cross-section and the beam current density j. The cross-section depends on the energy of the beam and the specimen material.

In insulators with localized electrons, electronic excitations may lead to ionization, bond breaking, or other effects causing damage of the specimen. In metals, on the other hand, delocalized conductions electrons quench electronic excitations so that persistent damage can only be created when the atoms are displaced as a cause of electron-nucleus scattering. Generally, the cross-section for electronic excitation decreases with increasing beam energy whereas the cross-section for atom displacements increases slightly when displacement cascades occur. It is important to note that no atom displacements are possible below a certain threshold energy of the electron beam. Hence, electronic excitations (in non-metallic specimens) are important at low beam energies whereas atom displacements dominate at high energies.

Another important point (which is often misjudged) is the total energy dissipation in the specimen. The range of the beam electrons in typical specimen materials is mostly of the order of millimeters. Considering that TEM specimens for lattice resolution imaging have thicknesses of approximately 10 nm or even less, only a very small fraction of the electron energy dissipates in the specimen, even when atoms are displaced. As a consequence, the overall heating of nanoparticles under the beam is typically a few degrees only (fully focused beams might lead to somewhat more heating). Therefore, thermal effects can mostly be neglected when radiation-induced alterations in elemental or inorganic nanoparticles are considered. The situation may be different, however, in sensitive organic materials with low thermal conductivity, e.g. polymers, where even moderate electron irradiation can lead to severe damage which is in part due to heating. Nanoparticles can often be considered as "zero-dimensional" objects, but they have to be held on a grid in the TEM. Therefore, heat dissipation occurs by heat conduction through the contact area and by thermal radiation. In extended one-dimensional systems such as nanowires or nanotubes or in two-dimensional systems such as planar specimens, heating is further reduced by lateral heat conduction.

2.1.1. *Electronic excitations*

The scattering of the energetic beam electrons at the electrons in the specimen can lead to several types of excitations. The most important mechanisms are ionization, bond breaking, excitation of electrons into higher levels, the generation of electron-hole pairs or excitons, the emission of secondary electrons, or collective excitations such as plasmons. Each of these contributions can be quantified by the respective scattering cross-section.[10] In most materials, plasmon excitation has the highest cross-section, followed by ionization. Plasmons are collective vibrations of the electron system in the sample and cause no or little damage. But the dissipation of plasmons which occurs on a short time scale generates phonons and is the main source of specimen heating.

An ionization event in solids with conduction electrons (metals, graphite) is followed by immediate recombination so that the structure remains unchanged. In insulators, however, ionization can lead to the

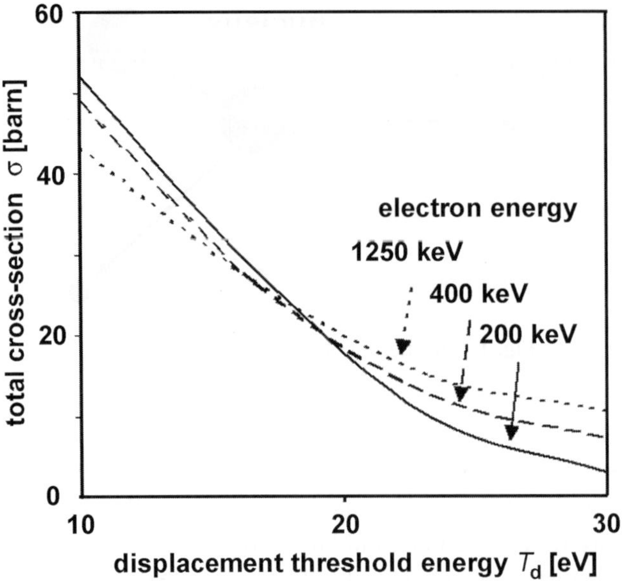

Fig. 2. Total cross-section σ as a function of the displacement threshold energy T_d for the displacement of a carbon atom by an electron. The curves are plotted for three electron energies.

The cross-section increases with the electron energy and the atomic number and decreases with the threshold energy. This is shown for the example of carbon in Fig. 2. When the beam energy is just slightly above the displacement threshold E_{thr}, only head-on collisions (forward scattering) are possible while for higher beam energies large-angle scattering dominates. This is an important point when the crystal orientation (anisotropy) is considered.

When the energy of the electron exceeds twice the displacement threshold, some of the displaced atoms may have enough energy to cause further displacements. Such displacement cascades may dominate in larger objects. In the center of a displacement cascade, a thermal spike on the scale of a few interatomic distances develops and dissipates on the time scale of picoseconds.[13]

In nanoparticles the fraction of surface atoms is high. Due to the open space above the surface it is generally easier to displace a surface atom than an atom inside the bulk. The displacement threshold for surface atoms is accordingly lower but has hardly been measured to date. Surface atoms may leave the specimen completely (this is known as sputtering[14]) or migrate on the surface without getting lost. For sputtering the atoms need enough energy to leave the surface of the specimen from where they are attracted. Due to the fact that most scattering events occur with large scattering angle and, thus, low energy transfer, most displaced atoms at the surface do not leave the specimen.

The product of the displacement cross-section σ and the beam current density j tells us how often each atom is displaced per second. Typical beam current densities in TEMs range between 10 A/cm^2 during normal inspection in the high-resolution mode and 10^4 A/cm^2 in the fully focused beam of a field emission microscope. As an example we take a displacement cross-section of 20 barn (graphite with $T_{\text{thr}} \approx 17$ eV[6]) and so obtain values between 10^{-3} displacements per second ($j = 10$ A/cm^2) and 1 displacement per second ($j = 10^4$ A/cm^2) for each atom under the beam.

2.2. *Defects generated under electron irradiation*

The displacement of atoms and the successive dissipation of energy happens on a time scale of less than 10^{-11} seconds. After that period, the migration of point defects and annealing take place. Interstitials and vacancies can recombine with each other and anneal radiation damage. On the other hand, mutual shielding of defects of the same type can lead to the formation of stable interstitial or vacancy agglomerates.

The diffusivity D of defects follows an Arrhenius law

$$D = D_0 \exp\left(-\frac{E_m}{kT}\right), \tag{3}$$

where D_0 is a pre-exponential factor, E_{m} the migration energy, T the temperature, and k Boltzmann's constant. With increasing temperature interstitials and vacancies are getting mobile whereas agglomerates are generally rather immobile. For example, in graphite single vacancies have a migration energy in the ab-plane of 1.0–1.6 eV [15,16] and are much more mobile than divacancies with a migration energy of 7 eV.[16] Once defect agglomerates have formed, annealing of damage is difficult and requires very high temperatures for dispersing the agglomerate.

In most materials interstitials are more mobile than vacancies. This is due to the fact that the migration of vacancies requires a rearrangement of the lattice by site exchange of atoms. The energy barrier for site exchange is particularly high in covalently bonded crystals because several bonds have to be broken during each exchange. Interstitials, on the other hand, are located between the lattice sites and often do not form bonds to adjacent atoms. This results in a lower energy barrier for their migration and makes them mobile at lower temperatures. The migration of vacancies and interstitials is generally anisotropic due to the particular symmetry of the crystal, making some migration pathways easier than others. This has to be considered in highly anisotropic lattices such as graphite.

During irradiation of a crystal with electrons, interstitial-vacancy pairs ('Frenkel pairs') permanently appear, migrate, and anneal or agglomerate. Therefore, a dynamic evolution of the defect structure takes place and depends sensitively on the irradiation intensity (defect production rate) and on the temperature (migration dynamics and annealing). Due to the agglomeration of point defects, larger defect structures such as dislocation loops, voids, or other extended crystal defects may successively appear during irradiation. An interesting example that will be treated in more detail in the following sections is graphite where divacancies as the smallest possible agglomerates lead to the formation of non-hexagonal rings and curvature in the basal planes.

3. The Experimental Techniques of *In-Situ* Electron Microscopy in the Study of Irradiation Effects

When irradiation effects in solids are studied by *in-situ* transmission electron microscopy, not much experimental periphery is required and the experiments can be done in almost every standard TEM without technical modifications. However, a heating specimen stage which is available commercially is generally of advantage. The appeal of such experiments is that irradiation and imaging are carried out with the same electron beam. But it has to be noted that the settings for irradiation and imaging are in most cases not optimal at the same time. The optimum for high-resolution imaging is the coherent illumination of the specimen with a parallel beam of moderate intensity. The optimum beam settings for irradiation, however, depend on the problem that has to be studied. In most cases a beam current density higher than for normal imaging is needed. For achieving a high current density, the largest condenser aperture and the lowest excitation of the first condenser lens (large spot size) are needed, but these settings do not allow coherent illumination of the specimen. The highest possible beam intensity is achieved when the beam is fully focused onto a spot. The diameter of the spot may be less than 1 nm in field emission TEMs, so the illuminated specimen area is much too small for imaging. Inspecting the result of irradiation under these conditions requires the spreading of the beam periodically. Some of the most spectacular irradiation experiments showed processes on the atomic scale that were recorded in real time, for example, the restructuring of lattice planes or the migration of individual atoms. However, the irradiation intensity in these experiments was moderate and the irradiation effects occurred under the conditions of normal imaging.

An interesting aspect of irradiation is the possibility of structuring the specimen on the atomic scale. In standard field emission TEMs, beam spots of 0.2–0.3 nm in diameter are feasible. In such beams, which have approximately a Gaussian intensity profile, current densities of more than 10^4 A/cm^2 can be achieved although the total beam current from a field emission gun is low. Therefore, specimen alterations on the scale of single lattice spacings can be achieved. A new generation of field emission TEMs is now available with an aberration-corrected condenser system. In these instruments, the beam diameter can be reduced to less than 0.1 nm and the total beam current density is accordingly higher (up to 4×10^5 A/cm^2 might be achieved). Standard TEMs with LaB_6 emitters have high total beam currents but do not allow beam spot diameters of less than approximately 10 nm (smaller spots are possible but at the expense of beam current density). Therefore, no manipulation on the nanometer scale can be achieved. But for the intense irradiation of larger specimen areas, LaB_6 systems are better suited than field emission guns. The maximum beam current density in LaB_6 systems is approximately 100–300 A/cm^2.

As a recipe for obtaining maximum beam brightness, the condenser aperture should be retracted and the largest spot size should be selected.

It is also indispensable that the tilt adjustment of the gun is optimized so as to obtain the highest brightness. In field emission TEMs, an appropriate gun lens setting should be used whereas in LaB_6 TEMs the filament heating and Wehnelt bias should be increased as much as can be tolerated.

The recording in real time by videotaping has been carried out since a long time by using image intensifiers and TV cameras below the column of the microscope. However, the videos taken in the high-resolution mode of the TEM were mostly affected by image noise. Nowadays, multi-scan CCD cameras allow real time recording with less noise and typically 20 frames per second. The output of the camera is converted by a computer to a standard video format and saved on a hard disc. Computer-assisted video processing can be carried out later to extract single frames, reduce noise, enhance contrast, or apply other functions of video processing. For very slow processes in the specimen that have to be viewed in time-lapse, it is often advisable to take conventional images with a slow-scan CCD camera in certain time intervals and generate a video later by linking the images together with an appropriate software. Since exposure times of typically 0.1–0.5 seconds can be used for each CCD image, noise is considerably lower than in real-time recording.

As mentioned above, the temperature of the specimen has a major influence on the evolution of radiation defects. Since heating due to electron irradiation is low, it is important that the specimen temperature can be adjusted and measured in a dedicated specimen stage. These stages are commercially available and allow resistive heating of the holder up to 1000–1300°C and tilting the specimen around one or two axes. The temperature of the stage is measured by a thermocouple. It is generally difficult to estimate the temperature of the specimen detail under the beam precisely because heat loss occurs locally by radiation, and the local temperature might differ from the temperature of the stage. Although the stage is thermally isolated against the rod of the holder, temperature changes of the whole setup are unavoidable and lead to thermal expansion and drift of the specimen. The drift has to be minimized which has been realized by water cooling of the rod. However, new problems arose due to mechanical vibrations of the rod caused by water turbulence, air bubbles, or vibrations of the water pumps. Several home-made solutions have been presented, for example by a gentle, gravity-driven flow of cooling water through the holder. The heating of nanoparticles up to very high temperatures has been achieved by depositing the particles onto a tungsten wire and heating the wire by an electric current in a specially designed specimen stage.[17,18]

The preparation of nanoparticles for *in-situ* electron irradiation experiments is in most cases much less time-consuming than the preparation of TEM specimens from bulk materials. The particles are often deposited on electron-transparent films, for example, holey or lacey amorphous carbon. The films are held by metal grids and can be purchased or made in the lab in a carbon evaporation apparatus. A general disadvantage of carbon films is that there is always an overlap of the image contrast of the specimen with the image of the film which can be quite distracting in high-resolution images. Although some carbon films can be used up to specimen temperatures of 800°C, the films have a tendency of warping or rolling when heated which can limit their applicability. An alternative preparation technique is the deposition of the nanoparticles directly onto pure metal grids. If the density of the particles is high enough, some are visible at the edge of the grids without overlap by other objects. It is important that the metal grids are stable up to the maximum desired temperature, do not evaporate, and that no welding, alloying, or reaction with the nanoparticles or the material of the stage occurs. Commercially available molybdenum grids fulfill these conditions for most applications. Another possibility is the deposition of carbon nanotubes on a metal grid so that the tubes span over the open squares and then to deposit the nanoparticles onto the tubes. The particles can be found sitting on the tubes and can easily be inspected at high temperature. However, shape transformations of the tubes such as bending or curling happen under electron irradiation and limit the applicability of this technique in irradiation studies.

An unwelcome artefact of electron irradiation in the TEM is the deposition of amorphous hydrocarbon contamination onto the specimen under the beam.[19] Although this effect can be applied for lithography[20] or contacting of nanowires[21,22] and shows some interesting phenomena, for example, the ramified growth of tree-like structures,[23] it should generally be avoided. Contamination under irradiation is dominant around room temperature whereas at high or low specimen temperatures contamination is negligible.

4. Electron Irradiation Effects in Nanoparticles

4.1. *Basic principles of irradiation of small objects*

Transmission electron microscopy has always been the most important technique to study the size, morphology, and structure of nanoparticles. Many irradiation phenomena in nanoparticles were observed accidentally. The different behaviour of nanoparticles and bulk materials is generally due to the proximity of surfaces.[24] The threshold energy for atom displacements is smaller for surface atoms, accordingly radiation effects are expected at electron energies below the bulk displacement threshold.

Migrating interstitial atoms or vacancies that have been generated by displacements in the bulk of the particle arrive at the surface after a short diffusion path. The surface is a perfect annealing site for both interstitials and vacancies. Hence, annealing of defects is faster when the particle size is smaller, giving nanoparticles a clearly higher stability against defect agglomeration than bulk materials. This is often observed in high-resolution electron micrographs when, for example, crystals of a few nanometers in size are irradiated. Then the particles shrink due to sputtering, but no lattice defects appear.

Another reason for the lack of defects in nanometer-sized crystals is the instability of some defect types in small crystals. For example, dislocations are hardly observed in nanocrystals because the operation of dislocation sources often needs a critical crystal size. Furthermore, dislocations are mobile and may anneal at surfaces or grain boundaries within a short time. Since the shortest timescale which is accessible to *in-situ* TEM is approximately 0.1 seconds, defects with a shorter lifetime cannot be detected.

In the following sections, examples of electron irradiation effects in some types of nanoparticles will be shown. The main focus will be on graphitic nanoparticles because graphite shows unique possibilities of lattice reconstruction after atom displacement, leading to the formation of novel structures under electron irradiation.

4.2. *Irradiation effects in nanometer-sized crystals*

As mentioned above, irradiation phenomena in nanometer-sized crystals were often observed as side effects during inspection in the electron microscope. Therefore, not much work was devoted to detailed studies. Some "trivial" phenomena such as amorphization or sputtering are known since a long time. Crystals that are very sensitive to irradiation or have a very low mobility of point defects lose their crystalline order and amorphize successively under the electron beam. Crystals that are more stable but whose surface displacement threshold is below the energy of the beam maintain their structure but shrink under sustained irradiation due to sputtering of surface atoms. The lower thermodynamical stability of small crystals as compared to bulk materials is reflected by the lowering of the melting point as the size of the crystal decreases. At high specimen temperatures this becomes apparent as an additional effect, and melting or evaporation of the crystals is promoted further by electron irradiation.[25]

Displacements with low energy transfer to surface atoms do not lead to sputtering. The atoms remain on the surface and migrate if their thermal energy exceeds the activation energy for migration. These mobile surface atoms can cause shape changes of the crystal under the beam.[26–28] Since the crystals are always in contact with another material, for example, the support film, displaced atoms from the surface can also migrate over the grid so that the crystal loses mass. In dense aggregates of crystals with different shapes Ostwald ripening is often observed because larger crystals are more stable and collect the migrating atoms. A very common phenomenon is the coalescence of nanometer-sized crystals under the beam[29] which reduces the total surface energy. In bulk materials composed of nanometer-sized crystallites (these materials are commonly denoted as "nanocrystalline materials"), irradiation can lead to grain growth or sintering.[30] Several other interesting irradiation phenomena have been observed by high-resolution TEM, for example, the thinning of metal nanowires under electron irradiation[31–34] or the drilling of holes into nanowires by a focused electron beam.[35]

4.3. *Irradiation effects in graphitic nanoparticles*

4.3.1. *Radiation defects in graphitic structures*

Radiation defects in the graphite lattice have been a subject of intense research for several decades because graphite is an important material in reactor technology. Several studies of bulk graphite samples have been carried out by *in-situ* electron microscopy. The early work on point defect formation and migration or displacement threshold energies is in part contradictory or out-dated and can be found in the literature.[6,36,37] Nowadays, the renewed interest in graphitic structures after the discovery of fullerenes and carbon nanotubes has initiated numerous electron microscopy studies. Again, as a by-product of electron microscopy, irradiation phenomena were observed in graphitic nanoparticles and some results were so striking and unexpected that a new field of irradiation studies has emerged. New insight into the atomistic processes was not only gained by the improved resolution of modern electron microscopes but also by first principles calculations with steadily increasing computing power that are meanwhile able to calculate the behaviour of structures with sizes of several nanometers.[16,38,39]

The displacement threshold energies in graphitic structures strongly depend on the direction of momentum transfer relative to the lattice. Atom displacements are easier when momentum is transferred normal to the basal plane of graphite. However, in most TEM studies the electron energy is much higher than the displacement threshold so that larger scattering angles dominate and knocks in different directions occur. When curved structures such as fullerenes or nanotubes are irradiated, a range of crystallographic directions appears relative to the beam so that anisotropy effects can often be neglected. Displacement threshold energies have been measured by gradually increasing the electron energy in the microscope and observing the onset of visible structural alterations. For single-wall carbon nanotubes which represent a cylindrically closed single layer of graphene, a threshold electron energy of $E_d = 86$ keV has been measured;[40] this corresponds to an energy of $T_d = 16$ eV which has to be transferred to a carbon atom for permanent displacement. A value of approximately $E_d \approx 100$ keV has been measured in multi-wall nanotubes[6] which might also be an appropriate value for graphite along the c-direction. Fullerene molecules (C_{60}) are more sensitive to electron irradiation than cylindrical structures. Above a threshold of somewhat less than 80 keV, fullerene molecules that are packed in a fullerite crystal tend to coalesce and form larger cages.[41]

The formation energy of a single vacancy in graphite is 7.3–7.5 eV as obtained from first principles calculations.[16,42,43] This energy is high compared to the corresponding values in typical metals so that almost no vacancies exist in thermal equilibrium even at high temperatures. The formation energy of a divacancy as calculated in simulations is 8.7 eV[16] which is surprisingly close to the formation energy of a monovacancy.

The migration behavior of vacancies and interstitials is reflected by their respective migration energies. The situation in graphite is complex because the migration of defects is highly anisotropic and depends on the number and curvature of the graphene sheets. (A recent review of defect formation and migration in graphitic structures is found in the literature.[9]) For example, a vacancy can easily migrate in the ab-plane of graphite (E_m = 1.0–1.6 eV) whereas migration in the c-direction is hindered by a high migration energy (E_m = 4.7 eV).[15,16,42] Interstitials, on the other hand, have quite similar migration energies in the ab-plane (1.5 eV) and in the c-direction (2.3 eV).[42,43] The situation can change dramatically when a graphene sheet has a high curvature; for example in a typical single-wall nanotube the migration of interstitials moving inside the tube (E_m = 0.3 eV) is much faster than the migration on the outer side of the tube (E_m = 0.8 eV).[44,45]

A graphene sheet has a unique ability to reconstruct after the generation of vacancies. As mentioned above, the formation energies of monovacancies and divacancies are close (7.5 eV and 8.7 eV, respectively) and monovacancies have a rather low migration energy in the ab-plane (1.0–1.6 eV) whereas divacancies are almost immobile (7 eV). Therefore, monovacancies migrate rapidly and tend to coalesce to form stable and immobile divacancies.

Figure 3 shows in a simple example that single vacancies cannot anneal by reconstruction of the lattice (whichever way new bonds form, there is always one open bond left). Divacancies, however, can vanish without any dangling bond by reconstruction of the lattice. Hexagonal rings can be converted to pentagonal rings (as shown in the example in Fig. 3) or, in more complicated cases, to heptagonal or octagonal rings. Such a reconstruction which is related to the Stone-Wales transformation[46] is always accompanied by a local change in curvature because the hexagonal lattice is flat whereas pentagons introduce positive, heptagons negative curvature. As an additional effect, the surface area of the graphene sheet is getting smaller upon removal of atoms and reconstruction. The

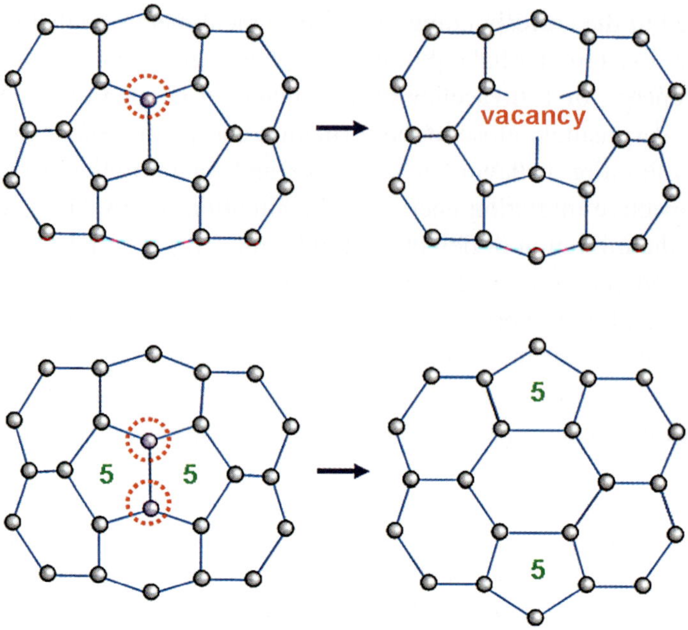

Fig. 3. The removal of one atom from a graphenic layer leaves a single vacancy (top). When two adjacent atoms are removed (bottom), the divacancy can close by reconstruction of the bonds. The number of hexagons decreases by one. For simplicity of the reconstruction a structure with pre-existing pentagons was chosen.

example in Fig. 3 shows that the number of hexagons decreases by one when two atoms are removed. As a consequence, cylindrically or spherically closed graphitic structures (nanotubes or fullerene-like cages) shrink when atoms are lost but may remain coherent due to the ability to close divacancies.

Of course, the reconstruction mechanisms as described here can only take place when each monovacancy is mobile enough to coalesce with another vacancy (there may be several activation barriers for the coalescence of vacancies that have to be surmounted). Therefore, the temperature of the specimen has to be high enough. At low temperature, monovacancies and interstitials would remain stationary until the concentration of vacancies is so high that the lattice loses its coherence. The minimum temperature which is necessary for the reconstruction of the lattice may depend on the defect production rate (this has not been studied carefully yet). Under the typical conditions of electron irradiation in the electron microscope, this temperature has been found to be in the range of 200–300°C.[47–49] These temperatures can be easily reached in heating stages inside the TEM so that it is advisable to heat every specimen containing graphitic carbon to higher temperature even when just inspection of the material is desired and no deliberate irradiation is planned. Artefacts due to defect agglomeration at room temperature irradiation are likely to appear and have sometimes been misinterpreted as structural details.

4.3.2. *Carbon nanotubes*

Carbon nanotubes are meanwhile the most intensely studied new form of graphitic carbon.[50–53] It has already been observed in early electron microscopy studies that single- or multi-wall nanotubes change their structure and collapse under electron irradiation.[54,55] However, controlled structural modifications of nanotubes by electron irradiation have not been achieved until heating stages were used.[56]

Single-wall carbon nanotubes (SWNTs) are generally less stable under the electron beam than multi-wall nanotubes (MWNTs). This is due to the fact that sputtered atoms from SWNTs are lost and cannot be taken up by other shells. Irradiation of SWNTs at temperatures above 300°C with moderate beam intensity leads to the successive loss of atoms and reconstruction of the graphene shell by the closure of divacancies. The surface of the initially perfect cylinder becomes wavy and the diameter of the tube decreases. A simulation of this process is shown in Fig. 4.[39] After the formation of vacancies and interstitial atoms, the lattice reconstructs via the transformation of hexagons to pentagons, heptagons, or octagons. Besides changing the curvature of the cylinder locally, the tube shrinks although the cylinder remains coherent. It has to be pointed out that the covalent sigma bonds between carbon atoms in graphene remain strong even around non-six-membered rings. As a consequence, the high mechanical stability of the tube is maintained to a large extent during shrinkage. Assuming a compressible material inside the hollow of a SWNT, irradiation would lead to an internal pressure of up to 40 GPa until the tube would break.[57,58]

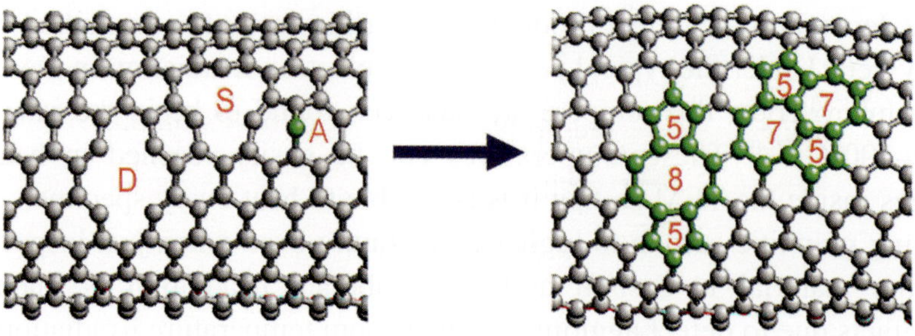

Fig. 4. Model of a single-wall nanotube with a single vacancy (S), a divacancy (D), and an interstitial as adatom (A). Reconstruction of the lattice as obtained by computer simulation[39] leads to the formation of 5-, 7-, and 8-membered rings. The decreasing surface area causes a shrinkage of the tube. (Courtesy of A. Krasheninnikov).

The shrinkage of the tube cannot continue when its diameter falls below 0.4 nm which is the lower stability limit of carbon nanotubes. Then the tube breaks and closes the ends by a fullerenic cap[45,59,60] (this is shown below in Fig. 5 where the innermost layer of a MWNTs has been cut by an electron beam).

It has first been believed that most displaced atoms leave the tube by sputtering. However, experimental evidence has shown that a substantial fraction of the displaced atoms is injected into the interior of the tubes where the atoms have a much higher mobility than on the outer surface.[45,60,61] Hence, carbon nanotubes can be considered as pipelines for the efficient transport of carbon atoms.

The injection of atoms into the interior of the tubes has several consequences. When a SWNT has a closed end, mobile atoms inside the tube cannot vanish in this direction and are trapped inside the tube. These atoms are available for annealing with new vacancies so that such a tube shows an increased stability under the beam. This becomes apparent when a SWNT is cut twice. A second cut close to the capped end (where the tube has been cut for the first time) requires a clearly higher electron dose than the first cut.[60] When multi-wall nanotubes are irradiated, the injection of carbon atoms into the interior leads to the agglomeration of these interstitial atoms and the formation of new graphene sheets. Figure 5 shows the evolution of a MWNT under a moderate electron beam. The shrinkage of

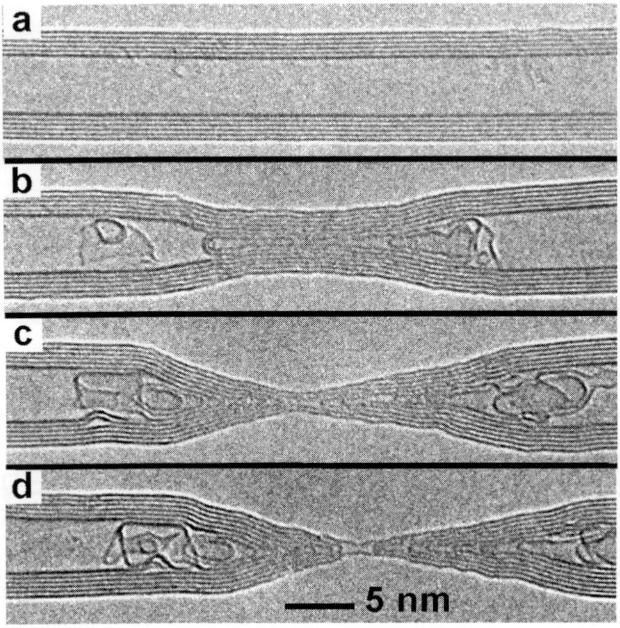

Fig. 5. Irradiation series showing the collapse of a multi-wall nanotube under the electron beam at 600°C. While the outermost layer remains intact, the inner layers break once their diameter falls below 0.4 nm. The aggregation of sputtered carbon atoms inside the tubes leads to the formation of disordered graphenic structures. (Beam current density: 450 A/cm^2, irradiation times: b: 160s; c: 340s; d: 820s).

all shells is apparent and the outermost shell is not destroyed by sputtering but instead remains intact. The innermost shells break successively when their diameter falls below the stability limit and close their ends by caps. It can also be seen how new disordered graphenic sheets form inside the tube by aggregation of the injected atoms.

When carbon nanotubes are filled with metallic or carbide nanowires[62] (this can be made by using metal-containing gases in the CVD production of the tubes), both the collapse of the tube and the injection of carbon atoms can lead to interesting phenomena. As pointed out above, the collapse occurs by exerting high pressure, in the radial directions, onto the encapsulated material. Under a pressure of the order of 40 GPa, the metal or carbide crystals can be deformed heavily and extruded as shown in Fig. 6

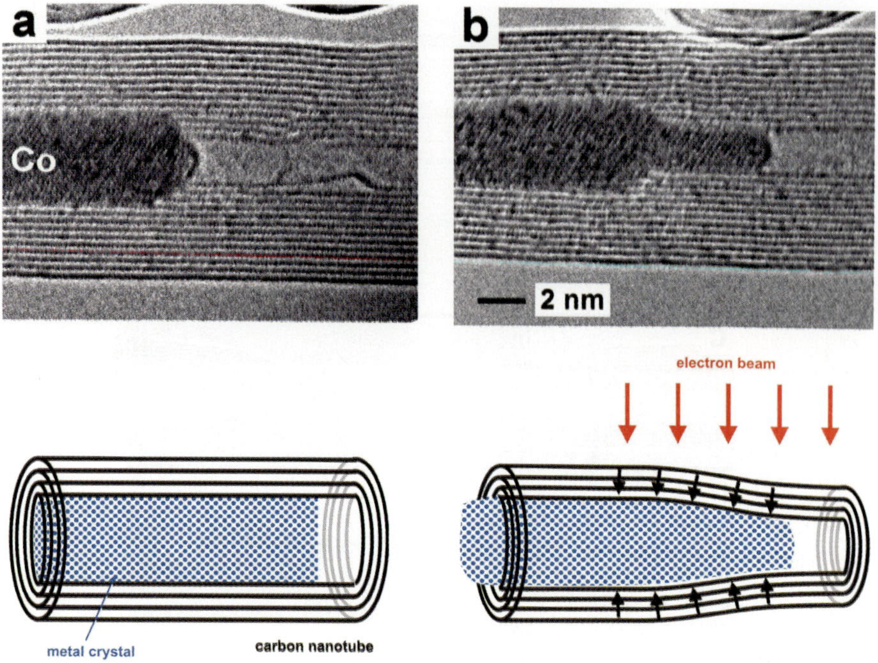

Fig. 6. Deformation of a cobalt crystal inside a collapsing carbon nanotube under electron irradiation. The Co crystal has a cylindrical shape before irradiation (a) and is thinned and extruded when the tube collapses locally (b).

for the example of a cobalt wire inside a MWNT under electron irradiation.[63] Since the crystal can be imaged with lattice resolution during such a deformation, this experiment offers us, for the first time, the possibility to carry out an *in-situ* study of the deformation of individual nanometer-sized crystals. Here carbon nanotubes act as robust jigs that deform even hard materials. Furthermore, the nanotubes are electron-transparent so that all structural changes in the filling such as defect formation become visible. The deformation rate in these experiments is generally low (less than 1 nm/sec) but can be controlled by the intensity of the electron beam.

When carbon nanotubes that are partly filled with catalytically active metal crystals (e.g., Fe or Co) are subjected to moderate electron irradiation, carbon atoms from the shells of the tube are injected into the metal crystal. As long as carbide formation does not occur, the carbon

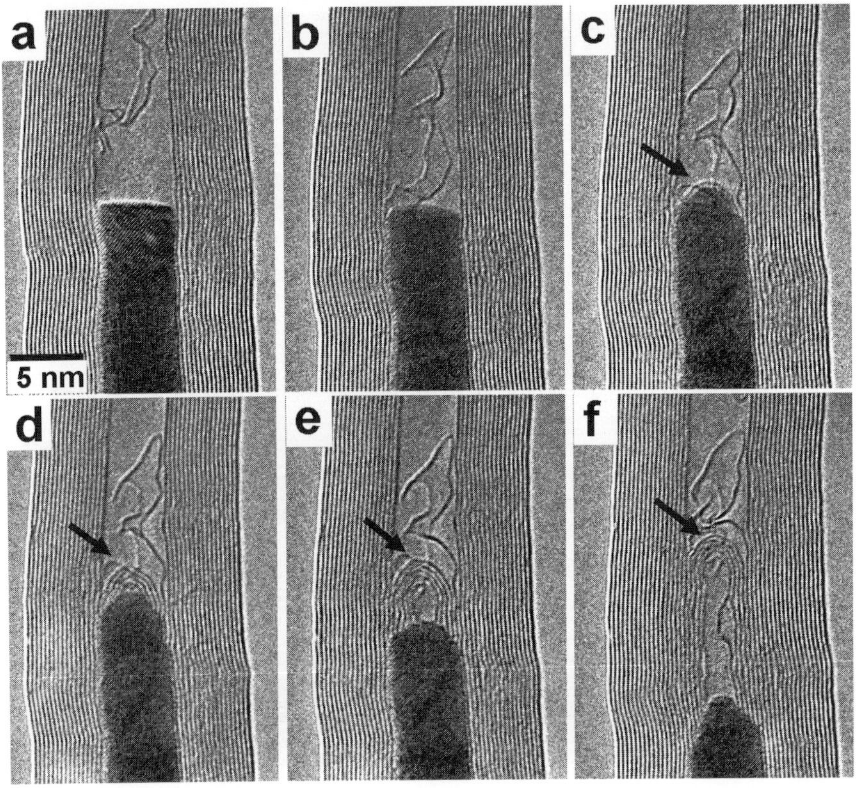

Fig. 7. Irradiation of a nanotube filled with a FeCo crystal. The injection of carbon atoms into the metal crystal leads to the growth of a new nanotube (arrowed) from the metal inside the host tube. (Temperature: 600°C, beam current density: 100 A/cm^2, irradiation times: b: 360s; c: 386s; d: 411s; e: 420s; e: 500s).

atoms diffuse through the metal crystal but cannot remain in the crystal because the solubility of carbon in Fe or Co is low. Therefore, new graphenic layers precipitate at the end face of the metal crystals, and these layers can grow as new nanotubes as shown in Fig. 7. In such an experiment, nanotube growth can be monitored *in-situ* at high spatial resolution in the electron microscope.[58] Useful information about the growth of carbon nanotubes can be obtained from such experiments.

Selective irradiation of MWNTs with a focused electron beam can induce changes of their shape. Two examples are shown in Fig. 8.

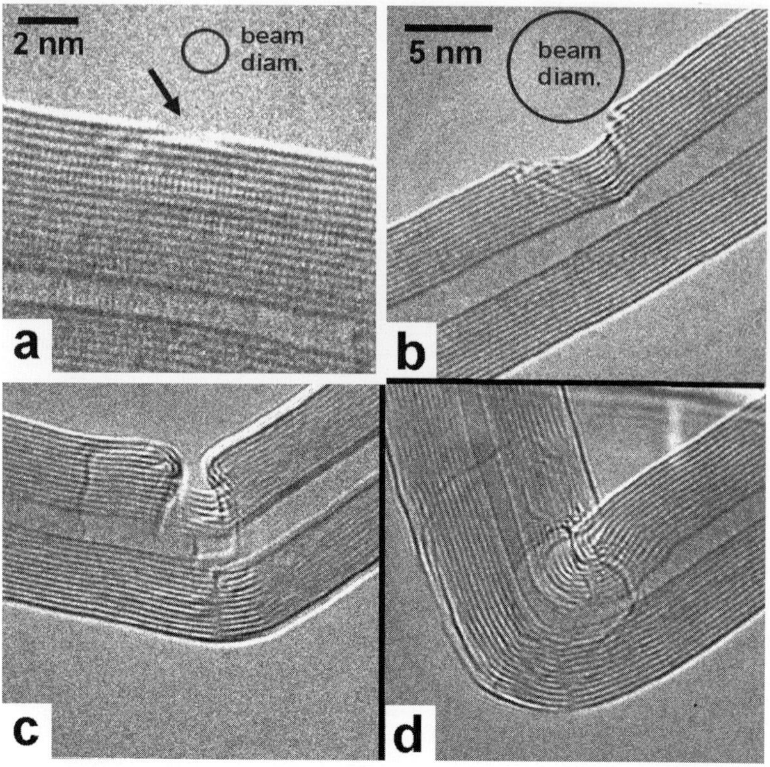

Fig. 8. Electron beam-induced structuring of a multi-wall carbon nanotube. A fully focused beam can be used to remove a few atoms from the outermost layer (a). Irradiation with a beam diameter of half the diameter of the tube leads to the gradual bending of the tube (b–d). (Temperature: 600°C, irradiation intensity: a: 10^4 A/cm^2; b-d: 10^3 A/cm^2).

Irradiation of the outermost layer by a fully focused electron spot (1-2 nm) can remove just a few atoms so that a "perforation" of the outer layer is possible (Fig. 8(a)). In Fig 8(b)–(d) a MWNT has been bent by predefined angles by irradiation of one half of the tube's diameter.[56] Sustained irradiation of nanotubes with an intense electron beam transforms the tubes to spherical carbon onions as described in Sec. 4.3.3.

The irradiation of a bundle of single-wall nanotubes with an electron beam can lead to a collapse of the bundle which is followed by graphitization and formation of a multi-wall tube.[56] Irradiation of a SWNT bundle with a low-intensity electron beam can cause the coalescence of two

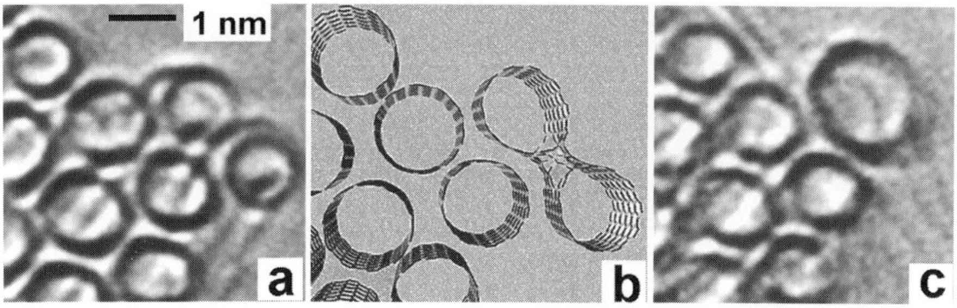

Fig. 9. Electron irradiation of a bundle of single-wall nanotubes (a), seen here in a cross-sectional view, can lead to the coalescence of two adjacent tubes (c). The model shows an intermediate stage (b). (Temperature: 800°C, the coalescence occurs within a few seconds).

parallel tubes so that a new tube of double circumference appears as shown in a cross-sectional view in Fig. 9.[64,65] However, as molecular dynamics simulations show, such a parallel coalescence, which proceeds in a zipper-like way in the axial direction of the tubes, can only take place when the two tubes have the same geometrical structure (n,m-type). Irradiation of nanotube bundles can also lead to an interlinking of the tubes[66] which increases the resistance of the bundle against sliding of the tubes relative to each other.

By irradiating two crossing SWNTs that touch at one point, the tubes can be welded together so that an X-junction is created (Fig. 10).[67] The simulation of this process showed that two perfect tubes would never join because the energy of an X-junction is much higher than the energy of two individual tubes. However, when a certain concentration of vacancies prevails in both tubes, the formation of a junction can be energetically favourable. Further irradiation of such an X-junction can result in the sputtering-induced loss of one of the arms so that a Y-junction is left.

Single-walled nanotubes can be filled with fullerene molecules, resulting in a structure resembling to peapods. Electron irradiation of such an arrangement leads to the coalescence of the fullerene molecules inside the host tubes so that a double-wall nanotube structure is obtained.[68] In an electron irradiation study of peapod structures where the fullerenes contained metal atoms, the dynamic behavior of even individual metal atoms has been monitored.[69,70]

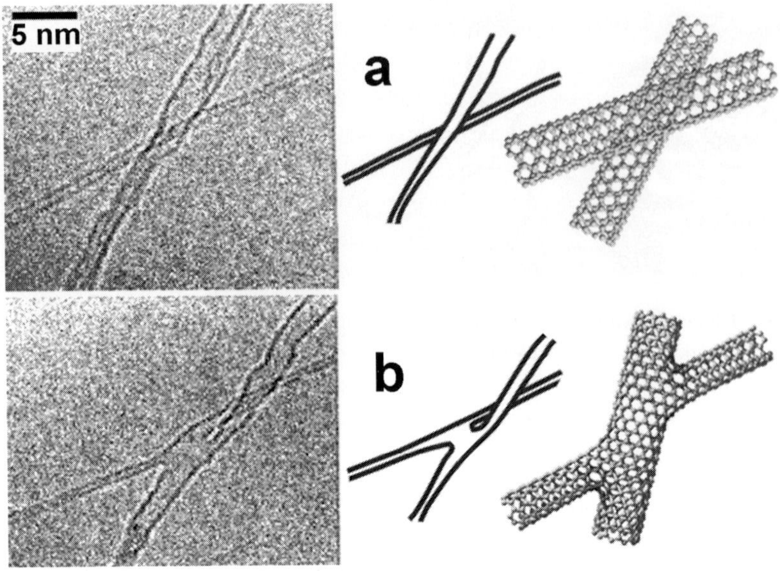

Fig. 10. Two crossing single-wall nanotubes (a) can be welded under electron irradiation (b). (Temperature: 800°C, irradiation intensity: 10 A/cm^2, irradiation time between a and b: 60s).

4.3.3. *Carbon onions*

Carbon onions have been the first example for the formation of novel carbon nanostructures under electron irradiation. In his pioneering work in 1992 Ugarte discovered that intense electron irradiation of graphitic soot leads to the transformation of polyhedral or irregular graphitic nanoparticles to spherical structures[71] which have been denoted as carbon onions and consist of concentric fullerene-like spherical shells. When these irradiation experiments have been repeated at higher specimen temperatures, carbon onions with perfectly spherical and coherent shells were obtained[48] as shown in Fig. 11 together with a possible structure model.[72] In the model, a spherical cage has been designed under the condition that all carbon-carbon bonds have approximately the same length as in unperturbed graphene. Hexagons and pentagons are not sufficient to build a structure with uniform curvature when the shell is much lager than a C_{60} molecule. It is also necessary to introduce larger rings, e.g. heptagons. The formation

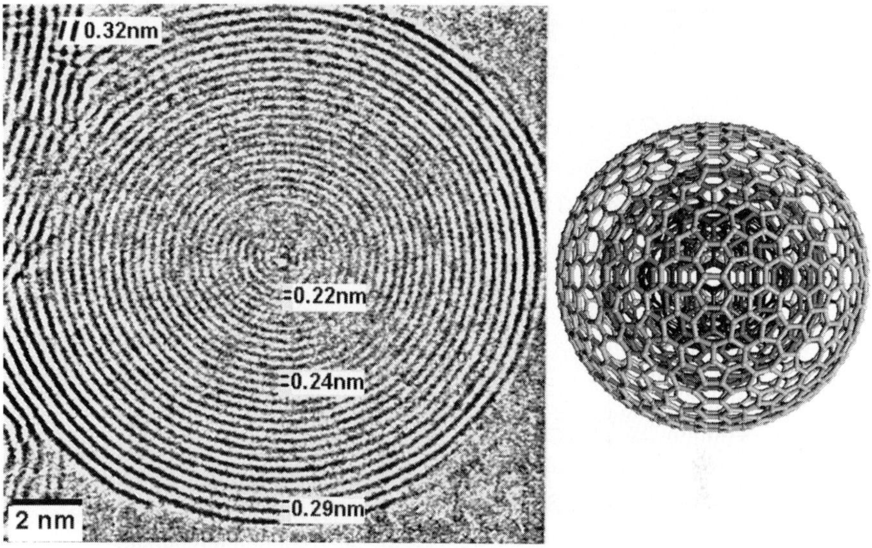

Fig. 11. A spherical carbon onion as obtained under electron irradiation of a polyhedral graphitic grain. The distance between the graphenic shells decreases towards the centre of the particle. The model on the right hand side (not to scale) shows a possible structure including pentagons and heptagons (courtesy of M. Terrones).

mechanism of carbon onions is meanwhile understood. Atom displacements in planar graphene sheets lead to the formation of non-hexagonal rings and, thus, curvature of the sheets. The sheets remain unstable as long as dangling bonds remain at their edges. But dangling bonds can close when the sheet is rolled up in two dimensions. When a closed cage exists, sustained loss of atoms under the electron beam makes the surface of the cage smaller and induces surface tension which tends to make the structure spherical.

The surface stress from the outer shells of a multi-shell onion has another important consequence. As in the case of carbon nanotubes, the collapsing shells of carbon onions induce high pressure in the interior. This was first seen as a decreasing spacing between the graphenic shells from the surface towards the centre (see Fig. 11).[48,73] Under sustained and intense electron irradiation of such a compressed carbon onion, the nucleation of a diamond crystal in the centre can be achieved as shown in Fig. 12.[74]

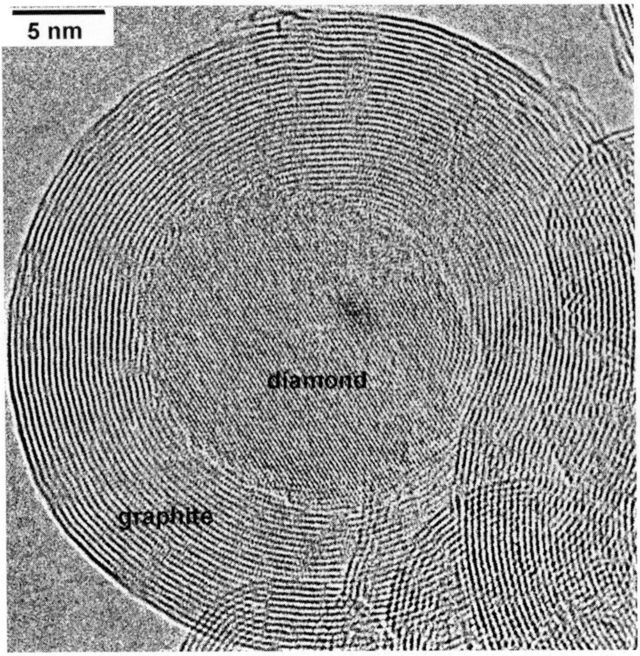

Fig. 12. Diamond crystal in the centre of a carbon onion. The nucleation of the diamond was observed after intense electron irradiation (1.25 MeV, 300 A/cm^2) of a carbon onion for 30 minutes. (Temperature: 700°C).

Two issues can explain the transformation of graphitic material to diamond inside carbon onions. The pressure in the centre of the collapsing onion is hydrostatic and may reach values even higher than inside collapsing nanotubes. Since the pressure in nanotubes can already be as high as 40 GPa (see Sec. 4.2.2), the pressure in the centre of carbon onions may be far inside the stability regime of diamond in the phase diagram of carbon.[58] As the second reason, the hybridization of carbon atoms in a graphene sheet is sp^2 when the sheet is flat but changes gradually towards sp^3 with increasing curvature of the sheet. In the centre of a compressed onion the curvature of the innermost shell may be even higher than the curvature of a C_{60} molecule (the s-character in the π-orbital of C_{60} is approximately 0.1) so that the transformation to diamond with sp^3 hybridization is facilitated.

If we consider the whole transformation process from an irregular graphitic particle via a spherical carbon onion to a diamond crystal, it is apparent that the structural order of the system increases considerably. But this is not a relaxation of the system towards thermal equilibrium because graphite has thermodynamically a lower free energy than diamond. Under the conditions of intense electron bombardment, the system is highly dissipative because only a small fraction of the energy which is transferred through the system is stored as the formation energy of defects. Thus, the transformation of graphitic material to diamond under electron irradiation can be considered as an example of self-ordered structure formation in a dissipative system.[75,76]

Interesting phenomena have been observed when carbon onions encapsulating crystalline non-carbon materials have been subjected to electron irradiation in the electron microscope.[62] Similar to nanotubes, carbon onions exert pressure onto the encapsulated materials, but the crystals inside onions have no freedom to evade the pressure. Metal crystals can be encapsulated in onion-like shells when carbon is evaporated together with the respective metal in an electric arc discharge. These core-shell particles have a polyhedral shape and transform towards a perfect sphere under the electron beam at specimen temperatures above 300–400°C. During this shape transformation, the encapsulated crystal is subjected to deformation so that the formation of deformation defects is accessible to direct observation. This has been demonstrated in an *in-situ* electron microscopy study of W and Mo crystals in graphitic shells and is shown in Fig. 13.[77]

It has been observed that metal atoms, e.g. Co, Fe, Ni, or Au, are able to migrate through the graphite lattice when metal nanocrystals are compressed inside a carbon onion.[78] This leads to a gradual shrinkage of the metal crystal while the atoms diffuse through the shells towards the surface. Nickel atoms form intermediate metal-carbon phases when diffusing through the graphitic layers.[79]

When the encapsulated metals tend to form carbides, the reaction can be induced by the pressure of the graphitic shells against the metal particle and by ballistic displacements of interface atoms. This has been observed for Fe crystals inside carbon onions and is shown in Fig. 14.[80]

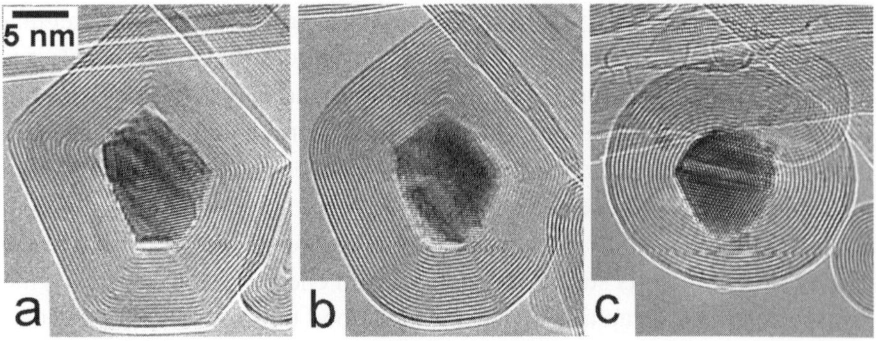

Fig. 13. Deformation of a molybdenum crystal inside graphitic shells. Irradiation leads to shape changes of the shells. Deformation defects (grain boundary in b, twin boundaries in c) can be recognized. (Temperature: 600°C).

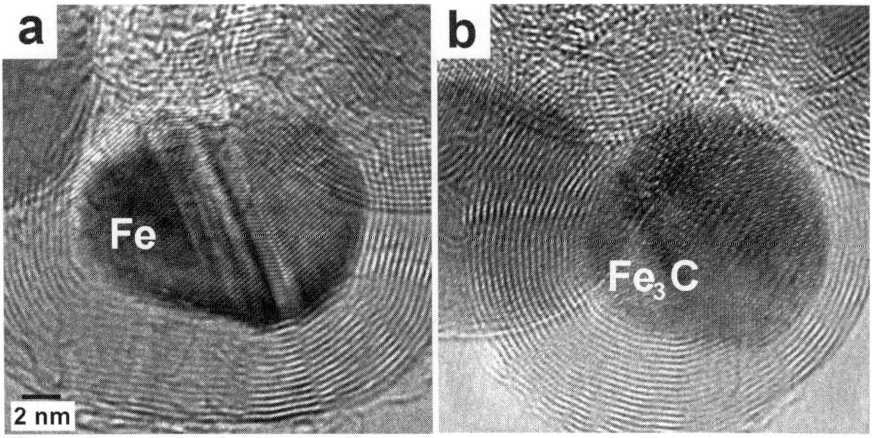

Fig. 14. Reaction between an iron crystal and carbon inside a carbon anion. The starting configuration is an fcc-Fe crystal encapsulated in graphitic shells (a). After intense electron irradiation, the crystal reacts with the shells by forming an Fe_3C crystal (b). (Temperature: 600°C, irradiation intensity: 300 A/cm^2, irradiation time between a and b: 63min).

4.4. *Phase transformations in nanoparticles under irradiation*

Electron irradiation can cause solid-solid or solid-liquid phase transformations in nanomaterials. It has often been observed that a system relaxes from a metastable to the equilibrium state under the electron beam. This is obvious because momentum transfer from the energetic electrons can help the atoms to overcome the kinetic barrier for a phase transformation towards equilibrium. An example is the crystallization of metastable amorphous material, e.g. Si, under the electron beam.[81] Here, electron irradiation causes a transition that could also be achieved by heating. On the other hand, when atoms are "mixed" at lower temperatures by the electron beam, this can cause disorder and the transformation of a stable crystalline to a metastable amorphous structure.[82,83]

When the atoms in a two-phase system are displaced at a certain rate, the equilibrium between the two phases can be shifted. This occurs when the two phases have a different "radiation hardness", i.e. when the displacement rate in the two phases is not the same. Then, the transformation of a thermodynamically more stable into a less stable phase can occur. However, such a transition generally needs the presence of both phases, i.e. a nucleus of the thermodynamically less stable phase has to exist. The phase transformation happens at the interface between the two phases and is seen as a growth of the radiation-harder phase at the expense of the less radiation-hard phase. An example is the reversal of the graphite-diamond equilibrium under electron irradiation.[84] Although graphite is the thermodynamically stable phase of carbon at low pressure, electron irradiation can cause the transformation of graphite to diamond when a diamond nucleus is in a graphitic environment. Due to the low density of graphite and the open structure with much space between the basal planes, atoms are easily displaced in graphite whereas diamond with its densely packed lattice does not allow atom displacements at the same rate as in graphite. This is reflected by the different displacement threshold energies in graphite ($T_d \approx 17$–25 eV) and diamond ($T_d \approx 30$–48 eV). Figure 15 shows, as an example, the interface between graphite and diamond.[85] Electron irradiation in the TEM at high specimen temperature leads to the growth of the diamond crystal at the expense of graphite. When a carbon atom at the interface is displaced, it can aggregate to either of the two phases but its lifetime on a certain site is higher when it is bound in the diamond phase. Of course, the new equilibrium between the phases also depends on the temperature.

As an indirect cause of irradiation, phase transformations in nanoparticles can occur under the action of high pressure such as inside carbon

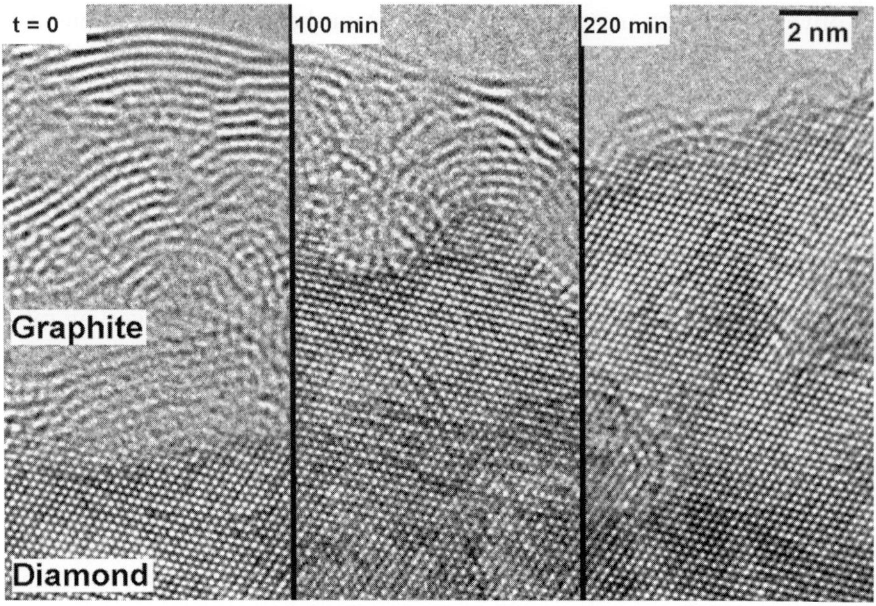

Fig. 15. Interface between graphite (top) and diamond under electron irradiation at 700°C. The diamond crystal grows slowly at the expense of the graphite layer.

onions. For example, solid-liquid transformations of metal crystals encapsulated in carbon onions depend on the pressure in the carbon shells that, in turn, depends on the irradiation conditions. Metals with low melting temperature such as Sn or Pb have been encapsulated in graphitic shells and show considerable superheating or supercooling under electron irradiation.[86] An example is shown in Fig. 16. A melting hysteresis with a width of almost 400 K has been measured in encapsulated Sn crystals. The unexpected superheating has been explained by the occurrence of pressure and, more importantly, by the suppression of surface melting of the Sn crystals by the firmly encapsulating graphitic shells.

5. Conclusions

The irradiation of nanomaterials with electrons shows a great variety of phenomena of fundamental importance that can only be studied by *in-situ* electron microscopy. The ability to both generate and investigate

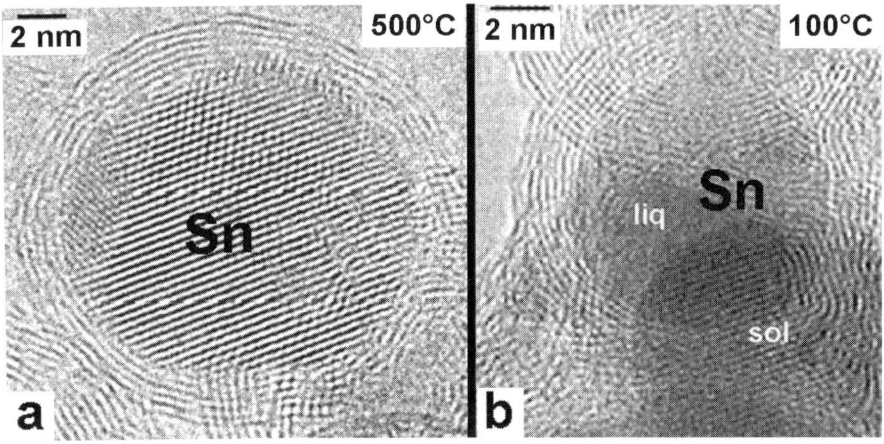

Fig. 16. Solid tin crystals can be superheated up to 500°C (a), and liquid Sn droplets can be supercooled down to 100°C (b) when encapsulated in graphitic shells. The melting temperature of bulk Sn is 232°C.

structural transformations at the atomic scale with the same electron beam and at the same time is a particular appeal of *in-situ* microscopy. With the development of specially designed heating stages it is now possible to study irradiation effects in a wide temperature range with lattice resolution.

Nanoparticles show many interesting beam-induced transformations that do not occur in bulk materials. In this chapter, results on graphitic particles were treated because they show the largest variety of structural transformations. Carbon nanotubes or carbon onions are not just destroyed under the electron beam but transformed into new nanoparticles that have not been obtained hitherto by other techniques. With the irradiation of composite nanoparticles, for example, nanotubes or carbon onions encapsulating metal crystals, experiments on the nanoscale can be carried out such as the exposure of nanometer-sized crystals to high pressure. In such a way, composite systems can act as nanolaboratories that are transparent to the electron beam and so allow the direct observation of the processes at high spatial resolution.

With the steadily improving image resolution of transmission electron microscopes it can be expected that this field will continue to expand.

46. A. J. Stone and D. J. Wales, *Chem. Phys. Lett.* **128**, 501 (1986).
47. A. P. Burden and J. Hutchsion, *J. Cryst. Growth* **158**, 185 (1996).
48. F. Banhart, T. Füller. Ph. Redlich, and P. M. Ajayan, *Chem. Phys. Lett.* **269**, 349 (1997).
49. K. Urita, K. Suenaga, T. Sugari, H. Shinohara, and S. Iijima, *Phys. Rev. Lett.* **94**, 155502 (2005).
50. A. Oberlin, M. Endo, and T. Koyama, *J. Crystal Growth* **32**, 335 (1976).
51. S. Iijima, *Nature* **354**, 56 (1991).
52. R. Saito, G. Dresselhaus, and M. Dresselhaus, *Physical Properties of Carbon Nanotubes* (Imperial College Press, London, 1999).
53. P. J. F. Harris, *Carbon Nanotubes and Related Structures* (Cambridge University Press, Cambridge, 1999).
54. P. M. Ajayan, V. Ravikumar, and J.-C. Charlier, *Phys. Rev. Lett.* **81**, 1437 (1998).
55. N. G. Chopra, F. M. Ross, and A. Zettl, *Chem. Phys. lett.* **256**, 241 (1996).
56. J. X. Li and F. Banhart, *Nano Lett.* **4**, 1143 (2004).
57. L. Sun, F. Banhart, A. Krasheninnikov, J. A. Rodriguez-Manzo, M. Terrones, and P. M. Ajayan, *Science* **312**, 1199 (2006).
58. L. Sun, J. A. Rodriguez-Manzo, and F. Banhart, *Appl. Phys. Lett.* **89**, 263104 (2006).
59. H. E. Troiani, M. Miki-Yoshida, G. A. Camacho-Bragado, M. A. L. Marques, A. Rubio, J. A. Ascencio, and M. Jose-Yacaman, *Nano Lett.* **3**, 751 (2003).
60. F. Banhart, J. X. Li, and M. Terrones, *Small* **1**, 953 (2005).
61. A. Krasheninnikov, F. Banhart, J. X. Li, A. S. Foster, and R.M. Nieminen, *Phys. Rev.* B **72**, 125428 (2005).
62. F. Banhart, N. Grobert, M. Terrones, J.-C. Charlier, and P. M. Ajayan, *Int. J. Mod. Phys.* B**15**, 1437 (2001).
63. J. A. Rodriguez-Manzo, M. Terrones, H. Terrones, L. Sun, F. Banhart, and H. W. Kroto, *Nature Nanotech.* **2**, 307 (2007).
64. M. Terrones, H. Terrones, F. Banhart, J.-C. Charlier, and P. M. Ajayan, *Science* **288**, 1226 (2000).
65. M. Yoon, S. Han, G. Kim, S. Lee, S. Berber, E. Osawa, J. Ihm, M. Terrones, F. Banhart, J.-C. Charlier, N. Grobert, H. Terrones, P.M. Ajayan, and D. Tomanek, *Phys. Rev. Lett.* **92**, 075504 (2004).
66. A. Kis, G. Csanyi, J. P. Salvetat, T.-N. Li, E. Couteau, A. J. Kulik, W. Benoit, J. Brugger, and L. Forro, *Nat. Mater.* **3**, 153 (2004).
67. M. Terrones, N. Grobert, F. Banhart, J.-C. Charlier, H. Terrones, and P. M. Ajayan, *Phys. Rev. Lett.* **89**, 075505 (2002).
68. E. Hernandez, V. Meunier, B. W. Smith, R. Rurali, H. Terrones, M. Buongiorno-Nardelli, M. Terrones, D. E. Luzzi, and J.-C. Charlier, *Nano Lett.* **3**, 1037 (2003).
69. K. Suenaga, R. Taniguchi, T. Shimada, T. Okazaki, H. Shinohara, and S. Iijima, *Nano Lett.* **3**, 1395 (2003).
70. K. Urita, Y. Sato, K. Suenaga, A. Gloter, A. Hashimoto, M. Ishida, T. Shimada, H. Shinohara, and S. Iijima, *Nano Lett.* **4**, 2451 (2004).
71. D. Ugarte, *Nature* **359**, 707 (1992).
72. M. Terrones and H. Terrones, *Fullerene Sci Technol.* **4**, 517 (1996).
73. M. Zaiser, *Mater. Res. Soc. Symp.Proc.* **540**, 243 (1999).
74. F. Banhart and P. M. Ajayan, *Nature* **382**, 433 (1996).

75. A. Seeger and W. Frank, *Solid State Phenomena* **3&4**, 125 (1988).
76. A. Seeger, N. Y. Jin, F. Phillipp, and M. Zaiser, *Ultramicroscopy* **39**, 342 (1991).
77. J. X. Li and F. Banhart, *Adv. Materials* **17**, 1539 (2005).
78. F. Banhart, Ph. Redlich, and P. M. Ajayan, *Chem. Phys. Lett.* **292**, 554 (1998).
79. F. Banhart, J.-C. Charlier, and P. M. Ajayan, *Phys. Rev. Lett.* **84**, 686 (2000).
80. L. Sun and F. Banhart, *Appl. Phys. Lett.* **88**, 193121 (2006).
81. G. Lulli, P. G. Merli, and M. V. Antisari, *Phys. Rev.* B **36**, 8038 (1987).
82. G. J. C. Carpenter, and E. M. Schulson, *J. Nucl. Mater.* **23**, 180 (1978).
83. H. Mori and H. Fujita, *Jap. J. Appl. Phys.* **21**, L494 (1982).
84. M. Zaiser and F. Banhart, *Phys. Rev. Lett.* **79**, 3680 (1997).
85. Y. Lyutovich and F. Banhart, *Appl. Phys. Lett.* **74**, 659 (1999).
86. F. Banhart, E. Hernandez, and M. Terrones, *Phys. Rev. Lett.* **90**, 185502 (2003).
87. A. Zobelli, C. P. Ewels, A. Gloter, G. Seifert, O. Stephan, S. Csillag, and C. Colliex, *Nano Lett.* **6**, 1599 (2006).

formation through the knock-on defect is first discussed for a single-wall carbon nanotube (SWNT).[2] Then the formation of the Frenkel-type defects (a pair defect of an interstitial and a vacancy) at the inter-layer of double-wall carbon nanotube (DWNT) and their annihilation are investigated at the elevated temperatures.[3] Finally the migration of the individual metal atoms through the atomic defects of fullerene cage is corroborated by the HR-TEM [4,5] and by the associated DFT calculations.[6]

2. Mono-vacancy Formation in SWNT

Imaging point defect in a graphen layer, such as atomic vacancy, has been challenging but indeed has a crucial importance since it directly refers to the physical and even the industrial properties of this material. Only scanning probe microscopy was used to investigate the structure of such defects at the surface of graphite.[7,8] In the case of using an aberration non-corrected HR-TEM, the acceleration voltage and the defocus value should be carefully adjusted to visualize the atomic defects in graphen.[2] Figure 1(a) shows a HR-TEM image of a graphen layer which consists of a SWNT, extracted after the irradiation time of ($t = 320$ s) from a series of hundreds of images. The defocus value is not at the Scherzer condition but set to $\Delta f \sim -38$ nm to enhance the contrast corresponding to the carbon zig-zag chain spacing (0.21 nm) in our microscope ($C_s = 0.45$ mm at 120 kV, JEM-2010F). The electron irradiation is estimated as 60,000 electrons/nm^2 per sec. Obviously a number of white spots appeared under electron irradiation and more than six spots can be obviously isolated from the noise level, leading to a crude estimation of around 0.3 point defects nm^{-2} in graphen layer. This results leads to a cross-section for atomic displacement at 120 kV electrons as 160 barn (160×10^{-10} nm^2) which well fits with a calculated value (180 barn) for 100 kV electrons.[1] HR-TEM simulations have been done to corroborate the contrast arising for some of the expected point defects on a single graphen layer. The HR-TEM contrast for one carbon atom vacancy V_1, two neighboring vacancies (or an intra-planar di-vacancy V_2), a pentagon-octagon-pentagon (5–8–5) defect and a carbon adatom are considered for the possible defect structures in Fig. 1(b). The atomic positions of the carbon atoms neighboring the V_1 and V_2 vacancies have not been relaxed. Nevertheless,

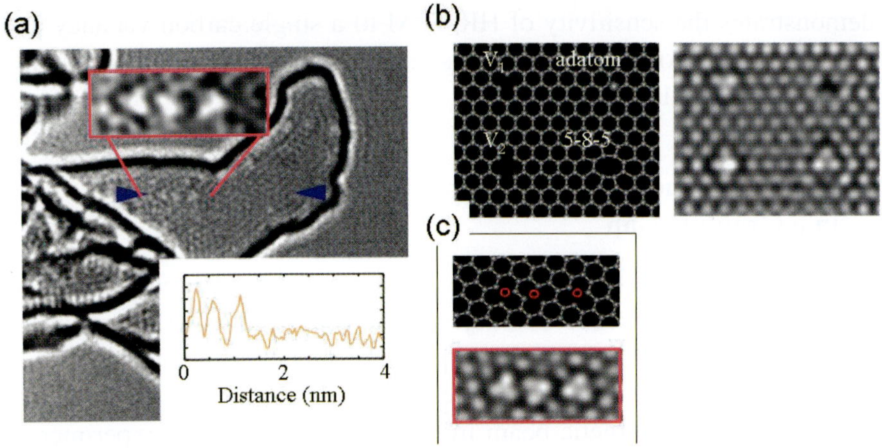

Fig. 1. *In-situ* observation of vacancy formation in a graphene layer. (a) A HR-TEM image of a single-walled carbon nanostructure recorded after 320 s of irradiation. Intensity profiles (insets) between the two arrows illustrate the contrast measured at the carbon vacancies. (b) Atomistic models and simulated HR-TEM image for a graphene layer with an adatom, unrelaxed vacancies (V1 and V2) and a 5–8–5 rearrangement. Squares indicate the location of the vacancy. (c) Atomic model and simulated HR-TEM image with three mono-vacancies, giving the best match to the experimentally observed defect structures. Scale bar, 2 nm. [See A. Hashimoto *et al.* (Ref. 2)].

theoretical works [9,10] have predicted that intra-planar relaxation is weak for a V_1 vacancy and indeed, no differences have been seen in our HR-TEM simulation using either relaxed or un-relaxed atomic positions. On the other hand, inter planar di-vacancy is supposed to strongly reconstruct and transform to an agglomeration of non-hexagonal rings such as the 5–8–5 defect presented (Fig. 1(b)). The adatom has also been positioned in the theoretically predicted equilibrium structure, i.e. to be in a bridge-like structure, between two surface atoms. The major issue of the HR-TEM simulations is that, even for a single vacancy, the contrast may be strongly enhanced at the center of the neighboring hexagons. This delocalization effect is particularly strong at the chosen defocalization. The three point-defects imaged at Fig. 1(a) can then remarkably be well simulated by the removal of only three carbon atoms from a graphen layer as shown in Fig. 1(c) and we were not able to obtain a better matching using various other types of (multi)-vacancies or non-hexagonal clusters. This demonstrates the sensitivity of HR-TEM to a single carbon vacancy and the stability of mono-vacancy at least a typical exposure time for the HR-TEM imaging (~1 sec).

3. Formation and Relaxation of Inter-Layer Defects in a Graphite Gap

Among various types of atomic defects expected in graphite, the inter-layer defect "in between" two adjacent layers is particularly crucial for graphite in practical use such as graphite moderator employed for a nuclear reactor. The inter-layer defect has long been believed to appear frequently under energetic beam irradiation[11] but no direct experimental evidence has been provided. Theoretical studies have suggested a metastable Frenkel-type defect, which consists of an interstitial carbon atom and a vacancy (an I-V pair) in between bi-layer graphite.[11,12] In order to provide a direct evidence for this type of inter-layer defect, which may be stable for a macroscopic time but likely to annihilate just after, we have used a DWNT as a simplest model of two layer graphite and investigated the defects formation during the observations. There are two major reasons for employing the DWNTs; (i) a DWNT consists of a bi-layer graphite and is the thinnest specimen for HR-TEM observation of the inter-layer defects. (It should be noted that a single-atomic defect would become more and more difficult to be imaged in a thicker specimen such as HOPG.) (ii) Both the top-wall and side-wall defects can be easily imaged in a single experiment and a reasonable atomic model for the defect structure can be more convincingly proposed. We should note, however, that the DWNT is representative for the turbostratic graphite but not for the HOPG with the regular c-axis stacking. The inter-layer orientational relationship is mostly random and the inter-layer distance is slightly larger in DWNTs (0.38 nm) than that for HOPG (0.335 nm).[13] The energetics of defect formations can also be affected by the curvature of the graphene layer in DWNT and should therefore be somehow different from that for the flat HOPG.

Since the temperature dependence of the formation and relaxation behavior of the inter-layer defects is extremely intriguing in terms of an energy barrier for the recombination of I-V pairs, a systematic study with

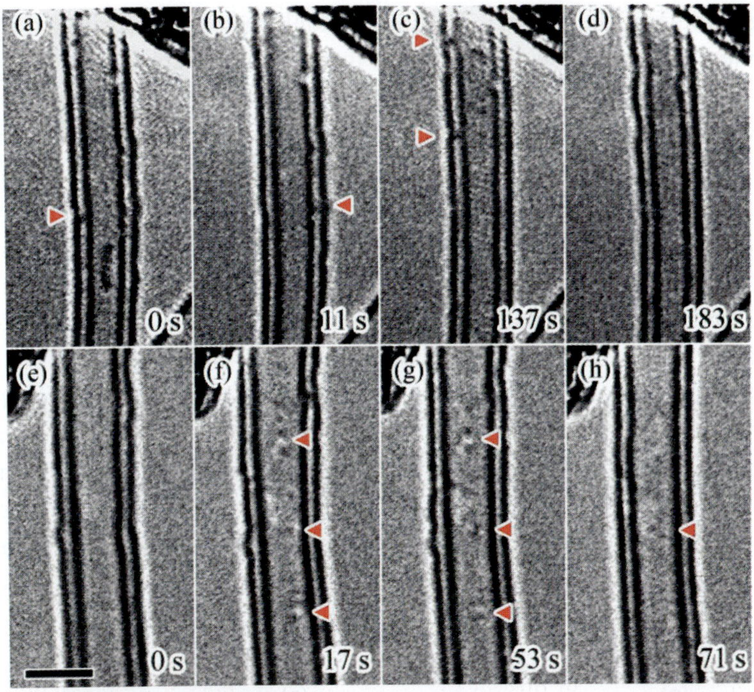

Fig. 2. Sequential HR-TEM images of the inter-layer defects which are appearing and disappearing in DWNTs recorded at 573 K. In the side-wall (a–d), several bridges connecting two graphene layers frequently appear in dark contrast (marked by red arrows) and disappear just after during the observation. Also in the top-wall (e–h), pairs of dark and bright contrast often appear and then vanish (marked by red arrows). Scale bar = 2 nm. [See K. Urita *et al.* (Ref. 3)].

a wide range of temperatures from 98 K to 573 K should be performed. We have used both a cold stage and a heating stage for the temperature elevated experiments.[4]

Figure 2 shows two sets of sequential HR-TEM images of the DWNTs recorded at 573 K. In the first sequential images (Fig. 2(a)–(d) from 0 to 183 sec), several bridges in dark contrast obviously appear in between two layers (indicated by red arrows). These inter-layer bridges often appear and disappear during the observation. Also in the other sequential images (Fig. 2 (e)–(h)), from 0 to 71 sec), pairs of dark and bright spots are appearing (Fig. 2(f) and (g)) and disappearing just after (Fig. 2(g) and (h)) in the middle of the DWNT. Both sequential images are attributed to the side-view and top-view of the similar inter-layer defects.

Anisotropy of the defect formation rate can be found between top- and side-wall[14] and is more prominent at lower temperature. In the irradiation experiments of DWNTs, the inner nanotube always suffers the damage faster than the outer nanotube, and the outer nanotube is more resistive. This is most probably because the inner nanotube exhibits a larger curvature (a smaller diameter) and more easily suffers the knock-on damage. Therefore we can reasonably assume that the vacancy is more likely to be generated in the inner nanotube.

HR-TEM image simulations of the theoretically predicted defect structures have been also performed. A model for the I-V pair in bi-layer graphite[11] (Fig. 3(a)) was relaxed to fit within the DWNT gap by using a semi-empirical potential. Vacancy has been assumed in the inner layer of DWNT. Fig. 3(c) and 3(d) show the simulated HR-TEM images calculated from the I-V pairs, involving a single and a couple of I-V pairs at the

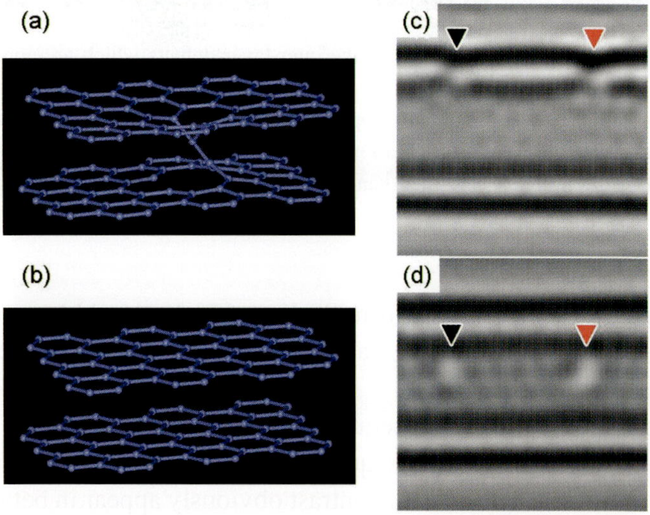

Fig. 3. Expected atomic configurations by using a semi-empirical potential and their image simulations of the inter-layer defects. (a) A single of I–V pair and (b) its annihilation. (c) (d) Image simulations for the side-view and top-view of the inter-layer defects of single I–V pair (black arrow) and a couple of I–V pairs (red arrow).

inter-layer space in side- and top-wall. The latter consists of an isolated interstitial pair and a di-vacancy. Although the detailed atomic structure of the defect cannot be experimentally derived, the calculated image does exhibit a good agreement with the observed images (Fig. 2(a)–(d)), i.e. the dark contrast bridging the two layers experimentally observed is clearly reproduced. Also as for the top-wall defects, the simulated image shows at least a qualitative agreement with the observed contrast, the dark and bright pair (Fig. 2(e)–(h)), in the middle of DWNT. It is hardly possible from these images to distinguish the number of I-V pairs (one or two in this case). This is mainly because the insufficient spatial resolution of the experiments at that time.

Then the temperature dependence of the defect formation in DWNTs is investigated to corroborate the stability of these defects. Figure 4

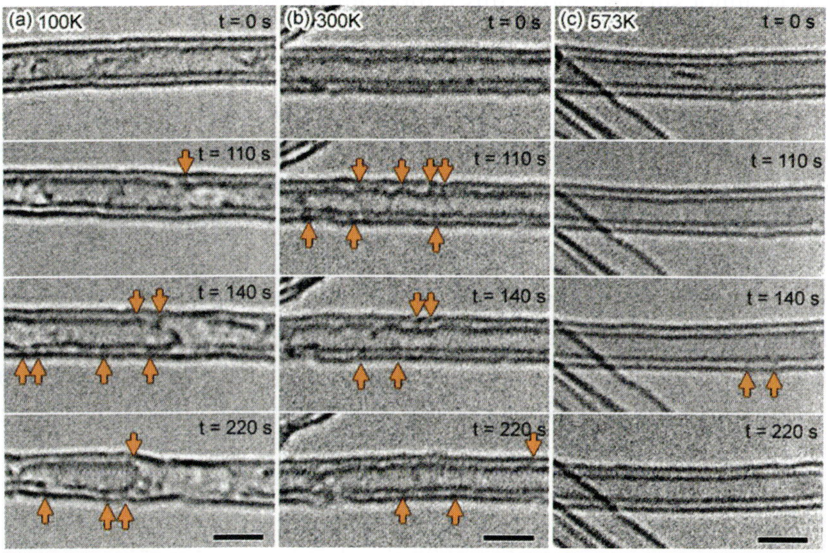

Fig. 4. Sequential HR-TEM images for the formation rates of the inter-layer defects at different temperatures with the same time scale (0 to 220 sec). (a) At 93 K, the defects due to electron irradiation are quite prolific and the nanotube inside quickly damages due to the complex defects. (b) At 300 K, the nanotubes are more resistive but the defects can be also found frequently. (c) At 573 K, the defect formation can be hardly seen and the DWNTs are completely resistive due to the electron beam irradiation. The arrows indicate possible inter-layer defects. Scale bar = 2 nm.

fullerene has been frequently detected and the examples are shown at $t =$ 0 sec and 51 sec (Fig. 5(a) and (b)). An opening pathway at the fullerene cage is unambiguously visible at the indicated region by arrows.

Thereafter the two defected fullerenes (probably turn around to face its once defected area to each other and then) start to coalesce ($t = 123$ sec). Such a coalescence of fullerene molecules has been already reported for C_{60} or C_{82} cages.[20,21] It should be mentioned that the presence of many pentagons (or a consequent higher curvature) of the fullerene cage is one of the important reasons why the defect is more likely to be induced at the fullerene site although the nanotube wall is more resistive.

In the case of metallofullerenes (the fullerene cages containing metal atoms inside), these atomic defects should surely act as pathways for migrations of metal atoms, through which the metal atoms can go out from the cage. Two different metallofullerenes with the same cage but with the different valence state have been chosen for this experiment. The $Ca@C_{82}$ contains a di-valent Ca^{2+} inside the C_{82} (C_{2v} symmetry) cage and the $Gd@C_{82}$ contains a tri-valent Gd^{3+} in the same cage. These two molecules show a very good contrast. The breakout of Gd atoms has been often observed, although the Ca atoms have never been observed to breakout from the cage. A DFT calculation supports the different interaction between the encaged metal atoms and the defective fullerenes.[6]

In Fig. 6, a comparison of the behavior of metal atoms with the defective fullerenes is summarized in the case of Ca and Gd metallofullerenes. The DFT calculations has been performed at the level of B3LYP/CEP-121G + 6 − 31G*. In the case of $Ca@C_{82}$ (Fig. 6(a)), the interaction between the metal atoms and the fullerene cage is quite small and the Ca atoms tend to stay inside the cage even the cage has a defect or a vacancy. On the contrary, the Gd atoms exhibit a strong interaction with the defects of fullerene cage and are very likely to break out through the vacancy. The experimentally observed HR-TEM images (middle) show excellent agreements with the simulations based on the calculations (right).

Although the valence state may not be the only factor to determine the behavior of metal atoms inside the fullerene cage, this result has an important implication about the possibilities to control the atomic migration rate by means of the number of defects in matrix.

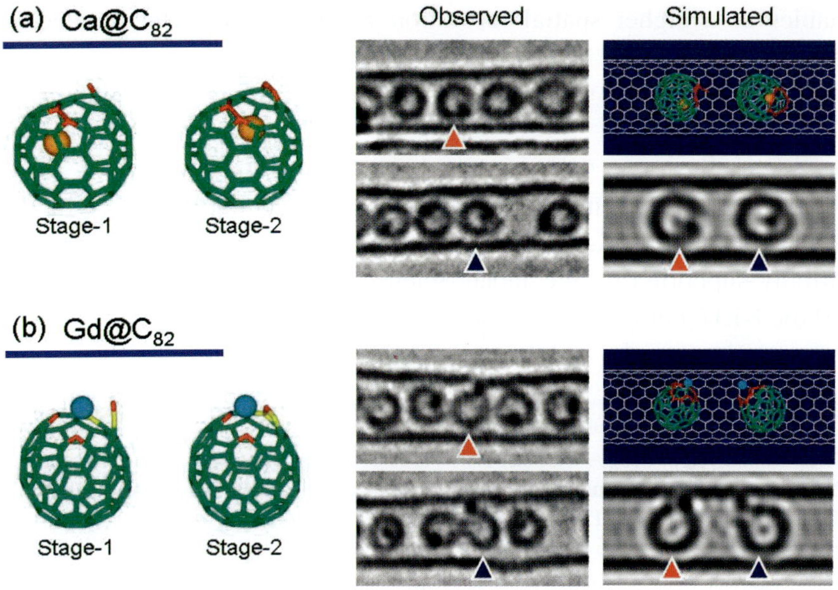

Fig. 6. A comparison with the DFT calculations (left) and the HR-TEM imaging (middle) in the case of Ca@C_{82} (a) and Gd@C_{82} (b). The simulations (right) based on the calculations show good agreements with the experimental images (middle). Scale bar, 1 nm. [See Y. Sato *et al.* (Ref. 5)].

5. Conclusions and Outlook

Formation and relaxation of lattice defects in crystal generally govern physical properties of a solid, and therefore have long been under intensive investigations in the field of physics. However this, kinetics of lattice disordering and thermal relaxation was so far discussed only in macroscopic viewpoint, and has never been investigated in atomic viewpoint where activities of individual defects can be monitored. As shown in this chapter, the progress of *in-situ* HR-TEM enables us to monitor the individual atomic defects during formation and annihilation.

Another important implication of the work demonstrated here is the fact that the HR-TEM imaging at individual molecular level now allows a direct comparison of the experiments with the DFT calculations.

Recent progress in the aberration-correction for HR-TEM imaging will lead to more precise analysis of the defect structures because it enables us a higher spatial resolution without raising the acceleration voltage.

Acknowledgments

The work presented in this chapter was done in collaborations with Drs. Y. Sato, T. Yumura, K. Urita, A. Hashimoto and S. Iijima. This work was partially supported by the Japan Science and Technology agency (JST) and the NEDO project for the nano-carbon technology.

References

1. F. Banhart, *Rep. Prog. Phys.* **62**, 1181 (1991).
2. A. Hashimoto, K. Suenaga, A. Gloter, K. Urita, and S. Iijima, *Nature* **430**, 870 (2004).
3. K. Urita, K. Suenaga, T. Sugai, H. Shinohara, and S. Iijima, *Phys. Rev. Lett.* **94** 155502 (2005).
4. K. Urita *et al.*, *Nano Lett.*, **4** 2451 (2004).
5. Y. Sato *et al.*, *Phys. Rev.* B **73** 233409 (2006)
6. T. Yumura, Y. Sato, K. Suenaga, K. Urita, and S. Iijima, *Nano Lett.*, **6** 1389 (2006).
7. J.R. Hahn, H. Kang, S. Song, and I.C. Jeon, *Phys. Rev. B* **53**, 1725 (1996).
8. D. Orlikowski, M. Buongiorno Nardelli, J. Bernholc, and C. Roland, *Phys. Rev. B* **61**, 14194 (2000).
9. A.A. El-Barbary, R.H. Telling, C.P. Ewels, M.I. Heggie, and P.R. Briddon, *Phys. Rev. B* **68** 144107 (2003).
10. L. Hjort and S. Stafström, *Phys. Rev. B* **61**, 14089 (2000).
11. R. H. Telling, C. Ewels, A. A. El-Barbary, and M. I. Heggie, *Nature Mater.* **2**, 333 (2003).
12. C. P. Ewels, R. H. Telling, A. A. El-Barbary, and M. I. Heggie, *Phys. Rev. Lett.* **91**, 25505 (2003).
13. A. Hashimoto *et al.*, *Phys. Rev. Lett.* **94,** 045504 (2005).
14. B. W. Smith and D.E. Luzzi *J. Appl. Phys.* **90,** 3509 (2001).
15. A. V. Krasheninnikov *et al.*, *Phys. Rev. B* **69,** 073402 (2004).
16. A. A. El-Barbary *et al.*, *Phys. Rev. B* **68,** 144107 (2003).
17. E. P. Wigner, *J. Appl. Phys.* **17,** 857 (1946).
18. E. W. Mitchell and M. R. Taylor, *Nature* **208**, 638 (1965).
19. K. Suenaga *et al.*, *Science* **290,** 2280 (2000).
20. T. Okazaki *et al.*, *J. Am. Chem. Soc.*, **123**, 9673 (2001).
21. H. Hernandez *et al.*, *Nano Lett.*, **3,** 1037 (2003).

INDEX